AF245588

PARTICULATES AND CONTINUUM

PARTICULATES AND CONTINUUM

Multiphase Fluid Dynamics

S. L. Soo
University of Illinois at Urbana–Champaign

HEMISPHERE PUBLISHING CORPORATION
A member of the Taylor & Francis Group

New York   Washington   Philadelphia   London

Library of Congress Cataloging-in-Publication Data

Soo, S. L. (Shao-lee), date.
  Particulates and continuum : multiphase fluid dynamics / S. L. Soo.
       p.   cm.
  Includes index.

  1. Fluid dynamics. 2. Two-phase flow. I. Title.
TA357.S62   1989
620.1′064—dc20   89-31741
ISBN 0-89116-918-0   CIP

CONTENTS

PREFACE

The term *Multiphase Flow* in the dynamic sense of phase interaction was introduced in 1965 (*Ind. Eng. Chem. Fund.*, Vol. 4, p. 426) to account for systems including particles, droplets, and bubbles in fluids, as well as general configuration of interfaces between the phases.

Although the engineering practice of processing mixtures of phases began much earlier, multiphase flow as a fluid science is identified with second half of the twentieth century. In this period, recognition of fundamental concepts, definitive experiments, and general formulations have undergone rapid progress. Studies of the dynamics of multiphase systems encompass broad lines of disciplines in engineering and sciences. Since the publication of the volume, *Fluid Dynamics of Multiphase Systems,* in 1967, many advances have been achieved in research on multiphase flow by a large number of contributors. A revision in 1983 led to the volume *Multiphase Fluid Dynamics.* While these volumes consisted of extensive surveys of various methods and have served as general references for research and development, the present volume treats multiphase systems with emphasis on the aspect of fluid dynamics and as an introduction to research in multiphase flow.

At this stage, multiphase flow can be formulated in terms of fundamentals of continuum mechanics. However, rigorous solutions are available only to particulate systems and stratified flow systems. The present volume treats definitive concepts, methods, and theories which have been validated by experimental results. Such a choice limits the application of the basic formulation to solving, principally, problems of dilute dispersed systems. These consist of multiphase systems in the form of suspensions of particulates in predominating continuous fluid media. The particulates may take the forms of bubbles, droplets, or solid particles. Logical extensions to dense suspensions are treated with some necessary empiricism.

This book is intended as a textbook for college seniors and graduate students and as a research reference. A coherent presentation facilitates the understanding of physical interactions. The principle concern is fluid dynamics, with extension to other transport processes of heat and mass transfer, and chemical reactions to illustrate applications of multiphase flow. The exercise problems at the end of each chapter assist the reader in formulating and solving physical problems and gaining a sense of magnitude of interacting effects and events. Extended details and corollaries are also included in these exercise problems. Some of the topics in the exercise problems may also be incorporated as topics for the lectures. For extensive empirical details, the reader is referred to several excellent handbooks which are cited along with the fundamentals. Analytical solutions are preferred because they facilitate discussions. Computer codes are not included within the scope of this book. In a developing field such as multiphase flow, there are gaps to be filled and significant research topics are many.

An exhaustive survey of references is not intended. References not cited include material which is too advanced for the chosen scope. Those cited tend to be in books and journals easily accessible to the readers. Citations of recent work include those which may lead to new directions. Selection of topics has the aim of coherence of concepts and methods. Empirical relations are included only to bridge the gaps in the understanding. The organization of the topics allows as instructor to introduce additional materials as well as conflicting views. A few of the exercise problems suggest reading of references; the text, however, is self-contained as an introduction to fundamentals. A solutions manual to the exercise problems is available to instructors of the course.

The chapters are arranged in a nearly logical sequence for treating problems in multiphase flow. Chapter 1 is an outline of physical behaviors of single particles. However, details of a single free particle in suspension and its diffusivity are given.

Chapter 2 presents the basic equations as derived from continuum mechanics with simplification to stratified flow systems and dispersed flow systems. A reader may prefer to bypass Sec. 2.3 to Sec. 2.6, thereby following the earlier intuitive formulations especially of dispersed flow systems (Soo 1967).

Chapter 3 deals with transport properties, including particle-particle and particle-surface interactions and other transport process and flow regimes. Chapter 4 treats interactions of multiphase flow with waves and boundary conditions unique with multiphase flow.

Chapter 5 through 7 deal principally with cases of dilute dispersed flows and applications to physical systems. Readers may select topics according to their personal interest in the applications.

Chapter 8 deals with dense systems where gaps in concepts and procedures exist and treats possible avenues of future research.

The accuracy and helpfulness of Mrs. June W. Kempka and Mrs. Tammy S. Lawhead of the Publication Office of the Department of Mechanical and Industrial Engineering of the University of Illinois in preparing the lecture notes leading to the present volume is hereby acknowledged.

This volume is dedicated to my mother, Yunchuan Aisinjioro Soo, in honor of her ninetieth birthday.

S. L. Soo

SYMBOLS

In this volume various formulations include a large number of physical quantities. Use of the same letters or symbols for different entities or concept has been unavoidable. Choice of notations was also affected by the desire to retain commonly used notations familiar in each field of study, while at the same time avoiding ambiguity by making a single choice where two or more symbols for the same quantity are known to exist in the literature.

In general, notations are explained along the narative. The convention is seen in the list below. Unless otherwise defined in the text, the following symbols, with their meaning, are understood:

A	area or constant or coefficient as defined
a	radius of a particle (bubble, or droplet)
B	coefficient as defined
b	coefficient as defined or dimension
C	concentration or capacitance or coefficient as defined
c	specific heat at constant pressure of a fluid phase
c_p	specific heat at constant pressure of particulate phase
D	diffusivity
d	diameter or characteristic dimension
E	electric field or energy or modulus of material
e	electronic charge
F	time constant for momentum transfer, function or force
f	distribution function or force per unit mass
f	Coulomb friction coefficient
G	time constant for energy transfer
g	gravitational acceleration
H	curvature or time constant for charge transfer
h	height or heat transfer coefficient or enthalpy
i	$\sqrt{-1}$
J	flux
J_E	heat source
K	particle-fluid interaction parameter, charge mobility, or constant or function as defined
k	Boltzmann constant (1.381×10^{-23} J/K) or material parameter
L	length

L_p	interaction length
ℓ	length or scale of turbulence
M	total mass or molecular weight
m	mass of a particle
N	total number of particles or number of moles
N_{ED}	electrodiffusion number
N_{es}	electrosurface number
N_{et}	electrothermal number
N_{ev}	electroviscous number
N_{Fr}	Froude number
N_{Im}	Impact number
N_{Ja}	Jakob number
N_{Kn}	Knudsen number
N_{Le}	Lewis number
N_M	Mach number
N_m	Momentum transfer number
N_{Nu}	Nusselt number
N_{Pe}	Peclet number
N_{Pr}	Prandtl number
N_{Rc}	Radiative conductivity number
N_{Re}	Reynolds number
$N_{Re})_e$	Electric Reynolds number
N_{Sc}	Schmidt number
N_{Sh}	Sherwood number
N_{S1}	Strouhal number
N_W	Weber number
n	number density or harmonics or directional normal
P	static pressure or force
p	position
Q	flow volume or electric charge
q	electric charge per particle
R	radius of pipe
$\bar{R}$	gas constant (8314/molecular weight, J/kg K)
R	correlation as defined
r	radius or position
r, ϕ, z	cylindrical coordinates
$\bar{r}$	reflectivity
r^*	impact ratio
S	entropy
s	path
T	temperature or time interval
t	time
U,V	velocity as defined

U,V,W	conjugate velocity components or Cartesian or cylindrical coordinates
u	internal energy
u,v	fluctuating velocity components
V_e	electric potential
v	volume
W	volumetric erosion rate per unit area
x,y,z	Cartesian coordinates
Z	number of charges
z	elevation

Greek

α	thermal diffusivity, volume fraction, or parameter as defined
Γ	rate of generation
γ	ratio of specific heats
Δ	range or deformation tensor
δ	a small quantity or thickness
ε	fraction void or permitivity or ultimate yield stress
ε_0	permitivity of free space, 8.854×10^{-12} farad/m, C^2/J m, C/V m, or s/Ohm m
ε_r	dielectric constant
ε	perturbation parameter or energy to remove a unit volume of material
ζ	zeta potential or Fermi energy
η	fraction impacted
Θ	dimensionless temperature
θ	angle or polar angle or momentum thickness
κ	thermal conductivity
$\bar{\kappa}$	thermal conductivity of material
κ_{km}	thermal conductivity of phase k in the mixture
Λ	Lagrangian microscale or parameter as defined
λ	mean free path, or wavelength
λ_D	Debye shielding distance
μ	viscosity
$\bar{\mu}$	viscosity of material
μ_{km}	viscosity of phase k in the mixture
ν	Poisson ratio of material
$\bar{\nu}$	kinematic viscosity of fluid material
ν	frequency
ξ	position or function as defined
ρ	density
$\bar{\rho}$	density of material

σ	surface tension or normal stress or sticking probability
σ_e	electrical conductivity
τ	time or shear stress
Φ	impaction parameter for non-stokesian behavior
ϕ	angle or azimuthal angle
ϕ	thermionic work function
χ	energy level
Ψ	inertia parameter for impaction
ψ	stream function or general physical quantity
Ω	solid angle
ω	circular frequency

Unsuscripted quantities for fluid phase

Bold face	vector or tensor
$<\ >$	average of a quantity
$i_{<\ >}$	intrinsic average
$t_{<\ >}$	time average
$(\bar{\ })$	material property or mean value
$(\dot{\ })$	time rate
$(\ddot{\ })$	second derivative with time

Subscripts

B	bubble
b	brittle ir bed
c	concentration or contact
D	drag
d	diffusion or ductile
E	electric field
e	evaporation or electron
f	fluid phase or saturated liquid
g	vapor or saturated vapor
fg	change due to vaporization
i,j,k	components
i	ion or incoming
k	phase k
km	of phase k in the mixture
L	lift
l	liquid
M	mass
m	mixture or momentum
N	number

n	terms in series
o	reference value
p	particulate phase
q	heat or charge
r	relative motion or reflected
s	saturation or surface or interface
t	terminal
v	at constant volume
w	wall
1,2	phases or particles

Superscripts

*	dimensionless, reduced, or characteristic quantities
+	dimensionless quantities

Chapter 1

INTRODUCTION

1.1 Scope and Applications

Physical phases of matter consist of vapor, liquid, and solid of various crystalline forms. In a flow system with a mixture of phases, multiple interactions occur at the interfaces of various configurations or geometries. Within a given physical phase in the mixture, each element of similar dynamic response constitutes a dynamic phase in the flowfield and the potential field of the system. Multiphase flow arises from a multiplicity of elements of different dynamic responses. An example is a gaseous suspension of solid particles of identical material; particles of different sizes constitute different phases because of their different inertial responses to a changing flowfield. This multiphase concept is also applicable to the case of a suspension of monodispersed (identical sized) particles with different electrostatic charges; in a given electric field, particles of different charges constitute different phases because of different forces acting on them [Soo, 1965]. Similar examples can be drawn from the cases of bubbles, droplets, and particles suspended in a fluid, be it a gas or liquid, as long as the phases are immiscible. In general, these phases can be separated by a variety of configurations of interfaces.

The dynamics of multiphase systems include momentum, energy, mass and charge transfers between the phases, whether or not they are in the presence of a potential field. There are many multiphase systems among engineering equipment and processes. Here are a few examples:

1. Gas-solid particle systems: pneumatic conveyors, dust collectors; fluidized beds, heterogeneous reactors; metallized propellant rockets, aerodynamic ablation; xerography; cosmic dusts, nuclear fallout problems.
2. Gas-liquid droplet systems: atomizers, scrubbers, dryers, absorbers, combustors; agglomeration, air pollution; gas cooling, evaporation; cryogenic pumping, desuperheating.
3. Liquid-gas bubble systems: absorbers, evaporators, scrubbers, air lift pumps, cavitation, flotation, aeration; loss of coolant accident in nuclear reactors.
4. Liquid-gas systems: Boiling and nuclear reactor safety; surface waves of air over water.
5. Liquid-liquid systems: extraction, homogenizing, emulsifying.
6. Liquid-solid particle systems: fluidized beds, flotation, sedimentation.

We note that other than the need to account for the general interface configuration in treating the nuclear reactor safety problem and surface wave problem, multiphase systems, in most cases, concern clouds of particulates in a fluid phase, liquid or vapor. The magnitudes of sizes in multiphase systems in relation to other physical quantities are shown in Fig. 1.1. Solid particles are, in general, nonspherical but often isometric; further approximations are needed for treating fibers or sticks. Small liquid droplets usually attain a spherical shape due to surface tension. When gravity or other field forces are significant, they attain shapes of smallest flow resistance such as falling rain drops or minimum potential in the case of charged droplets. Gas bubbles would retain spherical shapes due to surface tension until modified by field forces. Foams or large nonspherical bubbles are further exceptions.

To account for a large variety of combinations of components to form a mixture, we shall express its properties in terms of those of

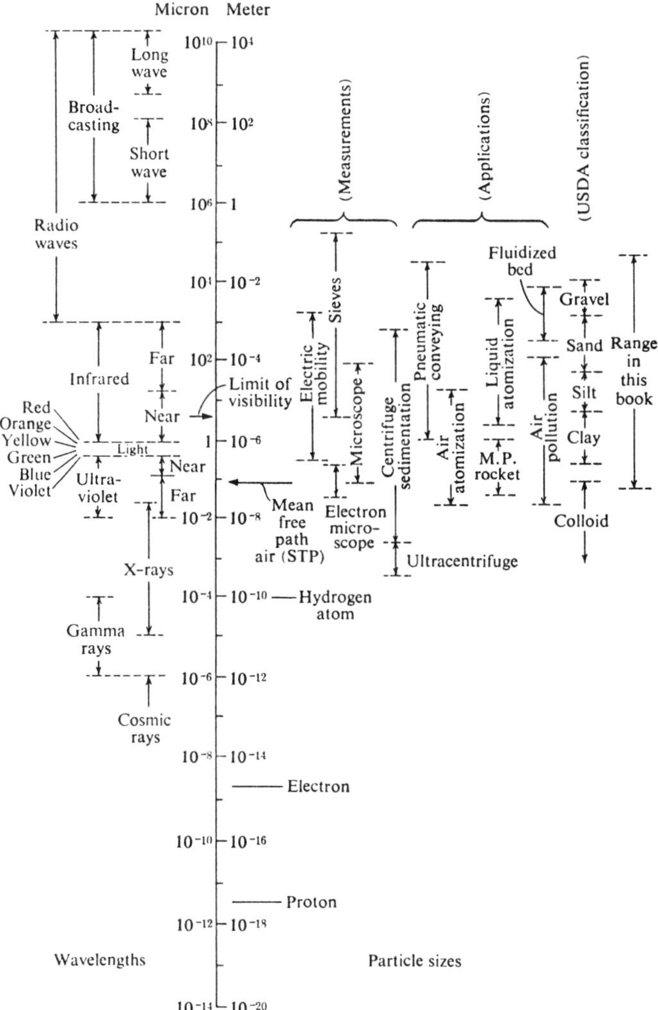

Figure 1.1　Magnitudes of particle sizes in multiphase systems in relation to other physical quantities.

the components.   In preparation for treating general mixtures, re-capitulation is made next on the elementary behaviors of a single particle in a fluid.   Several books treat single particle behaviors, generation, characterization, and instrumentation in detail (examples are Cadle [1965], Happel and Brenner [1965], Orr [1966]).

1.2 Elementary Particle-fluid Interactions

Studies of the magnitude of drag force on a sphere in steady motion of a viscous fluid began with Newton's experiment in 1710 which gave

$$F = 0.22 \, \pi \, a^2 \, \bar{\rho} \, U^2 \tag{1.1}$$

at a high relative velocity U where F is the force exerted on the sphere, a is the radius, and $\bar{\rho}$ is the density of the fluid material.   This relation mainly accounts for the inertia effect.

At very low relative velocities, Stokes [1891] suggested that the inertia effects were so small that they could be neglected from the Navier-Stokes equations.   His asymptotic approximation thus obtained gives a symmetric flowfield about a sphere.   The resulting drag force is

$$F = 6 \, \pi \, \bar{\mu} \, a \, U \tag{1.2}$$

where $\bar{\mu}$ is the viscosity of the fluid material, and two-third of F is the shear force (friction drag), and one-third of F is due to the pressure force (form drag) [Lamb, 1932] (Problem 1.1).

With the definition of the drag coefficient, C_D,

$$C_D = F/(\pi \, a^2)(\bar{\rho} \, U^2/2) \tag{1.3}$$

we get, for Newton's relation,

$$C_D = 0.44 \tag{1.4}$$

and for Stokes' law regime,

$$C_D = 24/(2\ a\ \bar{\rho}\ U/\bar{\mu}) = 24/N_{Re} \qquad\qquad (1.5)$$

where the Reynolds number, N_{Re}, is defined as

$$N_{Re} = 2\ a\ \bar{\rho}\ U/\bar{\mu} \qquad\qquad (1.6)$$

based on the diameter of the sphere.

Subsequent experiments yielded a so-called standard drag curve applicable to a fixed single solid sphere in an isothermal, incompressible fluid of infinite extent at constant velocity. The diagram in Fig. 1.2 shows that by comparison to the standard drag curve, the Stokes regime applies to $N_{Re} < 1$ nearly and Newton's regime holds nearly for $700 < N_{Re} < 2\times10^5$. Beyond $N_{Re} \sim 10^5$ (upper critical Reynolds number), a sharp reduction in drag coefficient occurs due to transition from laminar to turbulent boundary layers over the sphere.

One notes that above $N_{Re} \sim 1$, an analytical description of the flowfield around a sphere becomes complicated. Oseen's approximation includes inertial terms only in the flowfield away from the body, i.e.:

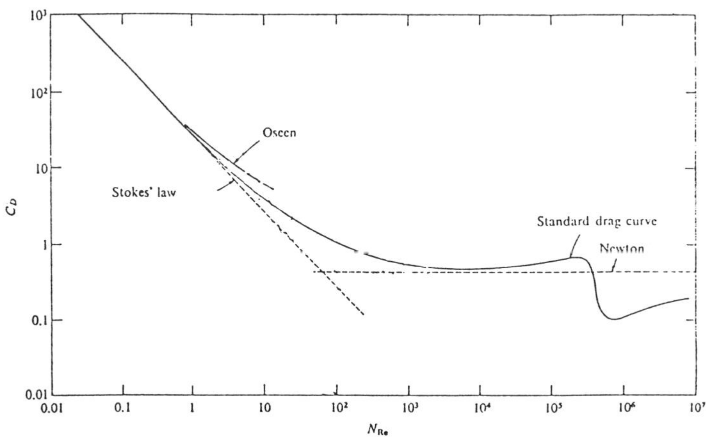

Figure 1.2 Drag coefficient of a sphere.

$$C_D = (24/N_{Re}) \, [1 + (3/16)N_{Re}] \tag{1.7}$$

which is applicable to $N_{Re} < 5$. As inertial forces become in-
creasingly significant, separation of laminar boundary layers occurs
at $N_{Re} \sim 10$; trailing vortices become prominent at $N_{Re} > 50$, with
growth of the vortices in size and strength as N_{Re} increases, with
oscillation of the vortex system. At $N_{Re} \sim 500$, called the lower
critical Reynolds number, the vortex system leaves the body and forms
a wake. The vortex rings continue to form and shed from the sphere,
giving rise to a periodic flowfield and instantaneous drag force.
This periodic flowfield gives rise to a transverse fluctuating force
[Schlichting, 1960]. In the case of a freely suspended particle in a
fluid, this fluctuating force gives rise to oscillatory motion of the
particle and local induced turbulence in the wake.

The turbulent intensity encountered in a practical situation may
amount to a few percent of the stream velocity [Laufer, 1951]. The
corresponding turbulent intensity of the particle flowfield decreases
with an increase of the ratio of the stream velocity to the relative
velocity between the particle and the fluid. The scale of turbulence, however, is unaltered. Moreover, turbulence may be generated
by the relative motion of the particle to the fluid. In the case of
a freely suspended sphere, the part of the turbulence spectrum having
an amplitude greater than the diameter of the sphere will contribute
to its oscillatory motion, while fluid elements having an amplitude
smaller than its diameter will contribute to the drag force.

<u>Rarefied Gas</u>. When the particles are small compared to the mean free
path of the gas λ, slip motion of gas molecules occurs at the surface
of the sphere, resulting in lowered drag force. The state of the gas
is known as a rarefied gas. The drag force can be give by a simple
modification suggested by Kundt and Warburg [1875]. The drag coefficient for slip flow is given by dividing the C_D obtained in
Eq. (1.5) by

$$C = 1 + A \, N_{Kn} + B \, N_{Kn} \, \exp(-b \, N_{Kn}^{-1}), \tag{1.8}$$

the Cunningham correction factor [Strauss, 1966], where the Knudsen number $N_{Kn} = \lambda/2a$, the empirical coefficients are A = 1.257, B = 0.4, and b = 1.10.   When λ is much greater than 2a, free molecule flow occurs; the drag coefficient is ideally 2.

<u>Inverse Relaxation Time For Momentum Transfer</u>. When a free spherical particle of mass m and initially at zero velocity is introduced into a fluid at a constant velocity U_o, it will be accelerated according to the relation

$$d\ U_p/dt = F(U_o - U_p) \tag{1.9}$$

where t is the time, F is the inverse relaxation time for momentum (linear) transfer from the fluid to the particle:

$$F = (\text{Drag force})/m \text{ (velocity difference)}$$

and has the dimension of $(\text{time})^{-1}$. Equation (1.9) has the solution

$$U_p = U_o(1 - e^{-Ft}) \tag{1.10}$$

which means that it takes a time interval of (1/F) for the velocity difference between the particle and the fluid to reduce to (1/e) of the initial value; (1/F) is called the relaxation time. The general value of F is given by

$$F = (3/8)\ C_D(\bar{\rho}/\bar{\rho}_p)|U - U_p|/a \tag{1.11}$$

and

$$F = 9\ \bar{\mu}/2a^2\ \bar{\rho}_p \tag{1.12}$$

for relative motion in the Stokes regime given by Eq. (1.5) (Problems 1.2 and 1.3)

1.3 Heat and Mass Transfer from a Sphere

When a rigid sphere at a given temperature is submerged in a completely quiescent fluid at another temperature, the rate of heat transfer due to conduction alone is readily shown to be given by the Nusselt number

$$N_{Nu} = 2\ a\ h/\bar{\kappa} = 2 \tag{1.13}$$

where 2a is the diameter, h is the heat transfer coefficient, and $\bar{\kappa}$ is the thermal conductivity of the fluid.

When relative motion of the fluid to the sphere exists, the local heat transfer coefficient depends on the local velocity and temperature distributions and flow separation. All the variables affecting the flowfield such as turbulence, rarefaction, variable fluid properties, and radiation also affect the heat transfer. Theoretical solutions are available only for Reynolds numbers where Stokes law applies and are typically given in the form

$$N_{Nu} = 2 + (1/2)\ N_{Re}N_{Pr} + (1/6)(N_{Re}\ N_{Pr})^2 + \ldots \tag{1.14}$$

where N_{Pr} is the Prandtl number of the fluid; $N_{Pr} = c\bar{\mu}/\bar{\kappa}$, and c is the specific heat of the fluid at constant pressure [Kronig and Bruijsten, 1951].

Experimental determinations were made for high Reynolds numbers. A good fit was proposed by Drake [1961]

$$N_{Nu} = 2 + 0.459\ N_{Re}^{0.55}\ N_{Pr}^{0.33} \tag{1.15}$$

for $1 < N_{Re} < 70,000$ and $0.6 < N_{Pr} < 400$, for a single particle in an incompressible fluid of infinite extent and in the limit of very small temperature difference. It was found that turbulence was most effective in increasing the heat transfer rate when the scale of turbulence was 1.5 times of the particle diameter [Van der Hegge Zijnen, 1958].

Rarefaction effects are expected for small particles in a gas. Imperfect accommodation and temperature jump reduce the heat transfer rate. With the Oseen approximation for the flowfield, Sauer [1951] gave

$$N_{Nu} = N_{Nu}^O \, [1 + 3.42 \, (N_M/N_{Re} \, N_{Pr}) \, N_{Nu}^O] \tag{1.16}$$

for an accommodation coefficient of nearly 0.8. An accommodation coefficient is defined as the ratio of the difference of the energy of incoming molecules and reflected molecules to that of the energy of incoming molecules and those at the wall temperature; this gives a temperature jump corresponding to the velocity slip in the above. In Eq. (1.16), N_M is the Mach number of the flow stream, while N_{Nu}^O is the Nusselt number based on Eq. (1.15) which is for $N_M = 0$. It is seen that N_M/N_{Re} is related to the Knudsen number, N_{Kn}, according to

$$N_{kn} = (\gamma\pi/2)^{1/2} \, N_M/N_{Re} = \lambda/2a \tag{1.17}$$

where γ is the ratio of specific heats of the gas.

<u>Inverse Relaxation Time for Heat Transfer</u>. In a manner analogous to the relaxation mechanism of momentum transfer, the inverse relaxation time for convective heat transfer from fluid to particle G is given by the relation

$$d \, T_p/dt = G(T_0 - T_p) \tag{1.18}$$

for a particle initially at T_{po} and fluid at constant temperature T_0; and

$$G = (\text{Heat transferred to particle})/(\text{its heat capacity}) \, |T - T_p|.$$

Equation (1.18) has the solution

$$(T_p - T_{po}) = (T_0 - T_{po}) \, (1 - e^{-Gt}) \tag{1.19}$$

and G is given by, for a spherical particle,

$$G = 4\pi a^2 h/m \, c_p = (3/2) \, N_{Nu} \, \bar{\kappa}/c_p \, \bar{\rho}_p a^2 \tag{1.20}$$

where c_p is the specific heat of the particle material.

Mass Transfer. Analogous to heat transfer, mass transfer to or from a sphere placed in a stagnant fluid is given by Langmuir [1918] as

$$N_{Sh} = 2 \, a \, k_c/D = 2 \qquad\qquad (1.21)$$

where N_{Sh} denotes the Sherwood number, D is the diffusivity of molecular species being transferred, k_c is the mass transfer coefficient given in (number of kmol)/(unit area)(unit time)(unit concentration in kmol/unit volume) such that the rate of mass transfer $\dot{N}$ in (number of kmol/area, time) is given by

$$\dot{N} = k_c(C_i - C) \qquad\qquad (1.22)$$

where C is the concentration of the diffusing component and subscript i denote equilibrium values based on the temperature of the surface (Problem 1.4). When relative motion exists, the effect of convection is given by Froessling [Sherwood and Pigford, 1952] from data on evaporation of small droplets of liquids and solid spheres of naphthalene as

$$N_{Sh} = 2ak_c/D = 2[1 + 0.276 \, N_{Re}^{1/2} \, N_{Sc}^{1/3}]. \qquad\qquad (1.23)$$

The quantity $\mu/\rho D$ has been denoted as the Schmidt number N_{Sc}. Analogous correlations have been made on adsorption [Brunauer, 1943].

1.4 Deformable Particles

The drag force of a fluid medium on a fluid sphere in relative motion is influenced by velocity distributions in both phases. Earlier theoretical studies were confined to creeping motion in the Stokes law regime (Section 1.2). Hadamard [1911] and Rybczynski [1911] gave the drag force as

$$F = 6\pi a\bar{\mu} \, U_o(2\bar{\mu} + 3\bar{\mu}_p)/3(\bar{\mu} + \bar{\mu}_p) \qquad\qquad (1.24)$$

where $\bar{\mu}$ and $\bar{\mu}_p$ are the viscosities of fluids outside and inside the sphere, U_0 is the velocity of the center of the sphere relative to the fluid medium infinite in extent. Chao [1962] applied the thin boundary layer approximation to motion inside and outside the sphere giving the drag coefficient C_D in the form

$$C_D = (32/N_{Re}) \ [1 + 2 \ \bar{\mu}_p/\bar{\mu} - 0.314 \ (1 + 4 \ \bar{\mu}_p/\bar{\mu}) \ N_{Re}^{-1/2}] \qquad (1.25)$$

where $N_{Re} = 2aU_0/\mu > 10$. Experimental studies showed that a bubble or droplet is nearly spherical even at relatively large Reynolds numbers, provided that the Weber number ($N_{We} = 2a\bar{\rho}U_0^2/\sigma$; σ is the surface tension at the interface) remains small [Moore, 1959]. Haberman and Morton [1953] proposed a drag curve for fluid spheres or, more precisely, gas bubbles by drawing the lower envelope of all experimental curves available. Figure 1.3 shows this drag curve in comparison with theoretical relations; ellipsoidal and spherical cap bubbles are also included. The motion of small bubbles below 1.2 mm in diameter is controlled by viscosity and surface tension; induced turbulence appears to be one of the controlling factors in the motion of large nonspherical bubbles.

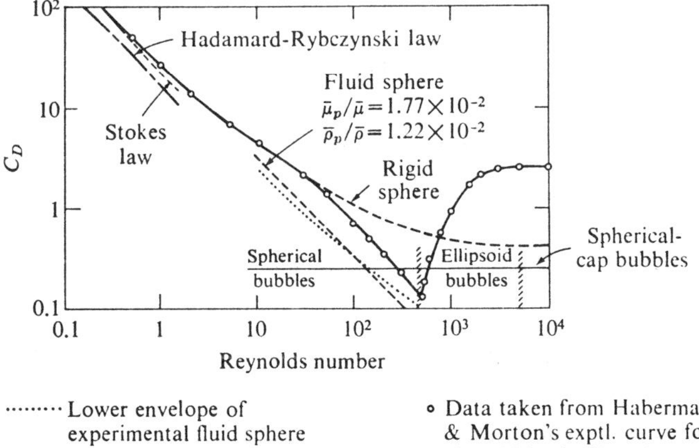

.... Lower envelope of experimental fluid sphere (gas bubble) data proposed by Haberman & Morton

o Data taken from Haberman & Morton's exptl. curve for air bubbles in filtered water at 19° C

Figure 1.3 Drag coefficients of air bubbles rising at their terminal velocity in filtered or distilled water [Chao, 1962].

Mass Transfer. The heat and mass transfer of a fluid sphere moving in a continuous fluid medium are affected by the motion inside the sphere. Hammerton and Garner [1954] showed that the transfer rate from circulating bubbles of slightly soluble pure gas was about five times as large as from noncirculating bubbles. Bowman, et al. [1961], computed the effect of internal circulation according to Hadamard [1911] and gave the Sherwood number in the form

$$N_{Sh} = 2 + (9/16) \ N_{Pe} + (9/64) \ N_{Pe}^2 \qquad (1.26)$$

for a thick boundary layer or small Peclet number ($N_{Pe} = N_{Re} \ N_{Sc} = 2aU_o/D$). An expression similar to Eq. (1.14) was obtained for a solid sphere with the Sherwood number replacing the Nusselt number and with N_{Pe} replacing $N_{Re} \ N_{Pr}$. For a thin boundary layer or a large N_{Pe},

$$N_{sh} = 0.978 \ N_{Pe}^{1/3} \qquad (1.27)$$

with the coefficient replaced by 0.89 for a solid sphere. Note that contaminants at the interface have an important effect on the transfer rate by their influence on the internal circulation [Hammerton and Garner, 1954]. The rate of dissolution of a gas bubble was determined by Ledig and Weaver [1923] to be

$$N_{Sh} = (2/\pi)^{1/2} \ N_{Pe}^{1/2} \qquad (1.28)$$

A general relation for the evaporation of a droplet was given by Frössling [1938] as

$$d \ a^2/dt = (da^2/dt)_o \ [1 + 0.276 \ N_{Sc}^{1/3} \ N_{Re}^{1/2}] \qquad (1.29)$$

where $(da^2/dt)_o$ represents the evaporation rate in still air. This relation is also applicable to the burning of a fuel droplet [Faeth, 1967]. The burning rates of liquid fuels and metals such as aluminum and magnesium are often evaporation rate controlled.

The shape of a distorted droplet under the action of hydrodynamic forces was reported by Savic [1953] for liquid droplets in an air stream. Mason [1957] suggested that a water droplet below 1 mm in radius in free fall can be treated as a rigid sphere.

<u>Mass and Heat Transfer between Gas and Liquid</u>. Evaporation and condensation involving a bubble or a droplet involves simultaneous heat and mass transfer. Finite rate of evaporation takes place from a superheated liquid, and rapid condensation occurs in a vapor which is at least slightly supersaturated. Phase change at a finite rate involves nonequilibrium states.

In a general form, the rate of evaporation or condensation of a bubble of a gas k in a liquid ℓ can be given by

$$d\ m_k/dt = (m_k/\overline{\rho}_k)(3N_{Sh}\ D_k/2a_k^2)(\overline{\rho}_g - \overline{\rho}_k) \qquad (1.30)$$

where $\overline{\rho}_k = \overline{\rho}_k(T_k, P_k)$ T_k is the temperature and pressure P_k includes the effect of curvature of the interface, and $\overline{\rho}_g$ is the density of a saturated vapor at T_g. The pressure in the phases are related by

$$P_k = P_\ell\ (2\sigma/a_k) + 4\overline{\mu}_\ell(\dot{a}_k/a_k) \qquad (1.31)$$

where a_k is the radius and $\dot{a}_k$ is the rate of bubble growth. T_g in the above equation is the saturation temperature at the surface corresponding to P. The liquid might be superheated and at a temperature T_ℓ. Among the above states, superheated liquid and supersaturated vapor are unstable equilibrium states while compressed liquid and superheated vapor are stable equilibrium states. Transition occurs when nonequilibrium occurs among these states. Figure 1.4 illustrates these states in corresponding pressure-volume and enthalpy-entropy diagrams using extrapolation by means of the van der Waals equation of states (Problems 1.5, 1.6).

Equation (1.30) is not often conveniently applied because N_{Sh} for this particular process is not readily determined for bubble growth by evaporation. Rohsenow [1961] gave

$$dm_{ek}/dt = 3(m_k/a_k^2)(\overline{k}_\ell/c_{p\ell}\ \overline{\rho}_\ell)\ N_{Ja}^2 \qquad (1.32)$$

where dm_{ek}/dt is the rate of change of bubble mass m_k, h_{fg} is the latent heat of evaporation at P_ℓ, and $\bar{k}_\ell/c_{p\ell}\bar{\rho}_\ell$ is the thermal diffusivity of the liquid and the Jakob number N_{Ja} is given by

$$N_{Ja} = c_{p\ell}(T_\ell - T_f)\,\bar{\rho}_f/h_{fg}\,\bar{\rho}_k \qquad (1.33)$$

where $T_f = T_f(P_\ell)$ is the saturation temperature.   When $T_\ell < T_f$, condensation occurs with loss from species k.   However, evaporation produced by superheated vapor in a bubble, say, due to adiabatic compression, is readily accounted for by the kinetic theory of effusion, or for gas constant $\bar{R}$, and $T_g = T_g(P_k)$,

$$dm_{ek}/dt = 3(m_k/\rho_k a_k)\,C(2\pi\bar{R})^{-\frac{1}{2}}(P_g\,T_g^{-\frac{1}{2}} - P_k T_k^{-\frac{1}{2}}) \qquad (1.34)$$

with the coefficient C less than 1 at ordinary pressures.   Droplet growth by mass transfer is, in many ways, analogous to the relations of bubble growth and is left for an exercise problem (Problem 1.7).

<u>Bubble Dynamics</u>.   Bubble growth also occurs in a liquid by adiabatic expansion or compression as the pressure of the system changes as

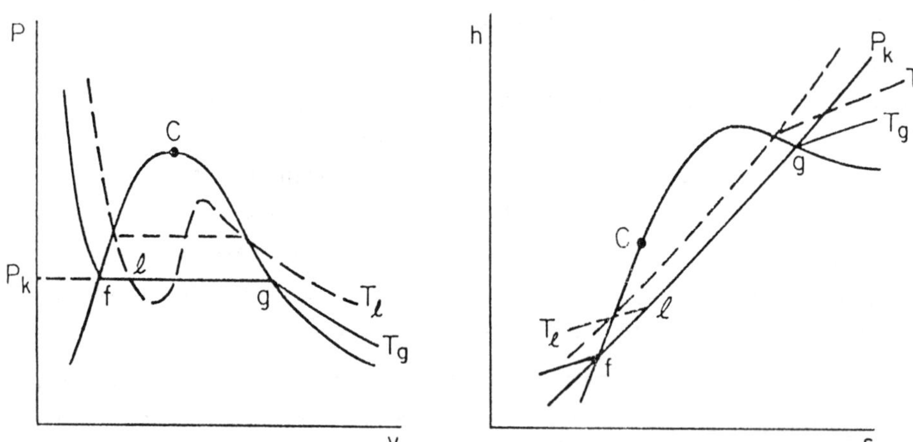

Figure 1.4   Nonequilibrium states ℓ in bubble growth as liquid is superheated at constant pressure as seen in corresponding pressure-volume and enthalpy-entropy diagrams--gives states represented by van der Waals equation of states.

well as via cooling or heating of the liquid. The initial phase of expansion of a free bubble is in the inertia controlled range. Rayleigh [1917] gave the rate of change of bubble radius as

$$da_k/dt = (2/3) \ [(P_{\ell O} - P_\ell)/\bar{\rho}_\ell]^{\frac{1}{2}} \tag{1.35}$$

where $P_{\ell O}$ is the initial liquid pressure which may change to $P_\ell(t)$ as a function of time. For this expansion from an initial equilibrium state, $T_{\ell O} = T_{fo} \ (P_{\ell O}) = T_{ko}$, T_{fo} being the saturation temperature at pressure $P_{\ell O}$. The initial expansion after the pressure drops to P_ℓ produces a non-equilibrium condition such that momentarily $T_\ell = T_{\ell O}$ and the liquid becomes superheated. Discussion of the trend is left for an exercise problem (Problem 1.6). Dynamics of bubble growth are also influenced by the effects of heat and mass transfer; the details, however, are beyond the scope of this text.

1.5 Momentum Transfer in a Nonuniform Fluid

Other factors which influence the motion and drag coefficient of a solid particle are velocity gradient, pressure gradient, temperature gradient, nonuniform radiation, and concentration gradient.

<u>Velocity Gradient</u>. Saffman [1965] computed the lift force acting on a sphere by various liquids at a low velocity U in simple shear at a rate dU/dy parallel to the streamlines and gave

$$F_L \sim K \ \bar{\mu} \ U(dU/dy)^{\frac{1}{2}} \ a^2/\bar{\nu}^{\frac{1}{2}} \tag{1.36}$$

normal to the flow direction; K = 6.46 was obtained from numerical integration for $U/(\bar{\nu} \ dU/dy) \ll 1$; $\bar{\nu} = \bar{\mu}/\bar{\rho}$ is the kinematic viscosity of the fluid. Saffman showed that unless the rotating speed is very much greater than the rate of shear for a freely rotating particle of angular velocity $\omega = (1/2)(dU/dy)$, the lift force due to particle rotation is less by an order of magnitude than that due to the shear. His conclusion is valid when the Reynolds number based on particle diameter is small. The dependence on viscosity is such that the drift velocity should be proportional to $N_{Re}^{2/3}$ if U is independent of $\bar{v}$. For a given spinning particle, the Magnus force could be significant [Rubinow and Keller, 1961] (Problem 1.8).

For the case of shear lift effect in a turbulent boundary layer, correlation [Soo and Tung, 1972] of experimental data of Graf and Acaroglu [1968] on erodible bed of sand in turbulent fluid gave

$$F_L = 25 \; \pi \; \bar{\mu} \; |(a/U_r)(dU/dy)]^{\frac{1}{2}} \; U_r \; a \qquad (1.37)$$

for relative velocity U_r between the fluid and the particle.

Pressure Gradient. It is readily shown that for a sphere in a static pressure gradient, the resultant force acting on the sphere is in the opposite direction of the pressure gradient; i.e.,

$$F = (4\pi a^3/3) \; (\partial P/\partial x). \qquad (1.38)$$

Note that one has to be careful in applying this relation, as a pressure gradient cannot exist arbitrarily in a fluid (in the absence of a potential field) without a corresponding velocity distribution which tends to counteract the above force [Soo, 1975,1976] (Problem 1.9). Passage of a particle through a shock wave with increase in temperature and viscosity of the gas may constitute an exception.

Temperature Gradient. Temperature gradient in a fluid may contribute to the variation of viscosity of the fluid. This variation may also arise due to heat transfer and the corresponding temperature distribution.

The forces acting on a particle due to a temperature gradient in a gas (thermophoresis) and nonuniform radiation (photophoresis) are called radiometric forces. For $a \ll \lambda$, that is, in the free-molecule flow range, this arises from the collision of gas molecules with different mean velocities from opposite directions and is given by [Cawood, 1936]

$$F_{Ti} = -(\pi P \; \lambda \; a^2/2T)(\partial T/\partial x_i)$$

$$= -(\pi/2) \; \rho \; \lambda \; a \; \bar{R} \; (\partial T/\partial x_i) \qquad (1.39)$$

for an accommodation coefficient α of 1; P is the pressure, $\bar{R}$ is the gas constant of the gaseous phase. Nonuniform radiation gives rise to a temperature gradient inside the particle of $\partial T_p/\partial x_i$. In the free-molecule flow range, Rubinowitz [1920] showed that

$$F_{Ti} = -(\pi\, \alpha\, P\, a^3/3T)\ (\partial T_p/\partial x_i) \qquad (1.40)$$

and the temperature gradient inside a spherical particle is given by [Epstein, 1929] as

$$\partial T_p/\partial x_i = 3\,\bar{\kappa}\ (2\bar{\kappa} + \bar{\kappa}_p)^{-1}\ (\partial T/\partial x_i) \qquad (1.41)$$

where $\bar{\kappa}$ is the thermal conductivity of the gas and $\bar{\kappa}_p$ is that of the solid material.

For the case $a \gg \lambda$, the velocity in this case rises to a maximum at a distance λ from the surface and then decreases with distance. The force is directed toward the lower temperature. Hydrodynamic considerations by Hettner [1926] lead to

$$\mathcal{F}_{Ti} = -(3\pi\, \bar{\mu}^2\, a\, \bar{R}/P)\ (\partial T_p/\partial x_i) \qquad (1.42)$$

Concentration Gradient. When a small particle or droplet is in a concentration gradient of a vapor such as having its surface wetted by a volatile liquid or condensation of a supersaturated vapor on a droplet, a relative motion, known as Stefan flow, was found [Fuchs, 1959]. For a spherical droplet or particle, the velocity of relative motion is given by

$$U = D\ (C_s - C_\infty)\ a\ M/C\ r^2\ M' \qquad (1.43)$$

where D is the diffusivity of the gas in the gas-vapor mixture, M is the molecular weight of the gas, M' that of the vapor; r is measured from the center of the droplet; C_s is the concentration of the vapor at the surface, C_∞ is that at $r = \infty$, and C is the concentration of the gas. This phenomenon is called diffusion phoresis. Thus an evaporating wall will tend to drive a droplet away.

For a $\ll \lambda$, Waldmann [1959] gave the terminal velocity U_t of an aerosol due to concentration gradient in a binary mixture of components 1 and 2 as

$$U_t = \frac{[(M_1/M_2)^{\frac{1}{2}} \delta_1 - \delta_2] D}{[C_1 (M_1/M_2)^{\frac{1}{2}} \delta_1 + C_2 \delta_2]} \nabla C_1 \qquad (1.44)$$

where $\delta = 1 - (\pi a/8)$, where a is the fraction of gas molecules reflected diffusely by the particle.

In air at room condition, the effect of thermophoresis and diffusion phoresis are significant only for particles in the submicron size range.

<u>Electrophoresis</u>.   Displacement of a dielectric particle in a dielectric fluid in an electric field E is known as electrophoresis.   Smoluchowski equation gives a velocity of

$$U_t = \zeta E[\bar{\mu} + (Q_d^2 d/\sigma_e a)]^{-1} \qquad (1.45)$$

where ζ is the zeta-potential across the mobile portion of the electric double layer of thickness d, $\zeta = Q_d d$, Q_d is the charge of the double layer in C/m^2, μ is the viscosity of the fluid, and σ_e is the electrical conductivity [Levich, 1962].   The viscous effect predominates for $Q_d^2 d/\sigma_e \bar{\mu}$ a $\ll 1$ while the effect of conductivity is significant for the same quantity $\gg 1$; the above velocity has a maximum at $Q_d = (\sigma_e a \bar{\mu}/d)^{\frac{1}{2}}$.

1.6 Motion of a Particle Suspended in a Fluid

Brownian motion of particles in a liquid suspension was studied by Einstein [1906] and Lorentz, [1896].   This interaction of a particle with its surrounding fluid is due to thermal random motion and statistical equipartition of energy.   The average kinetic energy of a particle per degree of freedom is given by (1/2)kT where k is the Boltzmann constant and T is the absolute temperature.   This random motion was given by Einstein [1956] in terms of root-mean-squared displacement $<(\Delta x)^2>$ over a long time τ as

$$<(\Delta x)^2>/\tau = 2kT/6 \pi \bar{\mu} a = D \qquad (1.46)$$

where D is the diffusivity of the particle in the fluid of viscosity $\bar{\mu}$. Equation (1.46) is valid for random motion with drag force in the Stokes law regime. Brownian motion is significant only for particles in the micrometer range or smaller.

Studies of relative motion of a solid sphere suspended in a viscous fluid began with that of Boussinesq [1885]. The sphere is large enough such that Brownian motion is negligible. The resistant forces consist of that of the effect of virtual mass (Problem 1.10), the Stokes drag, and that due to an unsteady flowfield often referred to as the Basset [1888] force. These forces are given by corresponding terms on the right-hand side of the following equation:

$$m_p \, d \, U_p/dt = \frac{1}{2} v_p \, \bar{\rho}(d/dt)(U - U_p) + 6 \, \pi\mu a \, (U - U_p) + m_p \, f$$

$$+ 6(\pi \, \bar{\rho} \, \bar{\mu})^{\frac{1}{2}} \, a^2 \int_{-\infty}^{t} (d/d\tau) \, (U - U_p) \, (\tau - t)^{-\frac{1}{2}} \, d\tau \qquad (1.47)$$

causing the sphere of mass m_p to accelerate or decelerate along its path at a rate given by dU_p/dt of creeping motion. In Eq. (1.47), v_p is the volume of the sphere of radius a, f is the field force per unit mass, $\bar{\rho}$ is the density of the fluid, $\bar{\mu}$ its viscosity, U its velocity; U_p is the velocity of the sphere, and t is the time. These terms are cumulative because of the linearization afforded by specifying creeping motion.

Similar terms as in Eq. (1.47) are included in the formulation by Tchen [1947] for the general motion of a small sphere suspended in a turbulent fluid, assuming the same cumulative relation of these forces, with those due to pressure gradient (see Section 1.5) and external field in addition. It is further assumed that:

1. The particle is small when compared to the smallest wavelength of the turbulence; the effect of particle motion due to shear flow (Eq. (1.36)) is negligible.
2. The flowfield is not perturbed by the presence of the solid particle.
3. During the motion of the solid particle, the same fluid

element remains in its neighborhood. This assumption is rather restrictive and makes small particle size necessary. One has to visualize that the fluid element continues to contain the particle P as both proceed along a meandering path as shown in Fig. 1.5, as time t progresses, although the fluid element may undergo distortions. This specification makes time derivatives along the streamline of the fluid and the path of the particle identical.

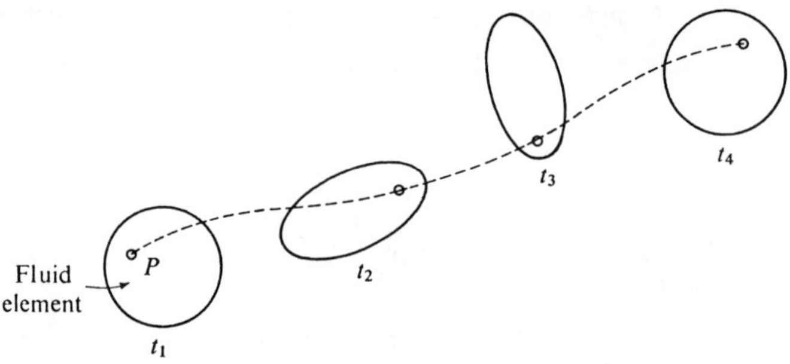

Figure 1.5 Tchen's assumption of transport of a particle in a fluid element.

The inherent inaccuracy of such a formulation is readily seen. It nevertheless marked the beginning of treating multiphase flow as different from molecular mixtures [Soo, 1970]. Advancements have been made from this step.

Particle-fluid Interaction. Some understanding of the basic interactions can be achieved via a simplified formulation and solution. Neglecting the Basset force term, the motion of the particle p is given by its velocity component U_p and that of the fluid U

$$d\, U_p/dt = F(U - U_p) \qquad (1.48)$$

and the virtual mass force is negligible for a particle suspended in a gas. Applying the concept of spectrum of turbulence [Taylor, 1937] with isotropy, the component u relative to mean velocity U of the stream can be represented by the velocity spectrum

$$u = \sum_n A_n \sin 2\pi nt \tag{1.49}$$

where A_n is the amplitude of the n-th harmonic and n is the frequency. Its mean squared value $<u^2>$ is given by

$$<u^2> = \lim_{T \to \infty} \frac{1}{2T} \int_{-T}^{T} u^2 dt = \frac{1}{2} \sum A_n^2 = <u^2> \int_0^\infty f_\eta(n) \, dn \tag{1.50}$$

where $<u^2>^{\frac{1}{2}}$ is the intensity of turbulence, $f_\eta(n)$ is the Lagrangian
statistical frequency function, and

$$\int_0^\infty f_\eta(n) dn = 1. \tag{1.51}$$

The correlation with time R_η is now

$$R_\eta = <u(t) \, u(t-\xi)>/<u^2> = \sum_n A_n^2 \cos 2\pi n\xi / \sum_n A_n^2 \tag{1.52}$$

and, by the definition of $f_\eta(n)$, can be expressed in terms of a
Fourier integral

$$R_\eta = \int_0^\infty f_\eta(\xi) \cos 2\pi n \, \xi \, dn = \int_0^\infty f_\eta(n) \cos 2\pi n\eta <u^2>^{-\frac{1}{2}} \, dn \tag{1.53}$$

where $dn = <u^2>^{\frac{1}{2}} dt$ and, for non-decaying turbulence, $\eta = <u^2>^{\frac{1}{2}}t$.
Hence R_η and $f_\eta(n)$ are Fourier transforms to each other. The concept
of diffusion by continuous movement gives [Taylor, 1922]

$$(1/2) \, d <x^2>/dt = <u^2> \int_0^T R_\xi \, d\xi = \int_0^\eta R_\eta \, d\eta \tag{1.54}$$

where $<x^2>$ is the average displacement. Further, the scale of turbulence of the fluid is given by

$$\ell_1 = \int_0^T R_\eta \, dn \tag{1.55}$$

and the microscale Λ as defined by the parabola in Fig. 1.6 is defined by [Taylor, 1935]

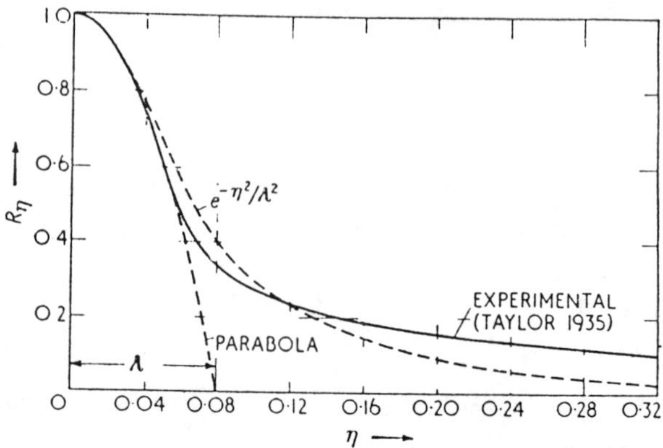

Figure 1.6 Approximations to Lagrangian correlation curve.

$$\lambda^{-2} = \lim_{\eta \to 0} (1 - R_\eta) \, \eta^{-2} \tag{1.56}$$

where λ is a measure of the size of the smallest eddies responsible for the dissipation of energy.

Based on assumption 3 in the above, the component u_p of U_p due to random excitation by the fluid is given by

$$d \, u_p/dt = F (u - u_p). \tag{1.57}$$

For u given by Eq. (1.49), Eq. (1.57) has the solution

$$u_p = \sum_n A_n [1 + (2\pi n/F)^2]^{-1} [\sin 2\pi nt - (2n/F) \cos 2\pi nt]$$

$$= \sum_n A_n [1 + (2\pi n/F)^2]^{-1} \sin (2\pi nt - \theta_n) \tag{1.58}$$

where θ_n is the phase lag and $\theta_n = \tan^{-1}(2\pi n/F)$. The intensity of turbulence of the particle $\langle u_p^2 \rangle^{1/2}$ is given by

$$\langle u_p^2 \rangle = \frac{1}{2} \sum A_n^2 [1 + (2\pi n/F)^2]^{-1} \tag{1.59}$$

and substitution of Eq. (1.50) gives

$$\langle u_p^2 \rangle = \langle u^2 \rangle \int_0^\infty [1 + (2\pi n/F)^2]^{-1} \, f_\eta(n) \, dn. \tag{1.60}$$

Further, it is readily shown that for relative velocity given by $\Delta u = u - u_p$, the relative intensity of motion of the particle to the fluid is given by substitution of Eqs. (1.49) and (1.58) and averaging

$$\langle (\Delta u)^2 \rangle = \langle u^2 \rangle - \langle u_p^2 \rangle. \tag{1.61}$$

The Lagrangian correlation of the particle motion is now

$$R_t = \frac{\langle u_p(t) \, u_p(t-\xi) \rangle}{\langle u_p^2 \rangle} = \frac{\sum_n A_n^2 [1 - (2\pi n/F)^2]^{-1} \cos 2\pi n\xi}{\sum_n A^2 [1 + (2\pi n/F)^2]^{-1}}. \tag{1.62}$$

Substitution of $\eta/\langle u^2 \rangle^{\frac{1}{2}}$ for ξ in Eq. (1.62) gives

$$R_\eta'(n) = \frac{\int_0^\infty [1 + (2\pi n/F)^2]^{-1} \, f_\eta(n) \, \cos(2\pi n\eta \langle u^2 \rangle^{\frac{1}{2}}) \, dn}{\int_0^\infty [1 + (2\pi n/F)^2]^{-1} \, f_\eta(n) \, dn} \tag{1.63}$$

from which the scale of turbulence can be computed for given $f_\eta(n)$.

As an approximation, we note that the curve of

$$R_\eta = \exp[-\eta^2/\Lambda^2] \tag{1.64}$$

as an approximation of the correlation function R has the same curvature at the apex of the parabola (Eq. (1.56))

$$R_\eta = 1 - (\eta^2/\Lambda^2) \tag{1.65}$$

in Fig. 1.6. Using the approximation in Eq. (1.64), Eq. (1.53) gives

$$f_\eta(n) = (2\sqrt{\pi} \, \Lambda/\langle u^2 \rangle^{\frac{1}{2}}) \, \exp(-\Lambda^2 \pi^2 n^2/\langle u^2 \rangle) \tag{1.66}$$

and

$$\ell_1 = (\sqrt{\pi}/2) \, \Lambda \, . \tag{1.67}$$

Equation (1.60) can be integrated in terms of a Laplace transform [Churchill, 1944] to give

$$\langle u_p^2 \rangle / \langle u^2 \rangle = \sqrt{\pi}\ \exp(K^{-2})\ \mathrm{erfc}(K^{-1})/K \tag{1.68}$$

where the particle-fluid interaction parameter K is given by

$$K = 2\langle u^2 \rangle^{1/2}/\Lambda F = (\sqrt{\pi}/18)\ \langle N_{Re} \rangle\ (2a/\ell_1)\ (\tfrac{1}{2} + \tfrac{\overline{\rho_p}}{\overline{\rho}}) \tag{1.69}$$

where $\langle N_{Re} \rangle$ is the characteristic Reynolds number of the particle based on random motion in the fluid. Figure 1.7 shows the relation in Eq. (1.68) in terms of a difference in intensities between the fluid and the particle versus K. Experimental data shown in Fig. 1.7 were obtained by Soo, et al. [1967], from measurements made on a 76 mm x 76 mm square duct with air flow at 15 m/s to 26 m/s suspending glass spheres with 105 to 125 and 210 to 250 micrometer size range. Determination of intensities and scales of motion were made on photo-graphic records of successive exposure with an 8000 cycles per seconds stroboscope with direct measurements on their enlargements [Soo, et al., 1960] and optical autocorrelation [Soo, et al., 1959]. Data shown were those under a small influence of gravity. The intensity of turbulence of the fluid was given by Laufer [1954] for core flow at velocity U_o in a pipe with $\langle u^2 \rangle^{1/2}/U_o = 0.026$ nearly.

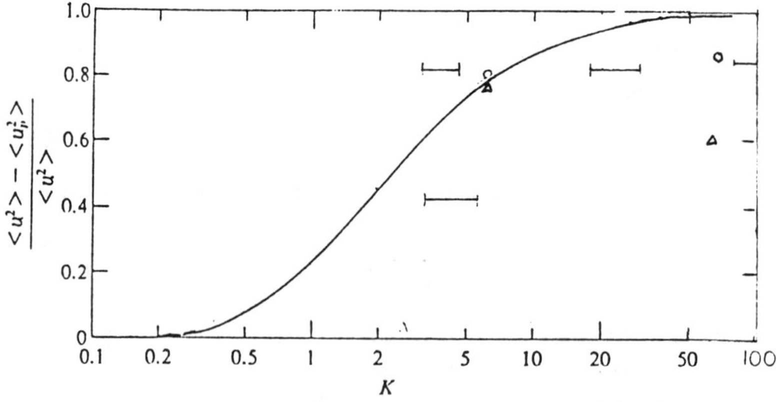

Figure 1.7   Deviation of local intensity of motion
             of a particle from that of the stream.

Using a time-dependent form of Eq. (1.66), Reeks [1977] obtained a transient dispersion relation in the form

$$<u_p^2>/<u^2> = \sqrt{\pi}\; K^{-1}\; \exp(K^{-2} + Ft)\; \mathrm{erfc}(K^{-1} + KFt) \qquad (1.70)$$

The above approximation as delineated in Fig. 1.5 gives identical diffusivities of the particle and the fluid, or $D_p = D$, with

$$D = <u^2>^{\frac{1}{2}} \ell_1 = (\sqrt{\pi}/2)\; \Lambda <u^2>^{\frac{1}{2}}. \qquad (1.71)$$

Determination of particle diffusivity calls for considerations of transfer of a particle from one bulk of fluid into another such that its direction may be significantly altered [Peskin, 1960]. By analogy to the concept of mean free path in the kinetic theory, we introduce an interaction length L_p such that

$$L_p = <(\Delta u)^2>^{\frac{1}{2}}/F \qquad (1.72)$$

where the intensity of relative motion is as given in Eq. (1.61) and L_p has the meaning of a mean free path of particle-fluid interaction [Soo, 1969]. The diffusivity of a particle in a fluid is then given by

$$D_{pf} \sim <(\Delta u)^2>^{\frac{1}{2}} L_p \qquad (1.73)$$

for subsequent development in the next section.

In the 1950's, attempts were made by many to solve Eq. (1.47) as the basic equation of multiphase flow. One readily notes that Eq. (1.47) is, in reality, a transport equation with position and velocity as independent variables. Rigorous solution of this equation is complicated by our lack of knowledge of the relation between the timewise (Lagrangian) correlation and the spacewise (Eulerian) correlation to account for the probability of encounter of a particle with various fluid elements [Peskin, 1960; Soo, 1967]. Except for the case of a particle much smaller than the size of the smallest of turbulent eddies, fluid motion of wave lengths greater than the particle diameter will give rise to random motion of a particle while those smaller than the particle will contribute to the drag on a particle in this fluid.   In the early 1960's, interest turned to the formulation and solution of problems involving a cloud of particles by extending continuum mechanics through some form of averaging with the introduction of transport properties (Chapter 2).   These transport properties can be experimentally determined or from solution of, for instance, Eq. (1.47) as illustrated next.

1.7 Particle Diffusivity

Determination of particle diffusivity via relations such as Eq. (1.47) is hampered by the lack of knowledge of a rigorous relation between correlation with respect to time (Lagrangian) and that with respect to space (Eulerian) for the fluid turbulence [Lumley, 1962].   An approximation procedure is needed at present because of the usefulness of this knowledge.

As a matter of fact, detailed knowledge on the diffusivity of the fluid phase is itself incomplete.   Fluid diffusivity in turbulent

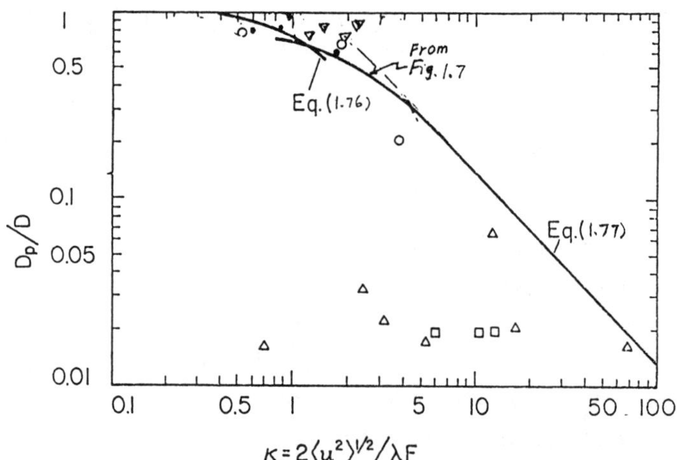

Figure 1.8   Diffusivity ratio as a function of a parameter K.
o from Peskin [1970], from Cheng, et al. [1970],
△ from Soo, et al. [1960], ▽ Nieh, et al. [1985],
• Vames and Hanratty [1986].

pipeflow appears to be best correlated from the postulate of mixing length according to van Driest [1956] to give

$$D = 0.01231 \ N_{Re}^{-1/8} \ [U_o(2R)] \tag{1.74}$$

where $N_{Re} = U_o \ (2R)/\bar{\nu}$, U_o being the mean velocity, $2R$ is the pipe diameter, and $\bar{\nu}$ the kinematic viscosity of the fluid. Equation (1.74) is in general, agreement with the results of measurements made by Kada and Hanratty [1960], Baldwin and Walsh [1961], Towle and Sherwood [1939], and Soo, et al., [1960], for air and water flow through round and square pipes and over a Reynolds number range of 10^4 to 6×10^5.

For a general situation of a free particle in a fluid with negligible wall effect, it is readily seen that for a very small particle or $K \rightarrow 0$, $D_p/D \rightarrow 1$. Deviation from this limiting condition is seen in that the mean squared displacement $<X_p^2>$ over a given time is

$$<X_p^2> = <X_f^2> - <X_{fp}^2> \tag{1.75}$$

where X_{fp} is the relative displacement of particles in fluid. Hence, for D given by Eq. (1.71) and D_p given by Eq. (1.73), we get

$$D_p/D \simeq 1 - [2<(\Delta u)^2>/F \ <u^2>^{\frac{1}{2}} \ \Lambda \ \sqrt{\pi}]$$

$$= 1 - [1 - (<u_p^2>/<u^2>)] \ K/\sqrt{\pi}$$

$$\simeq 1 - (K^3/2\sqrt{\pi}) \ + [0(K^5)]. \tag{1.76}$$

Since Eq. (1.68) gives, as $K \rightarrow 0$, $<u_p^2>/<u^2> \simeq 1 - (K^2/2) + [0(K^4)]$.

For a large particle or a large value of K, large relative motion of a particle to the fluid is expected. The diffusion result from displacements over a long time t, or $<u^2>t$ and $<u_p^2>t$ with $<u_p^2>$ given by Eq. (1.68). We thus have

$$D_p/D \gtrsim <u_p^2>/<u^2> \simeq \sqrt{\pi}/K \tag{1.77}$$

at large K. Another view is that the particle intensity is limited by the scale of random motion of the fluid, or $<u_p^2>^{\frac{1}{2}} \lesssim \Lambda F$, because the particle cannot be accelerated to a velocity beyond that permitted by the long relaxation time $(1/F)$ and the particle diffusivity $D_p \lesssim (\sqrt{\pi}/2)\Lambda(\Lambda F) \simeq (2/K)D$ at large K.

Equations (1.76) and (1.77) give the relation as shown in Fig. 1.8 together with the theoretical results of Peskin [1970] and measurements made on glass particles [Soo, et al., 1960], coal dust [Cheng, et al., 1970], and alumina particles [Nieh, et al., 1985] in dilute form in pipeflow. Also included are data on water droplets in air [Vames and Hanratty, 1986]. In the absence of further details, Eq. (1.77) is seen to be valid over a range of K.

Note that the above diffusivity relations at small relative motion are applicable to laminar flow where, besides Brownian motion, D could be induced by the perturbation of the flowfield of the particle such as via action and reaction of the wake of a sphere [Schlichting, 1960]. In these situations, D and D_p may become self-consistent, or D may be prominent only in the vicinity of a particle (Problems 1.11, 1.12).

1.8 Electrostatic Charging

Charge transfer occurs when there are contact and separation between two surfaces of different work functions (see, for instance, Montgomery [1959]). The energy-level diagram of the materials of two bodies before and after contact is shown in Fig. 1.9. In each, the Fermi energy is denoted by ζ, ν_o is the energy needed to remove an electron from the highest level of the uppermost nearly filled band, and χ_o is the energy released when an electron at rest outside the crystal is taken into the lowest level in the lower most nearly empty band. When the two surfaces are brought into contact, and after equilibrium is reached, the Fermi levels ζ_1 and ζ_2 equalize due to charge transferred q_o giving rise to modifications of χ_o's and ν_o's to χ_1, χ_2, ν_1 and ν_2 at the junction and a contact potential difference $|\Delta V_1| + |\Delta V_2|$. Rapid separation of these two bodies will leave surface 1 positively charged and surface 2 negatively charged. Further details will be given in Chapter 3.

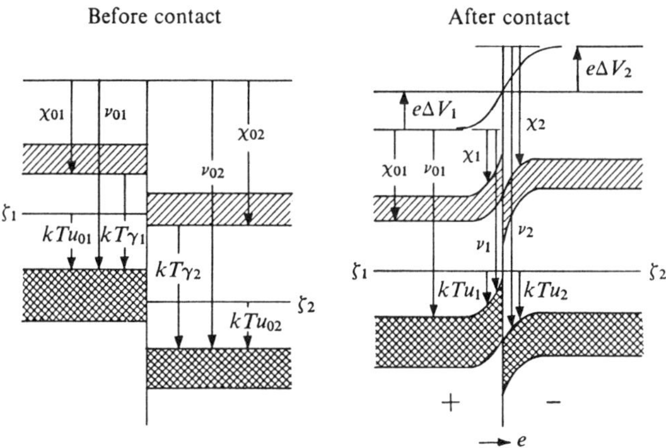

Figure 1.9   Potential diagram of charge transfer by contact
[Montgomery, 1959].

Charging of a Sphere by Gaseous Ions.   We first consider the cases
where the temperatures are moderate and the effect of emission from a
solid is negligible.   When a conducting sphere of radius a is placed
in a uniform electric field E_0 with an initially uniform unipolar ion
density n_z, the potential distribution is given by the Poisson
equation

$$\nabla^2 V = - q \; n_z/\varepsilon_0 \tag{1.78}$$

where ε_0 is the permittivity of free space, q is the charge per ion
(q = -e for an electron), and n_z the distributed ion density.   The
boundary conditions are $E = -\nabla V = E_0$ at infinity and $E = -Ze/4\pi\varepsilon_0 a^2$
for Z charges (Z > 0 for positive charges and Z < 0 for negative
charges) at the surface at a given time.   It is readily shown that at
any point on the sphere,

$$E_a = 3 \; E_0 \; \cos \; \theta - [Ze/4\pi\varepsilon_0 a^2) \tag{1.79}$$

where θ is the azimuthal angle in the spherical coordinate.

For particle radius much greater than the mean free paths of the
ions, the random motion of the ions can be neglected and the total

electric flux ψ entering the sphere is given by

$$\psi = \int_{0}^{\pi} 4\pi \ \varepsilon_o \ E_a \ 2\pi \ a^2 \ \sin \ \theta \ d \ \theta$$

$$= 12 \ \pi \ \varepsilon_o \ a^2 \ E_o \ [1 - (Ze/12 \ \pi \ \varepsilon_o \ E_o \ a^2)]^2. \qquad (1.80)$$

The rate of charging is given by the charging current i,

$$i = n_z \ q \ K_i \psi/4 \ \varepsilon_o = d(Ze)/dt, \qquad (1.81)$$

where K_i is the ion mobility. Integration gives

$$-n_z \ q \ K_i t/4\pi \ \varepsilon_o = (Z/Z_s)/[1 - (Z/Z_s)] \qquad (1.82)$$

where Z_s is the saturation charge at $\psi = 0$ or

$$Z_s = 12 \ \pi \ \varepsilon_o \ E_o \ a^2/e \qquad (1.83)$$

and, for a dielectric particle,

$$q \ Z_s = 4\pi \ \varepsilon_o \ E_o \ a^2 \ [1 + 2(\varepsilon_r - 1/\varepsilon_r + 2)] \qquad (1.84)$$

where ε_r is the dielectric constant, $\varepsilon_r = \varepsilon/\varepsilon_o$; ε is the permittivity of the material [White, 1963] (Problem 1.13).

When the radius a is of the order of mean free path or less, random motion of ions must be considered and the effect of the external field is less significant than in the above case of field charging. Based on an extensive study by Murphy, et al. [1959], on the diffusion charging of a sphere, Liu, et al. [1967], gave

$$N_{et} = Z \ e^2/4\pi \ \varepsilon_o \ a \ k \ T = \ln(1 + \tau_c) \qquad (1.85)$$

for charging by unipolar ions where $\tau_c = (n_{zo} \ e^2 a^2/4 \ \varepsilon_o \ kT) \ (\bar{v}t/a)$, $\bar{v} = (8kT/\pi m_z)^{\frac{1}{2}}$ is the mean thermal speed, n_{zo} is the unipolar ion density, and m_z is the mass of an ion. N_{et} is called an electro-thermal number [Soo, 1967] and is the ratio of the electrostatic to thermal energies. Thermal agitation tends to modify the charge acquisition of a particle. Dimick and Soo [1964] further showed that in an ionized gas, the flux of electrons and ions approaching a particle can be approximated by

$$dn_e/dt\Big|_{dA} = (8kT/\pi m_e)^{\frac{1}{2}} \, n_{eo} \, \exp(N_{et})$$

$$dn_i/dt\Big|_{dA} = (8kT/\pi m_i)^{\frac{1}{2}} \, n_{io} \, \exp(-N_{et}) \qquad (1.86)$$

with n_e, n_i denoting the electron and ion densities (Problems 1.14 and 1.15).

Emission charging. When applied to a charged spherical particle of radius a surrounded by an ionized gas, the distribution of electric potential $V(r)$ is given by the Poisson equation

$$\nabla^2 V = \frac{1}{r^2} \, \frac{\partial}{\partial r} \, r^2 \, \frac{\partial V}{\partial r} = - \frac{e}{\varepsilon_o} \, (n_e - n_z). \qquad (1.87)$$

Integration gives the potential at r

$$V(r) = (Ze/4\pi\varepsilon_o a) \, [1 - (a/r)] + (e/\varepsilon_o r) \int_a^r (n_e - n_z) \, r^2 \, dr$$

$$= (e/\varepsilon_o) \int_a^r (n_e - n_z) \, r \, dr \qquad (1.88)$$

where the first term on the right-hand side is due to charge Z (Z > 0 for positive charge) of the particle, the second term is due to shielding by the net charge density within radius r, and the third term is due to the image charge effect of these surrounding charges which is equal to the thermionic work function divided by the electronic charge e. Based on Eq. (1.87) and taking into account the work function of the solid and the ionization potential of the gas, the density distributions of species and potential are readily computed [Soo, 1967]. For an electron concentration of n_{cs} inside the surface of the solid, the concentration of electrons immediately outside the surface is given by

$$n_{ea} = n_{cs} \, \exp(-\phi/kT) \qquad (1.89)$$

where T is the absolute temperature. For V(r) given by Eq. (1.87) and $V_1(r) = V(r) - (\phi/e)$, the density distribution is given by

$$n_e(r) = n_{ea} \exp(-eV_1/kT). \tag{1.90}$$

Equation (1.87) is thus solved as a self-consistent problem with n determined from the Richardson-Dushman equation for thermionic current density J_e given by

$$J_e = A_o T^2 \exp(-\phi/kT) \tag{1.91}$$

where the coefficient A_o is related to the conduction electron concentration of the solid, and can be expressed in the form

$$n_{cs} = (A_o T^2/e)/(kT/2\pi m_e)^{\frac{1}{2}} (1 - \bar{r}) \tag{1.92}$$

by equating the thermionic current density J_e to the free-electron current density; r is the reflectivity of the surface for electrons.   For the surface of a general semi-conductor with an energy gap $\Delta\varepsilon$ from the conduction band [ter Haar, 1956], A_o can be expressed as [Soo, 1967]

$$A_o T^2 = e(2n_b)^{\frac{1}{2}} (2\pi m_e kT/h^2)^{1/4} (kT/h)(1-\bar{r}) \exp(-\Delta\varepsilon/2kT) \tag{1.93}$$

where n_b is the total electron (or impurity electron) concentration in the solid.   These relations will be applied in Chapter 4 dealing with the ionization of a suspension of particulates (Problem 1.16).

Charge Drag.   A drag force arises due to electrostatic interaction on a sphere moving through a highly ionized gas [Jastrow and Pearse, 1957].   A sphere of radius a acquires Z negative charges because of the greater velocities of electrons.   A positive ion sheath of thickness δ forms around the sphere due to repulsion of electrons. Assuming uniform δ and n_i ions or electrons, a surface potential ϕ_o, and for δ/a between 0.01 and 1, a semi-empirical relation was given by Chopra [1965]

$$\delta/a \sim 0.8[-e\, \phi_0/n_i a^2 e^2/2\varepsilon_0)]^{0.47}$$

$$= 0.8[-2N_{et}\, (\lambda_D^2/a^2)]^{0.47} \tag{1.94}$$

where λ_D is the Debye length given by $(\varepsilon_0 kT/n_i e^2)^{\frac{1}{2}}$, a characteristic dimension of a volume within which the ionized gas is electrically neutral. The additional resistant force F_q due to charge drag is given by

$$\frac{F_q}{F_{Di}} \sim \left(\frac{e\,\phi_0}{E}\right)\,\{1 - \exp[-2.4\left(-\frac{e\,\phi_0}{E}\right)\left(-\frac{2\varepsilon_0\, e\,\phi_0}{n_i a^2 e^2}\right)^{\frac{1}{2}}]\} \tag{1.95}$$

for $0 < (-e\phi_0/E) < 1$ where F_{Di} is that of an uncharged sphere and E is the kinetic energy of ions due to relative motion to the sphere, $(1/2)m_i U^2$. Note that

$$-\frac{e\,\phi_0}{E} = \left(-\frac{Ze^2}{4\pi\,\varepsilon_0\, akT}\right)\left(\frac{2\,kT}{m_i\, U^2}\right) = \left(\frac{2}{\gamma}\right)\left(-N_{et}\right)(N_M)^2 \tag{1.96}$$

where γ is the ratio of specific heats of the gas, N_M is the Mach number of the sphere based on a speed of sound given by $(\gamma RT)^{\frac{1}{2}} R$, and is the gas constant. Equation (1.95) is reduced to

$$\frac{F_q}{F_{Di}} \sim 2.4(-2N_{et})^{\frac{1}{2}}(\lambda_D/a) \tag{1.97}$$

for a low relative velocity. The additional drag coefficient due to charge drag is now

$$\Delta C_{Dq} = \frac{F_q}{(1/2)\bar{\rho}\, U^2 \pi a^2} \sim \left(\frac{14.4}{C}\right)\left(\frac{\lambda}{a}\right)\left(\frac{n_i}{n}\right)^{\frac{1}{2}}\left(\frac{\bar{v}_i}{U}\right)\left(-\frac{Z}{2\pi na^3}\right)^{\frac{1}{2}} \tag{1.98}$$

for F_{Di} given by $6\pi\mu a U/C$, C being the Cunningham correction factor given by Eq. (1.8), λ is the mean free path, $\mu = 0.499\, n_i m_i \bar{v}_i \lambda$, $\bar{v}_i = (8kT/\pi m_i)^{\frac{1}{2}}$; n is the total number of molecules (Problem 1.17).

Charged Liquid Droplet.   Charge collection by a liquid droplet follows similar relations as a solid sphere except that deformation of a spherical droplet may occur by the interaction of surface tension σ_s with electrostatic repulsion of charges. The ratio of these forces is called a electrosurface number

$$N_{es} = 2(q^2/2\epsilon_o a)/4\pi a^2 \sigma_s \qquad (1.99)$$

Rayleigh [1929] found that a conducting spherical droplet is stable for N_{es} < 4. Ailam and Gallily [1962] showed that stable forms change between oblate and prolate ellipsoids of revolution.

1.9 Multiphase Flow Systems

This chapter completes an outline of transport processes and transport properties such as relaxation times and has introduced diffusivities of a single particle in a flowfield for use in subsequent chapters dealing with multiple particle systems and systems consisting of general interfaces. The above discussion on the motion of a free particle in a randomized flowfield has shown the limitations of the kinetic equation (1.47) in predicting the induced motion of a single particle. In spite of its usefulness in determining the particle diffusivity in the fluid, difficulty of extending Eq. (1.47) to treating systems involving multiple particles and phases of general configurations has been recognized by many [Soo, 1971]. The reality is that, in most cases, one will deal not only with a multi-particle system but with a system with a distribution of particle sizes or with a system with interlaced phases of various combinations of solid, liquid, and vapor.

A system with a distribution of particle sizes may be represented in terms of a histogram or a distribution curve. A simple case of size distribution characterizing spheres or isometric particles is that in particle diameters. If the numbers of particles of a given specimen are counted between the size range of a and a + da, one has the number distribution function

$$d(N/N_o) = f_N(a) \, da \qquad (1.100)$$

for dN particles out of a total of N_o particles; $f_N(a)$ is called a number distribution function. Normalizing gives

$$\int_{a_1}^{a_2} f_N(a)\, da = 1 \qquad\qquad (1.101)$$

the overall range of sizes is between a_1 and a_2. When the mass of particles of a given size range is weighed, such as is done in sieving, we have a mass distribution function represented by $f_M(a)$, or

$$M_0\, d(M/M_0) = N_0\, m(a)\, d(N/N_0)$$

$$= f_M(a)\, da = M_0(m(a)/m_0)\, f_N(a)\, da \qquad (1.102)$$

where m_0 is the average mass of a particle based on the total number of particles. The mass distribution function $f_M(a)$ can be similarly

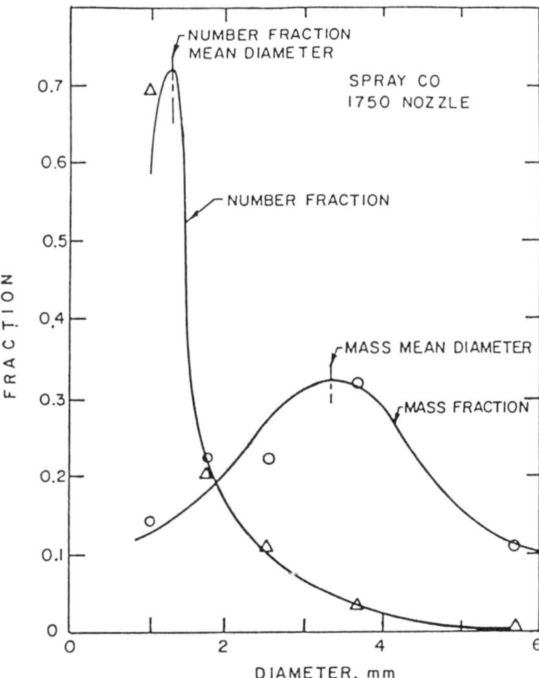

Figure 1.10 Typical spray distribution in size in number fractions and mass fractions largest number around 1 mm size.

normalized.   Equation (1.102) also gives the relation between the number distribution function and the mass distribution function. Figure 1.10 illustrates a case of particle sizes with a Gaussian ($f_M \sim \exp\ [-4(a-a_0)^2/\Delta^2]$, Δ is the "width" at $f_M = f_{Mo}/e$) mass distribution and the corresponding number distribution function (Problem 1.18). For other forms of distributions and equivalent mean diameters, see Orr (1966).

The above sections show that particulates of different sizes will have different dynamic response characteristics in mass, momentum, energy, and charge transfers even when they are constituted of similar materials. Each size range, for instance, can be identified as a dynamic phase, so also different temperature ranges, or charge ranges within each size range. Hence the term "multiphase flow" [Soo, 1965]. The general phase configurations in multiphase flow are illustrated by the vertical vapor-liquid pipe flow in Fig. 1.11 with their qualitative descriptions. The phase configurations between bubble flow and droplet flow remains to be quantified.

Basic formulations of multiphase flow begins in the next chapter with continuum mechanics of phases having discrete interfaces and their averages.

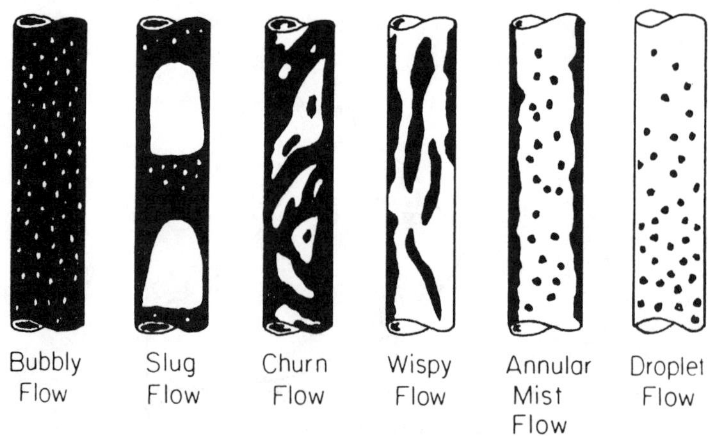

Figure 1.11 Flow regimes in liquid-vapor vertical pipe flow.

Exercise Problems

1.1 Determine the terminal velocities of free fall of a sphere in a
     fluid for the Stokesian range and the Newtonian range.

1.2 Calculate the distance over which a water droplet (assumed to be
     rigid) of 1 μm in diameter starting from rest is accelerated by
     an air stream of 1 m/s at room condition to a velocity of 1-
     (1/e) m/s. Calculate this distance also for a 100 μm droplet.
     Given: Densities of air and water are 1.18 kg/m^3 and 1000 kg/m^3
     respectively, viscosity of air 1.79 x 10^{-5} kg/m, sec

$$\text{Answer: } 1.142 \times 10^{-6} \text{m}, \ 5.107 \times 10^{-3} \text{m}.$$

1.3 Redo Problem 1.2 including motion in the water droplet. The
     latter is still assumed to be spherical. Given: Viscosity of
     water is 0.001 kg/m,s. Is there any conclusion that can be
     drawn from comparing the results?   Answer: 5.076 x 10^{-3} m.

1.4 Neglecting the effect of surface tension, compute the temper-
     ature of a water droplet evaporating in dry air at 100 C.
     Given: Vapor pressure of water = exp [11.93 - (5278/T)] MPa for
     T in K, latent heat of water 2.3 MJ/kg, diffusivity of water
     vapor in air is taken at 1.3 x 10^{-8} m^2/s, thermal conductivity
     is taken as that of air or 0.012 W/m K, assuming that the water
     vapor behaves as a perfect gas.   Answer: 98.43 C.

1.5 Indicate the states in corresponding P-v and h-s diagrams for
     the following cases:

     (a) When the pressure is suddenly lowered on a system consis-
          ting of a bubble in a liquid at equilibrium at a given T
          and P, the liquid becomes superheated and the vapor becomes
          supersaturated.

     (b) When the system consisting of a bubble in a liquid at
          equilibrium is suddenly compressed.

1.6 Indicate the states in corresponding P-v and h-s diagrams for
     the following cases of a droplet initially at equilibrium with
     its surrounding vapor:

     (a) When the pressure of the system is suddenly increased.

     (b) When the vapor is heated at constant pressure.

     (b) When the vapor is cooled at constant pressure.

1.7 The rate of change of bubble radius as the gas is dissolved in a
     liquid is given by [Liebermann, 1957]

$$da/dt = (D/\bar{\rho}) (C_s - C) [a^{-1} - (\pi Dt)^{-1/2}] \qquad (1.103)$$

where C_s is the gas concentration at saturation and C is the actual gas concentration. Determine the radius of the bubble after time t. Discuss the condition in which the last term of the above equation can be neglected.

1.8 In a fully developed turbulent pipe flow, the thickness of the laminar sublayer is given by

$$y_s = 60(U_o R/\bar{\nu})^{-7/8} R \qquad (1.104)$$

where U_o is the velocity at the center of the pipe of radius R. For velocity in the sublayer given by (U_s is the velocity at y_s):

$$U = 0.0225 \ U_o (U_o R/\bar{\nu})^{3/4} \ y/R \equiv U_s \ y/R. \qquad (1.105)$$

Compute the lift force acting on a solid particle or radius a and density $\bar{\rho}_p$ situated at one-half the thickness of the sublayer. Compare the results given by Eqs. (1.36) and (1.37). Determine the ratios of lift to drag forces.

1.9 Show that, for a spherical particle in a velocity gradient of a fluid due to pressure gradient, the net force of change of momentum of the fluid relative to the particle besides the viscous drag is equal to the volume of the sphere times the pressure gradient. Use a linear approximation for the deviation from the mean velocity over the sphere [Soo, 1976].

1.10 Determine the initial acceleration of free fall of a sphere of metal in air and that of rise of an air filled ballon in water. This exercise shows the meaning of virtual mass force.

Answer: g, -2g.

1.11 Given for a particle of alumina suspended in a gas in a pipe of 2R = 0.54 m, at a mean velocity of 6.6 m/s, viscosity of the gas is 40 micropoise (4 x 10^{-6} kg/m,s), density of gas 10.3 kg/m^3. The alumina particle has 2a = 50 μm, density 1370 kg/m. Compute: particle diffusivity D_p, intensity of particle motion for a fluid intensity of 0.05 of mean velocity, intensity of relative motion and interaction length.

Answer: 0.00356 m^2/s, 0.15 m/s, 0.0184 m.

1.12 For the above problem, compute the probable distance the particle will travel between collisions with the pipe wall.

Answer: 13.7 m.

1.13 For a space charge electron density of 5×10^{14} m^{-3}, and a mobility of -2.2 (cm/s)/(V/cm), compute the time to reach 1/2 of the saturation charge of a particle by field charging.

Answer: 0.002 s.

1.14 Express Eq. (1.85) in terms of number of charges per particle of radius a in micrometer, an electron concentration of n ions/cm and charging time in sec at a temperature of 300K. Compute for $Z = 100$, $n_0 = 10^{20}$ m^{-3}, and a = 1 μm.   Answer: 1.33×10^{-10} s.

1.15 10 μm diameter particle collect charges from unipolar ions in a corona discharge over a distance of 5 cm. For a flow velocity of 22.8 m/s, charge density of ions of 0.008 C/m^3. Taking the ions as electrons, compute the charge-to-mass ratio acquired for a density of 1370 kg/m^3 for the solid material.

Answer: 3.13×10^{-4} C/kg.

1.16 From Eq. (1.93), show that when the energy gap is zero and electron reflectivity of zero and electron concentration $n_e = n_0 = 2(2\pi m_e kT/h^2)^{3/2}$, $A_0 = 4\pi e m_e k^2/h^3$, the value for a pure metal.

1.17 Apply Eq. (1.98) as an approximation for a partially ionized gas and compute the increase in the drag coefficient due to charge drag for: (1) 10 μm fog particle, relative velocity, 0.01 m/s, $Z = -574$, $T = 300$ K, 10^{-6} fraction of ionization (n_i/n), $\lambda/a = 0.01$; (2) 10 μm dust particle, relative velocity 0.1 m/s, $Z = -31,000$, $T = 300$ K, 10^{-4} fraction of ionization, $\lambda/a = 0.01$.

Answer: (1) 0.0011, (2) 0.0085

1.18 For a number distribution function of

$$f_N(a) = A_N \exp[-4(a - a_0)^2/\Delta^2] \qquad (1.106)$$

and a narrow width such that normalization can be approximated by integration between infinities, compute the normalizing constant A_N and determine the normalized mass distribution function.

Chapter 2

BASIC EQUATIONS

2.1 Intraphase Equations and Balances at the Interfaces

The general multiphase flow system consists of phases of solid, liquid, or vapor separated by interfaces of any configurations. "Multiphase" arises from different dynamic phases because of different responses of different parts of the a physical phase of the same material due to their local configurations. While one phase may be a solid, the general mixture may consist of two or more immiscible fluids, each being a pure phase. Within a pure phase k, conservation principles of continuum mechanics give equations representing the conservation of mass, momentum, and energy.

<u>General transport theorem.</u> The basis of conservation in continuum mechanics is the general transport theorem. The rate of change of any volume integral can be determined by considering a function

$$F(t) = \int_{v(t)} f(r,t)dv \tag{2.1}$$

where f is integrable, noting that the volume $v(t)$ changes with time t. To calculate the material derivative of dF/dt, we integrate with respect to a volume in the ξ-space, where it would be possible to interchange differentiation and integration since d/dt is differentiation with respect to t, keeping ξ constant. The transformation $r = r(\xi,t)$ permits such an operation, and

$$dv = [\partial(r_1, r_2, r_3)/\partial(\xi_1, \xi_2, \xi_3)]d\xi_1 d\xi_2 d\xi_3 \equiv Jdv_0 \tag{2.2}$$

where J is the ratio of an elementary moving material volume $v(t)$ to its initial volume v_0 at $t = 0$. J is therefore a dilatation for $0 < J < \infty$ for surface velocity **U**,

$$(d/dt)\,(\partial r_i/\partial \xi_i) = (\partial/\partial \xi_j)\,(dr_i/dt) = (\partial U_i/\partial \xi_j) \qquad (2.3)$$

and d/dt is the differentiation with ξ kept constant so that

$$(\partial U_i/\partial \xi_j) = (\partial U_i/\partial r_k)\,(\partial r_k/\partial \xi_j) \qquad (2.4)$$

leading to the relation

$$dJ/dt = (\nabla \cdot \mathbf{U})\,(\partial r_k/\partial \xi_j) \qquad (2.5)$$

The above relations give dF/dt as follows:

$$\begin{aligned}
\frac{d}{dt} \int_{v(t)} f(\mathbf{r},\,t)dv &= \frac{d}{dt} \int_{v_0} f[r(\xi,t),t]J dv_0 \\
&= \int_{v_0} [\frac{df}{dt}\,J + f\,\frac{dJ}{dt}]\,dv_0 \\
&= \int_{v_0} [\frac{df}{dt} + f(\nabla \cdot \mathbf{U})]\,J dv_0 \\
&= \int_{v(t)} [\frac{df}{dt} + f(\nabla \cdot \mathbf{U})]\,dv \qquad (2.6)
\end{aligned}$$

combining with the Gauss theorem in the form

$$\int_A \mathbf{f} \cdot \mathbf{n}\, dA = \int_v \nabla \cdot \mathbf{f}\, dv \qquad (2.7)$$

where **n** is an outwardly directed normal and A is the surface area; and since $d/dt = (\partial/\partial t) + \mathbf{U}\nabla$, we get

$$\frac{d}{dt} \int_{v(t)} f(\mathbf{r},t)\, dv = \int_{v(t)} \frac{\partial f}{\partial t}\, dv + \int_{A(t)} f\,\mathbf{U} \cdot \mathbf{n}\, dA \qquad (2.8)$$

the general transport theorem. [Aris 1962, Slattery 1976].

<u>General conservation in a single phase of fluid.</u> For a flux $\mathbf{\Psi}$ of a property ψ per unit volume of a fluid element of volume v moving at velocity $\mathbf{U}$, the net flow across a closed surface is given by $\int_A \mathbf{\Psi} \cdot \mathbf{n} \, dA$, which is positive for net out flow and negative for net inflow. The time rate of change of the property ψ in this elemental volume is $(d/dt) \int_v \psi dv$ and the rate of generation of this property per unit volume is denoted as $\dot{\psi}_G$, and within this volume is $\int_v \dot{\psi}_G \, dv$. Since the change within this volume is a consequence of net flow across the surface and the generation, we get:

$$\frac{d}{dt} \int_V \psi \, dv = -\int_A \mathbf{\Psi} \cdot \mathbf{n} \, dA + \int_V \dot{\psi}_G \, dv \qquad (2.9)$$

Application of the relation given by Eq. (2.8) gives

$$\int_V \left(\frac{\partial \psi}{\partial t}\right) dv + \int_A \psi \mathbf{U} \cdot \mathbf{n} \, dA = -\int_A \mathbf{\Psi} \cdot \mathbf{n} \, dA + \int_V \dot{\psi}_G \, dv \qquad (2.10)$$

Application of the Gauss theorem further gives

$$\int_V \left[\frac{\partial \psi}{\partial t} + \nabla \cdot \psi \mathbf{U} + \nabla \cdot \mathbf{\Psi} - \dot{\psi}_G\right] dv = 0 \qquad (2.11)$$

Since the volume element is arbitrary, we have the general balance relation:

$$\frac{\partial \psi}{\partial t} + \nabla \cdot \psi \mathbf{U} + \nabla \cdot \mathbf{\Psi} - \dot{\psi}_G = 0 \qquad (2.12)$$

The conservation of mass is given by the continuity equation, with $\psi = \rho_k$: ($\mathbf{\Psi} = 0$, $\dot{\psi}_G = 0$)

$$(\partial \rho_k / \partial t) + \nabla \cdot (\rho_k \mathbf{U}_k) = 0 \qquad (2.13)$$

where ρ_k is the density of material of phase k, $\mathbf{U}_k$ is the vectorial velocity. The equation of conservation of momentum is given by having $\psi = \rho_k \mathbf{U}_k$, the momentum per unit volume, and the momentum flux is given by the stress tensor $\mathbf{\Psi} = P_k \mathbf{I} - \tau_k$, and $\dot{\psi}_G = \rho_k f_k$:

$$(\partial \rho_k \mathbf{U}_k / \partial t) + \nabla \cdot (\rho_k \mathbf{U}_k \mathbf{U}_k) = -\nabla P_k + \nabla \cdot \tau_k + \rho_k f_k \qquad (2.14)$$

where P_k is the static pressure and τ_k is the viscous stress tensor and $\mathbf{f}_k$ is the field force per unit mass. The equation of conservation of energy or the energy equation is given by, for $\psi = \rho_k E_k$, and other corresponding quantities:

$$(\partial_k \rho_k E_k / \partial t) + \nabla \cdot (\rho_k \mathbf{U}_k E_k) = - \nabla \cdot \mathbf{U}_k P_k$$

$$+ \nabla \cdot (\mathbf{U}_k \cdot \tau_k) + \rho_k \mathbf{U}_k \cdot \mathbf{f}_k - \nabla \cdot \mathbf{J}_{qk} + J_{Ek} \qquad (2.15)$$

where E_k is the total energy per unit mass, $\mathbf{J}_{qk}$ is the heat flux vector, and J_{Ek} is the heat source per unit volume inside phase k. By definition, $E_k = u_k + \mathbf{U}_k \cdot \mathbf{U}_k / 2$, with u_k being the internal energy per unit mass.

Alternately, the energy equation can be expressed in terms of u_k or enthalpy per unit mass h_k:

$$(\partial \rho_k u_k / \partial t) + \nabla \cdot (\rho_k \mathbf{U}_k u_k) = - P_k \nabla \cdot \mathbf{U}_k$$

$$+ \tau_k : \nabla \mathbf{U}_k - \nabla \cdot \mathbf{J}_{qk} + J_{Ek} \qquad (2.16)$$

The double dot in the last term denotes the scalar product of two second order tensors and is usually represented as Φ_k, the dissipation rate per unit volume of phase k. We further have:

$$(\partial \rho_k h_k / \partial t) + \nabla \cdot (\rho_k \mathbf{U}_k h_k) = (dP_k / dt)$$

$$+ \Phi_k - \nabla \cdot \mathbf{J}_{qk} + J_{Ek} \qquad (2.17)$$

in which the substantive derivative denotes:

$$(d/dt) = (\partial / \partial t) + \mathbf{U}_k \cdot \nabla \qquad (2.18)$$

Expressing Eq. (2.17) in terms of total enthalpy $H_k = h_k + \mathbf{U}_k \cdot \mathbf{U}_k / 2$ is left as an exercise. (Prob. 2.1)

<u>Local balance at a phase boundary</u>. The boundary conditions of the above equations of conservation of each phase, besides those at the boundaries of the physical system, are those at the interface or interfaces. The latter take into account balances of mass, momentum and energy at the interface represented by area A_k between phases k and f as shown in Fig. 2.1. Take the simple case of a fluid-fluid interface of zero thickness, the mass balance is given by:

$$\rho_k (U_k - U_s) \cdot n_k + \rho_f (U_f - U_s) \cdot n_f = \sum_{k,f} \dot{m}_k = 0 \qquad (2.19)$$

where the n's are direction normals pointing outward from respective phases. The balance is seen in terms of the mass flow of each phase relative to the interface, taking into account the transformation of one phase into the other at the interface. The momentum balance at the interface is given by, when the effect of change in mean curvature is ignored [Sha et al. 1988],

$$\sum_{k,f} (\dot{m}_k U_s + P_k n_k - \tau_k \cdot n_k) + F_s = 0 \qquad (2.20)$$

where F_s is the surface force:

$$F_s = \nabla_s \sigma - 2H_{kf} \, \sigma \, n_k$$

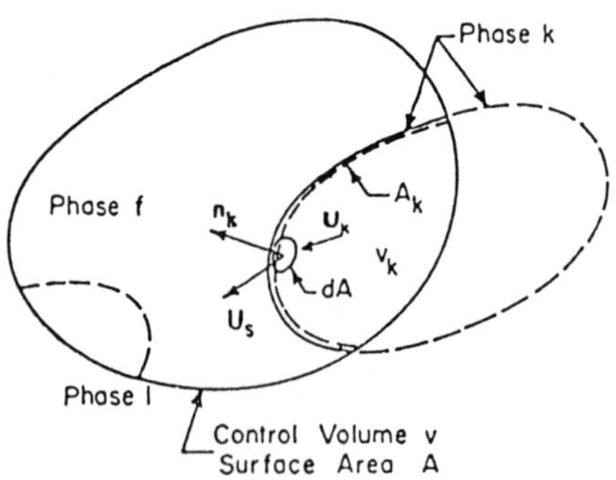

Fig. 2.1   Control Volume and Interacting Phases; v_k is the
            Volume of Phase k Inside of v and A_k is the Area
            of the Interface between Phases k and f inside of v.

where the mean curvature H_{kf} is along the direction normal $\mathbf{n}_k$, σ_{kf} is the interfacial tension, ∇_s is the surface gradient operator, and I is the unitary tensor; the interfacial velocity is given by $\mathbf{U}_s$ and H_{kf} is positive when the associated radius curvature is pointing outward, noting that for the interface between phase k and phase f, $A_k = A_f$, $\mathbf{n}_k = -\mathbf{n}_f$, and $H_{kf} = -H_{fk}$. The first term is the force due to gradient of surface tension between the phases; the second term is the force of surface tension due to the curvature; the third and fourth term is the force due to rate of change of momentum of phases at the interface, an effect of varying normal velocity; the fifth and sixth terms consist of the normal pressure force and the force due to shear and dilatation. In formulating the total energy balance at the interface, we neglect the capillary energy to get:

$$\rho_k E_k (\mathbf{U}_k - \mathbf{U}_s) \cdot \mathbf{n}_k + \mathbf{J}_{qk} \cdot \mathbf{n}_k + \rho_f E_f (\mathbf{U}_f - \mathbf{U}_s) \cdot \mathbf{n}_f$$

$$+ \mathbf{J}_{qf} \cdot \mathbf{n}_f - \mathbf{U}_k (-P_k I + \tau_k) \cdot \mathbf{n}_k - \mathbf{U}_f (-P_f I + \tau_f) \cdot \mathbf{n}_f = 0 \quad (2.21)$$

The balances include the relative energy flux, the heat flux, and the rate of work by pressure and the rate of dissipation by viscous stress. The interfacial energy balances in terms of internal energy and enthalpy can be expressed as:

$$\rho_k u_k (\mathbf{U}_k - \mathbf{U}_s) \cdot \mathbf{n}_k + \mathbf{J}_{qk} \cdot \mathbf{n}_k + \rho_f u_f (\mathbf{U}_f - \mathbf{U}_s) \cdot \mathbf{n}_f$$

$$+ \mathbf{J}_{qf} \cdot \mathbf{n}_f = 0 \quad (2.22)$$

and

$$\rho_k h_k (\mathbf{U}_k - \mathbf{U}_s) \mathbf{n}_k - P_k (\mathbf{U}_k - \mathbf{U}_s) \cdot \mathbf{n}_k + \mathbf{J}_{qk} \cdot \mathbf{n}_k$$

$$+ \rho_f h_f (\mathbf{U}_f - \mathbf{U}_s) \cdot \mathbf{n}_f - P_f (\mathbf{U}_f - \mathbf{U}_s) \cdot \mathbf{n}_f + \mathbf{J}_{qf} \cdot \mathbf{n}_f = 0 \quad (2.23)$$

One notes that only one of the Eqs. (2.21) to (2.23) is independent; in these three equations all the properties are those at the interface of area A_k. For further details of conservation equations see Aris [1962] or Brodkey [1967].

One readily notes that, when Eq. (1.47) is applied to the motion of a solid particle in a turbulent flow field, it is, in reality a kinetic equation with velocities as independent variables in addition to position and time coordinates; that is, it describes motion in the phase space. Attempts toward treating problems involving a cloud of particles via integrating equations such as given by Eq. (1.47) were abandoned around 1960 in favor of an approach via continuum mechanics. Formulations via continuum mechanics are in the configuration space, with position and time as independent variables through the introduction of transport properties to account for viscous stresses, heat fluxes, and mass generations. These transport properties can be determined experimentally, thus making accurate formulation and solution possible even when detailed interactions are not known.

In principle, the coupled equations of phases (2.13) to (2.15) or 3n equations for n phases, together with 3(n-1) interface balance equations (2.19) to (2.21), and for given initial and boundary conditions and physical properties can be solved deterministically. However, in the general case, the configuration and location of the interfaces cannot be rigorously defined even initially, direct solution of these equations of continuum mechanics is only possible for the simplest cases [Happel and Brenner 1965]. Therefore, either from the point of view of extending the single particle equation to a general multiphase system in the dynamic sense [Soo 1965], or from the point of view of intermingled phases as shown in Fig. 2.1, a proper averaging procedure is needed. This is in order to take into account the effect of the configurations of the interface. The significance of the latter is shown next, before we proceed with the averaging procedure.

2.2 Significance of Phase Configurations

The significance of the configurations of the interfaces of phases is readily seen by considering the simple cases represented by the Bernoulli equation of phases and their mixture. In this illustration, averaged quantities are assumed to exist, and according to the barycentric frame of reference, the mixture density (subscript m) and momentum are given by:

$$\rho_m = \sum_k \rho_k, \quad \rho_m U_m = \sum_k \rho_k U_k \tag{2.24}$$

to these we add $P_m = \sum_k P_k$. We note that if the Bernoulli equation for the steady flow of a phase k of incompressible fluid is written as

$$(1/2) \, \rho_k U_k^2 + P_k + \rho_k gz = \text{constant} \tag{2.25}$$

where g is the appropriate component of gravitational acceleration and z is the elevation. Then, according to Eqs. (2.24), the mixture will have the relation:

$$(1/2) \, \rho_m U_m^2 + \sum_k (1/2) \, \rho_k (U_k - U_m)^2 + P_m + \rho_m gz = \text{constant} \tag{2.26}$$

Formulations in terms of Eqs. (2.25) and (2.26) represents the case ignoring the transfer of inertia force across the interfaces, that is, the case of pure stratified flow. An example is air flowing over water without generating surface waves. (Prob. 2.2)

However, if one considers a highly dispersed system that can be represented by a macroscopically homogeneous mixture, the Bernoulli equation should take the form

$$(1/2) \, \rho_m U_m^2 + P_m + \rho_m gz = \text{constant} \tag{2.27}$$

For the same frame of reference, however, the Bernoulli equation of a phase k will have to take the form

$$(1/2) \, \rho_k U_k^2 - (1/2) \, \rho_k (U_k - U_m)^2 + P_k + \rho_k gz = \text{constant} \tag{2.28}$$

Eqs. (2.27) and (2.28) are exact for highly dispersed flow [Soo 1981]. Ideally, a highly dispersed system is defined as a mixture in which the characteristic dimension of the dispersed phase is less than the mean free path of the suspending fluid. An example is smoke particles smaller than 0.1 micrometer in air at room condition.

It is also well known that gradient of concentration of smoke, for instance, will lead to diffusion of smoke particles. yet a gradient in the amount of liquid in a slanted (with respect to gravity) can partially filled with air will not lead to diffusion. Another phenomenon is the propagation of sound. In a dispersed system, a common speed of sound exists [Richardson 1962]. However,

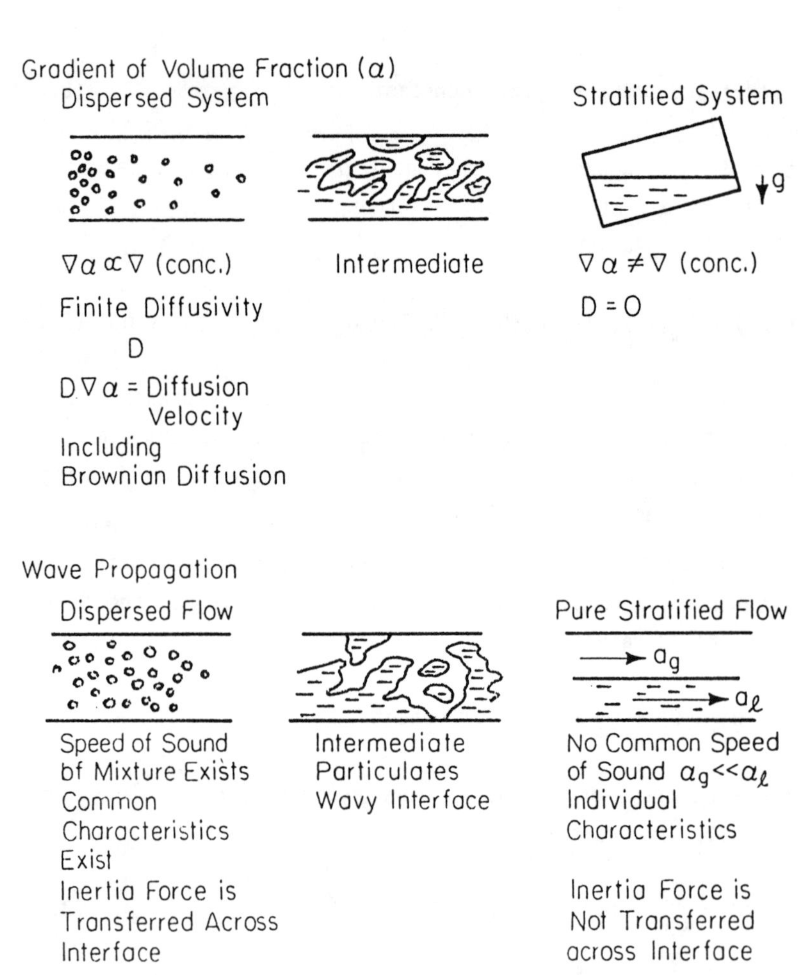

Fig 2.2 Significance of Configuration of Interface

in a pure stratified system, sound waves propagate through each phase at their own characteristic speeds. Other flow conditions, depending on configuration of phases, will have to be represented, in this illustration, by Bernoulli relations in forms intermediate between these two limiting cases. Clearly, a proper averaging procedure is needed to account for the real physical conditions. These ranges are depicted in Fig. 2.2; aspects relating to "characteristics" are for discussions in Chapter 5.

2.3 Averages and Averaging Theorems

For a general multiphase system consisting of interacting phases dispersed randomly in space and time, detailed solutions are not feasible nor needed in many applications. A realistic approach is to express the essential physical and dynamic quantities of such a system in terms of averages. Various averaging procedures have been proposed [Slattery 1967, Whitaker 1969, Delhaye & Archard 1976]. It is noted that volume averaging is convenient in expressing dynamic phases [Soo 1965] in terms of volume fractions of phases, while a priori time averaging [Ishii 1975] yields fraction residence times of phases. One notes that dynamic and thermodynamic properties of a mixture are not cumulative with fraction residence time but are cumulative with volume fractions, and fraction residence time is equal to volume fraction only in the instance of one-dimensional uniform motion of a mixture. Therefore, time and volume averaging operations are not commutative. A convenient procedure is to carry out time averaging after volume averaging to account for the high frequency fluctuations retained by instantaneous volume averaging.

Consider phase k in Fig. 2.1 which is partially included in a fixed control volume in space with the fluid phases passing through it. This part of phase k has a variable volume v_k with the interface area A_k separating phase k from the rest of the volume v of a fluid f and may contain another phase ℓ, $\mathbf{n}_k$ is the outwardly directed normal from phase k on A_k. The volume averaging, when applied to any scalar, vector, or tensor ψ_k associated with a unit volume or area of phase k is defined by an extensive average:

$$<\psi_k> = v^{-1} \int_{v_k} \psi_k dv \qquad (2.29)$$

when averaged over volume v, and when averaged over v_k,

$$\overset{i}{<}\psi_k> = v_k^{-1} \int_{v_k} \psi_k dv \qquad (2.30)$$

is called an intrinsic average. These two averages are related by the volume fraction of phase k, or

$$<\psi_k> = (v_k/v) \; v_k^{-1} \int_{v_k} \psi_k dv = \alpha_k \overset{i}{<}\psi_k> \qquad (2.31)$$

noting that α_k is an instantaneous quantity.

When these averaging relations are applied to a quantity per unit volume such as density, we have according to Eqs. (2.29) and (2.30):

$$\overset{i}{<}\rho_k> = v_k^{-1} \int_{v_k} \rho_k dv = \overline{\rho}_k \qquad (2.32)$$

$$<\rho_k> = v^{-1} \int_{v_k} \rho_k dv = \alpha_k \overset{i}{<}\rho_k> = \alpha_k \overline{\rho}_k \qquad (2.33)$$

The latter equalities are for a uniform density of material constituting phase k. Further, the densities in Eqs. (2.32) to (2.33) are related to v_k where the density of phase k is the material density, or $\rho_k = \overline{\rho}_k$. When applied to the mixture in volume v, the average density becomes the density of the phase averaged over that volume, or $\rho_k = <\rho_k>$. $<\rho_k>$ is the density of phase k over volume v such as in the case of a dispersed particle cloud or the average density of a phase in a system of stratified flow of two immisible fluid layers. The situation is seen in Fig. 2.3, on the equivalence due to averaging, and details due to the presence of interfaces are lost via averaging.

Figure 2.3 shows that, with non-uniform distribution within phase k included in volume v, there are distributions in density (indicated by distributed dots), velocity, as well as a general function ψ_k. Volume averaging may proceed along two routes:

(a) Intrinsic averaging gives rise to quantities inside v_k: average $^i\langle \rho_k \rangle$ is indicated by the uniform dots throughout v_k.

(b) Control volume average will spread phase k over the whole volume v, superposing on phase f which is correspondingly averaged in the same manner (in the way of a double exposed photographic plate). So the density $\langle \rho_k \rangle$ is averaged over v and is equal to $\alpha_k{}^i\langle \rho_k \rangle$.

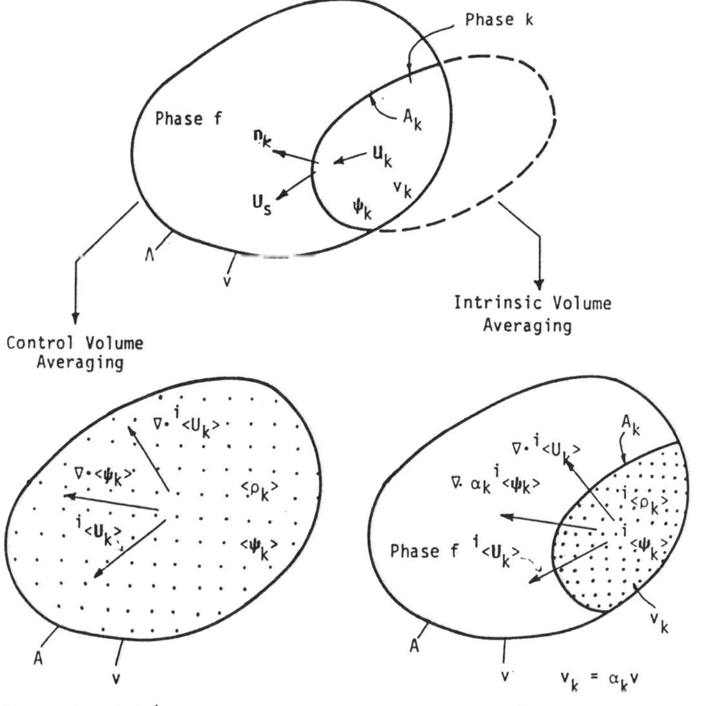

Phases k and f become distributed

$\langle \rho_k \rangle$ distributed over v
All functions and derivatives in v become point functions

Phase k averaged and distributed

$^i\langle \rho_k \rangle$ distributed over v_k
All functions and derivatives in v_k are treated as point functions in v via α_k

Fig. 2.3 Meaning and Consequences of Volume Averaging

(c) Volume averaging can only be applied to quantities per unit volume or area. These include density, momentum per unit volume, energy per unit volume, and gradients of stresses and fluxes for ψ_k in Eqs. (2.29) and (2.30). Therefore, $\overset{i}{<}U_k>$ is given by:

$$\overset{i}{<}U_k> = v_k^{-1} \int_{v_k} \rho_k U_k dv_k / \overset{i}{<}\rho_k> = v^{-1} \int_{v_k} \rho_k U_k dv_k / <\rho_k> \tag{2.34}$$

The same is applicable to internal energy per unit mass, u_k or enthalpy per unit mass h_k. One sees that only $\overset{i}{<}U_k>$ and $\nabla \cdot \overset{i}{<}U_k>$ are meaningful.

(d) Stresses and fluxes in a formulation can be expressed as $<\psi_k>$ = $\alpha_k \overset{i}{<}\psi_k>$, the physical meaning is represented in the "control volume average" where all interactions are represented.

As we shall see $<\tau_k>$, the viscous stress, is not necessarily contributed by the viscous stress inside phase k. It may represent the resistance to transfer of momentum by bodily displacement one phase through another. The same applies to $<J_{qk}>$, the heat flux, so also $<P_k>$.

Restrictions of scales. One further notes that in order that an extensive average be representative, it must be taken over a control volume whose characteristic dimension is greater than the characteristic dimension of a phase, such as the bubble size of k in Fig. 2.1. This is seen in a two-dimensional system as illustrated in Fig. 2.4, with a regular repeating pattern having phase k of characteristic dimension d, and control volume represented circles of diameter ℓ. We note that when its center is arbitrarily located at a in Fig. 2.4 (a), for ℓ of various sizes, the volume fraction α_k (2-dimensional) is given by the solid line in Fig. 2.4 (b) and the value of α_k is close to an asymptotic value when ℓ is large enough. If we take the center of the averaging volume at point b in Fig. 2.4 (a), the values of α_k follow the dotted line in Fig. 2.4 (b). When ℓ is large enough, the asymptotic value of α_k is independent of the location of the control volume. Hence a meaningful average of a physical system is obtained when $\ell \gg d$, in principle.

Yet in order that the average be representative of the local variations, the control volume must be small such that its characteristic dimension ℓ is smaller than that of the physical system under consideration L. The latter relation was shown by Whitaker [1969] by taking a Taylor series expansion of the average of an average quantity $<\psi_k>$, such that

$$<<\psi_k>> = <\psi_k>_o + O[<\psi_k>_o(\ell/L)^2] \tag{2.35}$$

The intrinsic average is not influenced by the above size restriction, but the size of v_k is, and $\overset{i}{<}\psi_k>$ should be representative of the phase k in the mixture. We note that in general, the averaging procedure favors a dispersed system. In general, for averaged physical quantities to be meaningful, it is required that:

$$d < \ell < L \tag{2.36}$$

(Prob. 2.3)

<u>Local volume averaging theorems</u>. The local volume average of a gradient of a quantity ψ_k, be it a scalar, a vector, or a tensor, has to be derived to apply averaging to the equations of conservation of a multiphase system. To apply the general transport theorem in Eq.

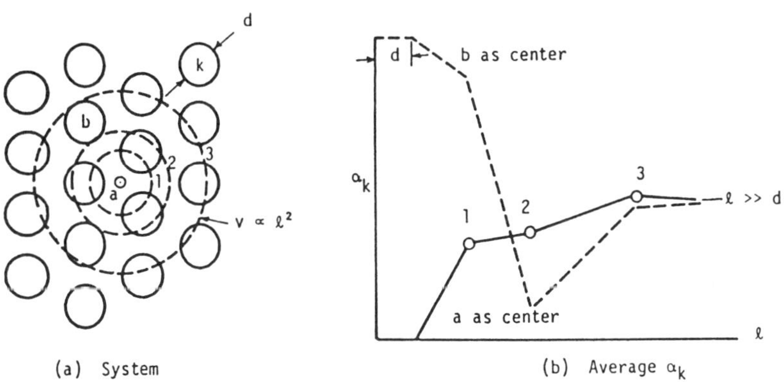

Fig. 2.4 Illustration showing Relation of Averaging Volume
($\propto \ell^2$) to Characteristic Volume of Phase ($\propto d^2$)

(2.8), we consider a point in the multiphase mixture located on an arbitrary, continuous curve s as shown in Fig. 2.5, the arc length measured along this curve being Δs. With each point on the curve as its center, we associate a spherical volume v(s) bounded by a surface A(s). Of this volume a portion is occupied by phase k; we denote this portion by $v_k(s)$ and its bounding surface by $A_{kc}(s)$, the total surface area of phase k included by v(s). The local volume fraction of phase k is $\alpha_k = v_k(s)/v(s)$. In considering the change of the integral over v_k (s) of a quantity as a function of s. Eq. (2.8) may be rewritten as

$$\frac{d}{ds} \int_{v_k(s)} \psi_k dv = \int_{v_k(s)} \left(\frac{\partial \psi_k}{\partial s}\right) dv + \int_{A_{kc}(s)} \psi_k \left(\frac{dr}{ds}\right) \cdot n_k dA \qquad (2.37)$$

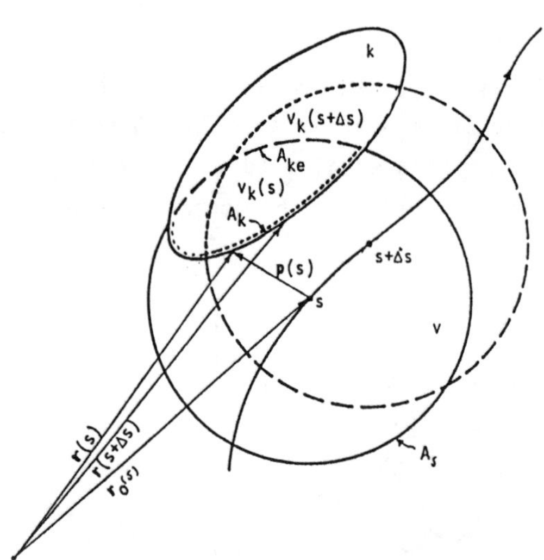

Fig. 2.5 Physical Meaning of Derivative of
Volume Integral with Respect to s

by replacing t with s, where n_k is the outwardly directed normal unit vector for $A_k(s)$.   This means that we assume a continuous and invertible mapping exists as in Sec. 2.1, with $r = r(\xi,s)$, with $\xi = r$ at $s = 0$ which maps $v_k(s = 0)$ into $v_k(s)$.   For quantities ψ_k which are explicit function of time and space coordinates only, or $\partial\psi_k/\partial s = 0$, giving

$$\frac{d}{ds} \int_{v_k(s)} \psi_k ds = \int_{A_{kc}(s)} \psi_k \left(\frac{dr}{ds}\right) \cdot n_k dA \qquad (2.38)$$

The functional dependence is such that, in terms of component x_i of r,

$$\psi_k = \psi_k(x_i, t) = \psi_k[x_i(s), t] \qquad (2.39)$$

and the total derivative of ψ_k with respect to s is, in general, non-zero and

$$\frac{d\psi_k}{ds} = \left(\frac{\partial\psi_k}{\partial x_i}\right) \left(\frac{dx_i}{ds}\right) \qquad (2.40)$$

while the partial derivative is zero as given before.   As shown in Fig. 2.5, dr/ds over the interface is a tangent vector to the interface, or $(dr/ds) \cdot n_k = 0$ on the interface.   The enclosing surface area $A(s)$ can be expressed as

$$A_{kc}(s) = A_k(s) + A_{ke}(s) \qquad (2.41)$$

where $A_{ke}(s)$ represents the area of the entrances and exits on the surface $A(s)$ and $A_k(s)$ represents the area of the interface within $v(s)$.   These considerations permit us to write Eq. (2.38) as

$$\frac{d}{ds} \int_{v_k(s)} \psi_k dv = \int_{A_{ke}(s)} \psi_k \left(\frac{dr}{ds}\right) \cdot n_k dA \qquad (2.42)$$

Let $r_0(s)$ be the position vector locating the reference point on the arbitrary curve and $p(s)$ the position vector locating points on $A_{kc}(s)$ relative to the reference point (the center of the sphere in Fig. 2.5), we have

$$r(s) = r_0(s) + p(s) \tag{2.43}$$

The directional derivative can be expressed in terms of $r_0(s)$ as

$$\frac{d}{ds} = \left(\frac{dr_0}{ds}\right) \cdot A \tag{2.44}$$

Substitution of Eqs. (2.43) and (2.44) into (2.37) yields

$$\left(\frac{dr_0}{ds}\right) \cdot \nabla \int_{V_k(s)} c_k dv = \int_{A_{ke}(s)} \psi_k \left(\frac{dr_0}{ds}\right) \cdot n_k dA + \int_{A_{ke}(s)} \psi \left(\frac{dp}{ds}\right) \cdot n_k dA \tag{2.45}$$

Since dr_0/ds is not a function of the limits of integration for a fixed value of s, dr_0/ds can be taken outside of the integral

$$\left(\frac{dr_0}{ds}\right) \cdot \left(\nabla \int_{V_k(s)} \psi_k dv - \int_{A_{ke}(s)} \psi_k n_k dA\right) = \int_{A_{ke}(s)} \psi_k \left(\frac{dp}{ds}\right) \cdot n_k dA \tag{2.46}$$

With the understanding that $v(s)$ is translated without rotation along the arbitrary curve as shown in Fig. 2.5, any differential variation in p, that is, dp/ds, is a unit tangent vector to the surface $A_k(s)$ of the volume. Thus dp/ds and n_k are orthogonal and the right-hand side of Eq. (2.46) is zero. In addition dr_0/ds is an arbitrary vector, one gets [Slattery 1967]:

$$\nabla \int_{V_k} \psi_k dv = \int_{A_{ke}} \psi_k n_k dA \tag{2.47}$$

Expressing the Gauss theorem in terms of

$$\int_{V_k} \nabla \psi_k dv = \int_{A_{ke}} \psi_k n_k dA + \int_{A_k} \psi_k n_k dA \tag{2.48}$$

Eq. (2.47) can be written as

$$\int_{V_k} \nabla \psi_k dv = \nabla \int_{V_k} \psi_k dv + \int_{A_k} \psi_k n_k dA \tag{2.49}$$

By the definition in Eq. (2.29), we get the averaging theorem of a gradient:

$$\langle \nabla \psi_k \rangle = \nabla \langle \psi_k \rangle + v^{-1} \int_{A_k} \psi_k \mathbf{n}_k dA \tag{2.50}$$

When Eq. (2.8) is applied to the same system in Fig. 2.5, with the position of k fixed in space and A_k as a function of time, or $d/dt = \partial/\partial t$, and $\mathbf{U} = \mathbf{U}_s$, which gives

$$v^{-1} \int_{v_k} \frac{\partial \psi_k}{\partial t} dv = \frac{\partial}{\partial t} (v^{-1} \int_{v_k} \psi_k dv) - v^{-1} \int_{A_k} \psi_k \mathbf{U}_s \cdot \mathbf{n}_k dA \tag{2.51}$$

The averaging theorems of Whitaker [1969] and Slattery [1967] of derivatives as applied to the system in Fig. 2.1 are summarized as:

$$\langle \partial \psi_k / \partial t \rangle = \partial \langle \psi_k \rangle / \partial t - v^{-1} \int_{A_k} \psi_k \mathbf{U}_s \cdot \mathbf{n}_k dA \tag{2.52}$$

$$\langle \nabla \psi_k \rangle = \nabla \langle \psi_k \rangle + v^{-1} \int_{A_k} \psi_k \mathbf{n}_k dA \tag{2.53}$$

$$\langle \nabla \cdot \boldsymbol{\psi}_k \rangle = \nabla \cdot \langle \boldsymbol{\psi}_k \rangle + v^{-1} \int_{A_k} \boldsymbol{\psi}_k \cdot \mathbf{n}_k dA \tag{2.54}$$

where $\mathbf{U}_s \cdot \mathbf{n}_k$ is the speed of displacement of the interface. For $\psi_k = 1$, Eqs. (2.29), Eq. (2.52) gives, for volume fraction α_k,

$$\partial \alpha_k / \partial t = v^{-1} \int_{A_k} \mathbf{U}_s \cdot \mathbf{n}_k dA \tag{2.55}$$

and Eq. (2.54) gives

$$\nabla \alpha_k = - v^{-1} \int_{A_k} \mathbf{n}_k dA \tag{2.56}$$

Note that within the averaging volume, just as depicted in Fig. 2.3, we have one averaged value of ψ_k, one averaged value of its gradient, and one averaged value of its time derivative now distributed over the volume v. (Prob. 2.4) We note that the averaging theorems, being mathematical theorems, are not restricted by the aforementioned restriction of characteristic dimensions. The physical meaning of the averaged quantities $\langle \psi \rangle$ are subject to that restriction, however.

Time averaging will be dealt with after considering the volume averaged equations of conservation.

2.4 Volume Averaged Equations of Conservation and Interface Balance Equations

Application of the volume averaging theorems given by Eqs. (2.52) to (2.54) to the equations of conservation given by Eqs. (2.13) to (2.17) leads to the volume-averaged conservation equations of multiphase flow.

The mass conservation equation is given by the continuity equation (2.13) as

$$(\partial/\partial t)<\rho_k> + \nabla \cdot <\rho_k U_k> = -v^{-1}\int_{A_k} \rho_k(U_k - U_s) \cdot n_k dA \qquad (2.57)$$

The integral on the right-hand side of Eq. (2.57) denotes the rate of total mass generation of phase k per unit volume v, the control volume. It is readily seen that if the interface A_k displaces in the way that the integral denoted by

$$\Gamma_k = -v^{-1}\int_{A_k} \rho_k(U_k - U_s) \cdot n_k dA \equiv -v^{-1}\int_{A_k} \dot{m}_k dA \qquad (2.58)$$

is greater than zero, there is a net generation of phase k in v. Note that Γ_k is related to the physical process of phase change and chemical reactions. (Prob. 2.5)

The linear momentum conservation equation is given by averaging Eq. (2.14):

$$(\partial/\partial t) <\rho_k U_k> + \nabla \cdot <\rho_k U_k U_k> = - \nabla<P_k> + \nabla \cdot <\tau_k> + <\rho_k>f_k$$

$$+ v^{-1}\int_{A_k} (-P_k I + \tau_k) \cdot n_k dA - v^{-1}\int_{A_k} \rho_k U_k(U_k - U_s) \cdot n_k dA \quad (2.59)$$

The integrals account for the transfer of pressure, viscous stresses, and inertia forces across the interface per unit volume. The field force per unit mass f_k is taken to be a constant. (Prob. 2.6)

The equation of energy conservation in terms of toral energy is given by averaging Eq. (2.15):

$$(\partial/\partial t) \; <\rho_k E_k> + \nabla \cdot <\rho_k U_k E_k> = - \nabla \cdot <U_k P_k> + \nabla \cdot <U_k \cdot \tau_k>$$

$$- \nabla \cdot <J_{qk}> + <\rho_k U_k> \cdot f_k + <J_{Ek}> - v^{-1} \int_{A_k} J_{qk} \cdot n_k dA$$

$$+ v^{-1} \int_{A_k} (-P_k U_k + \tau_k \cdot U_k) \cdot n_k dA - v^{-1} \int_{A_k} \rho_k E_k (U_k - U_s) \cdot n_k dA \qquad (2.60)$$

where the integrals account for the heat transfer across the interface, the rate of work done by the pressure force, the energy dissipated by viscous forces, and the energy transferred by the relative motion at the interface per unit volume v. In terms of internal energy, Eq. (2.16) gives

$$(\partial/\partial t) \; <\rho_k u_k> + \nabla \cdot <\rho_k U_k u_k> = - <P_k \nabla \cdot U_k> - \nabla \cdot <J_{qk}>$$

$$+ <J_{Ek}> + <\phi_k> - v^{-1} \int_{A_k} J_{qk} \cdot n_k dA$$

$$- v^{-1} \int_{A_k} \rho_k u_k (U_k - U_s) \cdot n_k dA \qquad (2.61)$$

where $<\phi_k>$ is the dissipation function or the dissipation rate per unit volume given by:

$$<\phi_k> = <\tau_k : \nabla U_k> \qquad (2.62)$$

where the double dot denotes the scalar product of two second-order tensors and the gradient is a dyadic operation. ϕ represents irreversible dissipation of mechanical work into thermal energy. In terms of enthalpy, Eq. (2.17) gives

$$(\partial/\partial t) \; <\rho_k h_k> + \nabla \cdot <\rho_k U_k h_k> = (\partial/\partial t) <P_k> + \nabla \cdot <U_k P_k>$$

$$- <P_k \nabla \cdot U_k> - \nabla \cdot <J_{qk}> + <J_{Ek}> + <\phi_k>$$

$$- v^{-1} \int_{A_k} J_{qk} \cdot n_k dA - v^{-1} \int_{A_k} \rho_k h_k (U_k - U_s) \cdot n_k dA \qquad (2.63)$$

We note that in these averaged equations, A_k includes all interfacial areas inside v, and all the terms inside the integrals are local values at the interface.

The volume averaged interface balance equations are given by applying the averaging theorems to Eqs. (2.19)to (2.23). The mass balance is now given by:

$$\Gamma_k = - v^{-1} \int_{A_k} \rho_k (U_k - U_s) \cdot n_k dA$$

$$= v^{-1} \int_{A_k} \rho_f (U_f - U_s) \cdot n_f dA = - \Gamma_f \qquad (2.64)$$

Note that the first and second equalities constitute two independent relations because Γ_k is given by the physical and phenomenological relations at the interface.

The linear momentum balance is given by:

$$v^{-1} \int_{A_k} (-\nabla_s \sigma_{kf} + 2\sigma_{kf} H_{kf} n_k) \, dA + v^{-1} \int_{A_k} (-P_k I + \tau_k) \cdot n_k dA$$

$$-v^{-1} \int_{A_k} \rho_k U_k (U_k - U_s) \cdot n_k dA = v^{-1} \int_{A_k} (-P_f I + \tau_f) \cdot n_k dA$$

$$+ v^{-1} \int_{A_k} \rho_k U_f (U_f - U_s) \cdot n_k dA \qquad (2.65)$$

For bubbles and droplets, the first integral gives the capillary pressure (subscript c) difference:

$$v^{-1} \int_{A_k} (-\nabla_s \sigma_{kf} + 2\sigma_{kf} H_{kf} n_k) dA = v^{-1} \int_{A_k} (P_{ck} - P_{cf}) n_k dA \qquad (2.66)$$

Ignoring the capillary energy, the total energy balance is given by:

$$v^{-1} \int_{A_k} (-P_k U_k + \tau_k \cdot U_k) \cdot n_k dA - v^{-1} \int_{A_k} \rho_k E_k (U_k - U_s) \cdot n_k dA$$

$$- v^{-1} \int_{A_k} J_{qk} \cdot n_k dA = - v^{-1} \int_{A_k} (-P_f U_f + \tau_f \cdot U_f) \cdot n_k dA$$

$$+ v^{-1} \int_{A_k} \rho_f E_f (U_f - U_s) \cdot n_f dA + v^{-1} \int_{A_k} J_{qf} \cdot n_f dA \qquad (2.67)$$

Alternately, the energy balance can be expressed in terms of internal energy or enthalpy as

$$-v^{-1} \int_{A_k} \rho_k u_k (U_k - U_s) \cdot n_k dA - v^{-1} \int_{A_k} J_{qk} \cdot n_k dA$$

$$= v^{-1} \int_{A_k} \rho_f u_f (U_f - U_s) \cdot n_f dA + v^{-1} \int_{A_k} J_{qf} \cdot n_f dA \qquad (2.68)$$

or

$$v^{-1} \int_{A_k} P_k (U_k - U_s) \cdot n_k dA - v^{-1} \int_{A_k} \rho_k h_k (U_k - U_s) \cdot n_k dA - v^{-1} \int_{A_k} J_{qk} \cdot n_k dA$$

$$= - v^{-1} \int_{A_k} P_f (U_f - U_s) \cdot n_f dA + v^{-1} \int_{A_k} \rho_f h_f (U_f - U_s) \cdot n_f dA$$

$$+ v^{-1} \int_{A_k} J_{qk} \cdot n_k dA \qquad (2.69)$$

The above equations include averages and averages of products as well as local values in the interphase transfer integrals. The configurations of the interface and its motion are given by U_s, n_k, and A_k. Solution of these equations calls for expressing averages of products in terms of products of averages and to express the integrals in terms of averaged dependent variables by introducing proper constitutive relations.

We note that, the volume averaged momentum of phase k is given by Eq. (2.34). Similarly the volume averaged rate of change of momentum flux of phase k per unit area is:

$$<\rho_k U_k U_k> = <\rho_k> {}^i\!U_k U_k> = \alpha_k {}^i\!<\rho_k> {}^i\!U_k U_k> \qquad (2.70)$$

and the volume averaged rate of change of internal energy per unit volume is given by:

$$<\rho_k u_k> = <\rho_k> {}^i\!u_k> = \alpha_k {}^i\!<\rho_k> {}^i\!u_k> \qquad (2.71)$$

and the rate of change of energy flux of phase k per unit area can be treated in an analogous manner.

Incorporation of a local deviation from each average has been suggested [Anderson and Jackson 1967] as a means for expressing the averages of products in terms of the products of averages but at the expense of introducing averages of products of deviations whose determination in general calls for solving the phasic equations with discrete interfaces; thus gaining no advantage of the averaging procedure. We shall show that local deviations can be neglected when treating dispersed systems (Sec. 2.7), the main object of the averaging theorems. (Prob. 2.7) However, useful details are gained by expressing averages of products such as $^i\langle U_k U_k \rangle$ and $^i\langle U_u u_k \rangle$ in terms of products of averages through the use of time averaging in the next section.

We further note that the above equations include extensive or control volume averages. They are related to the intrinsic averages via Eq. (2.31) by direct substitution subjecting to the consideration in item (c) of Sec. 2.3. Moreover, physical meaning is conveyed, in many cases, by the extensive averages, especially when viewing a dispersed phase. For instance, the density of a particle cloud is directly given by $\langle \rho_k \rangle$ for distribution over the entire control volume, while the equality in Eq. (2.33) is an alternate expression. Yet as has been noted previously, while the stress within a phase is given by $^i\langle \tau_k \rangle$, the resistance to shear motion of a particle cloud is pertinently represented by $\langle \tau_k \rangle$, because $\langle \tau_k \rangle$ here does not necessarily relate to the stress inside phase k. The same meaning is seen in the case of the heat flux $\langle J_{qk} \rangle$, as well as the quantities Γ_k, P_k, J_{Ek}, and ϕ_k. One sees, however, the physically meaningful averaged values of U_k, u_k, h_k, T_k, are their properly determined intrinsic averages.

2.5 Time Averaging and Basic Equations

Preference for a priori volume averaging to time averaging is seen for two reasons: (1) A priori time averaging eliminates the identity of different dynamic phases [Soo 1965]; (2) even when distinction can be made of dynamic phases via different durations of passage, time averaging gives residence time fraction of a phase, which is equal to volume fraction only for one-dimensional uniform motion [Sha et al. 1983]. This is in spite of the fact that a priori time averaging is valid for a single phase fluid such as in the Reynolds analysis. The purpose of time averaging after volume averaging is to express averages of products in terms of products of averages and to account for high frequency fluctuations. The time averaging theorems of Delhaye and Achard [1976] can no longer be applied because the interface disappears after volume averaging (Fig. 2.2). This step is therefore accomplished by identifying high frequency deviations within the volume averages and determining them by time averaging in the way of Reynolds analysis. While deviations can be identified for various frequencies, including the local deviations referred to in the above section, high frequency deviations alone are sufficiently illustrative.

The averaging time is chosen over a duration T such that

$$\tau_{HF} < T < \tau_{LF} \tag{2.72}$$

where τ_{HF} is the characteristic time of the high frequency component and is of the order of 1/(characteristic spectral frequency of fluctuations), while τ_{LF} is the characteristic time of the low frequency component and is of the order of (characteristic dimension of the physical system)/(characteristic low frequency speed variation at that location). While we speak of frequency, periodicity is not always implied. The local density can be expressed as

$$\rho_k \sim \rho_{kLF} + \rho_k' \tag{2.73}$$

where subscript LF denotes the low frequency component and the prime denotes the high frequency component. We note that its extensive average is given by

$$<\rho_k> = <\rho_k>_{LF} + <\rho_k'> \tag{2.74}$$

$<\rho_k'>$ here includes the fluctuations in the material density of the phase k and in its volume fraction. Velocity can be treated in a similar way, except only its intrinsic average is meaningful:

$${}^i<U_k> = {}^i<U_k>_{LF} + {}^i<U_k'> \tag{2.75}$$

so also u_k and temperature T_k. Time averaging now gives

$${}^t<<\rho_k>_{LF}> = T^{-1} \int_0^T <\rho_k>_{LF}\, dt = <\rho_k>_{LF} \tag{2.76}$$

while, for the high frequency term,

$${}^t<<\rho_k'>> = T^{-1} \int_0^T <\rho_k'>\, dt = 0 \tag{2.77}$$

That is, by a proper choice in the perturbation procedure,

$${}^t<<\rho_k>> = <\rho_k>_{LF} \tag{2.78}$$

and similarly for Γ_k, P_k, τ_k, J_{gk}, and J_{EK}. (Prob. 2.8)

When time averaging is applied to products of averages, we get

$${}^t<<\rho_k U_k>> = <\rho_k>_{LF}\ {}^i<U_k>_{LF} + {}^t<<\rho_k' U_k'>>$$

$$= <\rho_k>_{LF}\ {}^i<U_k>_{LF} - D_{km}^T \nabla <\rho_k>_{LF} \tag{2.79}$$

The last term in Eq. (2.79) represents the mass flux due to eddy diffusion of phase k in the mixture, originating from velocity fluctuations [Bousinesq 1877]. Similar procedure is applicable to $<\rho_k' u_k'>$ although $<\rho_k' u_k'>$ is simply the fluctuation in internal energy per unit volume.

The terms that arise from the time averaging of a volume averaged triple product is best illustrated by the term in the local volume averaged internal energy flux:

$$^t{<<\rho_k U_k u_k>>} = {<\rho_k>}_{LF} \, ^i{<U_k>}_{LF} \, ^i{<u_k>}_{LF} + {<\rho_k>}_{LF} \, ^t{_<}{^i{<u_k' U_k'>>}}$$

$$+ \, ^i{<u_k>}_{LF} + \, ^t{<<\rho_k' U_k'>>} + \, ^i{<U_k>}_{LF} \, ^t{<<\rho_k' u_k'>>} \qquad (2.80)$$

while the terms including triple correlations are neglected. The second term represents thermal energy transport due to fluctuation in velocity and internal energy (or temperature) and may be expressed as the eddy diffusivity for internal energy transfer D_{uk}^T. Therefore,

$$-{<\rho_k>}_{LF} \, ^t{_<}{^i{<U_k' u_k'>>}} = {<\rho_k>}_{LF} \, D_{uk}^T \, \nabla \, ^i{<u_k>}_{LF}$$

$$\equiv - \, J_{qk}^T = \kappa_k^T \, \nabla \, ^i{<T_k>}_{LF} \qquad (2.81)$$

Because $\nabla^i{<u_k>} = c_{vk} \nabla^i{<T>}$; when expressed in terms of the gradient of temperature, and κ_k is the eddy conductivity. The third term in Eq. (2.80) is simply $-^i{<u_k>}_{LF} D_{mk}^T \nabla{<\rho_k>}_{LF}$, the energy transport by diffusion. The average of momentum flux is given by replacing u_k with U_k in Eq. (2.80), and

$$^t{<<\rho_k U_k U_k>}} = {<\rho_k>}_{LF} \, ^i{<U_k>} \, ^i{<U_k>} + {<\rho_k>}_{LF} \, ^t{_<}{^i{<U_k' U_k'>>}}$$

$$+ \, 2 \, ^i{<U_k>} \, D_{km}^T \, \nabla{<\rho_k>}_{LF} \qquad (2.82)$$

The second term on the right hand side is the Reynolds stress, and the third term is the momentum flux arising from eddy mass diffusion.

The interface transfer integrals, after time averaging, also give transfer integrals with low and high frequency components. However, the behaviors of the interface must be accounted for; while the fluctuating part of v_k or v_k' varies from v_k with both positive and negative values, A_k' due to wrinkling of the interface A_k is always positive. An idealization is illustrated in Fig. 2.6, showing A_k consisting of A_{kLF} and A_k' and their coherent directional normals

n_{kLF} and n_k'. A quantity of interest is the volume fraction is an extensive averaged quantity, v_k/v. For α_k in the form

$$\alpha_k = \alpha_{kLF} + \alpha_k' \tag{2.83}$$

the meaning of α_k' is seen from Eq. (2.56), which gives coherent components of

$$\nabla \alpha_{kLF} = - v^{-1} \int_{A_{kLF}} n_{kLF} \, dA \tag{2.84}$$

and

$$\nabla \alpha_k' = -v^{-1} \int_{A_k'} n_k' \, dA \tag{2.85}$$

Similarly, from Eq. (2.55), we have

$$\partial \, \alpha_{kLF}/\partial t = v^{-1} \int_{A_{kLF}} U_{SLF} \cdot n_{kLF} \, dA \tag{2.86}$$

$$\partial \, \alpha_k'/\partial t = v^{-1} \int_{A_k'} U_s' \cdot n_k' \, dA \tag{2.87}$$

Hence α_k' exists only when the interface has high frequency disturbances such that U_s', n_k', and A_k' exist. The physical meaning of Eq. (2.87) is seen as a perturbation of velocity $U_s' \cdot n_k'$ on an advancing front (or oscillating bubbles) at velocity $U_s' \cdot n_k$ over area A_k, such

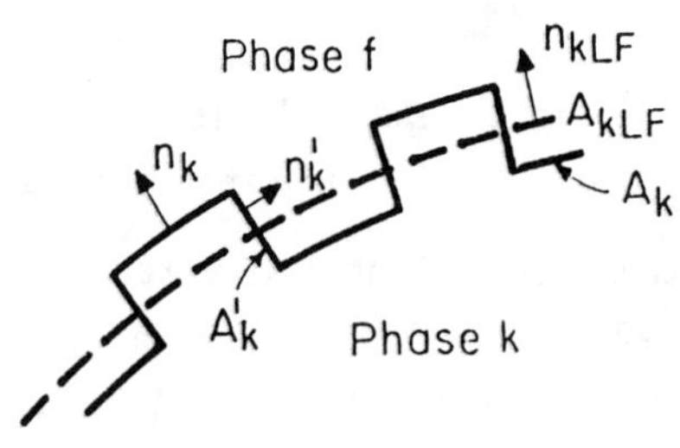

Fig. 2.6 Idealized Fluctuation of Interface

as in Fig. 2.1. Substitution of the above relations into Eqs. (2.57) to (2.69) gives rise to a set of time-volume averaged conservation equations in terms of averaged dependent variables with local values in the integrals. Neglecting all triple correlations, we have additional integrals with L'_{uk} in the energy equation given by

$$L_{uk} + L'_{uk} = {}^{t}\langle {}_{-v}{}^{-1} \int_{A_k} \rho_k\, u_k\, (U_k - U_s) \cdot n_k\; dA\rangle \qquad (2.88)$$

and

$$L'_{uk} = {}^{t}\langle {}_{-1}{}^{-1} \int_{A'_k} [\rho_k\, u'_k(U'_k - U'_s) + u_k\, \rho'_k(U'_k - U'_s)] \cdot n'_k\; dA\rangle$$

$$- v^{-1} \int_{A_{kLF}} \langle \rho'_k u'_k\rangle\, (U_k - U_s) \cdot n_k\; dA \qquad (2.89)$$

Similarly the interface momentum transfer integral can be expressed as

$$I_k + I'_k = {}^{t}\langle {}_{-v}{}^{-1} \int_{A_k} \rho_k\, U_k(U_k - U_s) \cdot n_k\; dA\rangle \qquad (2.90)$$

and I'_k is given by replacing u_k with U_k and u'_k by U'_k in Eq. (2.89). We note that volume-averaged equations with averages of products equal to products of averages are sufficient when high frequency fluctuations are absent. We note that not every quantity pertaining to high frequency can be related to known phenomenological relations or quantities identifiable with turbulent flow. This is especially true for those quantities inside the transfer integrals. Some of the details remains to be completed via future research, both theoretical and experimental.

By limiting the formulations to identifiable quantities, the volume averaged Eqs. (2.57) to (2.60), after time averaging, take the forms: (dropping subscript LF)

$$(\partial/\partial t)\langle\rho_k\rangle + \nabla\cdot\langle\rho_k\rangle {}^i\langle U_k\rangle - \nabla \cdot D^T_{mk}\, \nabla\langle\rho_k\rangle = \Gamma_k \qquad (2.91)$$

$$\frac{\partial}{\partial t}\left(<\rho_k>{}^i<U_k> - D^T_{mk}\,\nabla<\rho_k>\right) + \nabla\cdot<\rho_k>{}^i<U_k>{}^i<U_k> = -\nabla<P_k>$$

$$+ \nabla\cdot(<\tau_k> + \tau^T_k) + 2\nabla\cdot{}^i<U_k>\,D^T_{mk}\,\nabla<\rho_k> + <\rho_k>\,f_k$$

$$-v^{-1}\int_{A_k}\rho_k U_k(U_k-U_s)\cdot n_k\,dA + v^{-1}\int_{A_k}(-P_k I+\tau_k)\cdot n_k\,dA + I'_k \qquad (2.92)$$

$$\frac{\partial}{\partial t}(<\rho_k>{}^i<E_k>) + \nabla\cdot(<\rho_k>{}^i<U_k>{}^i<E_k>) = -\nabla\cdot(<\rho_k>{}^i<U_k> + {}^t<P'_k U'_k>)$$

$$+ \nabla\cdot({}^i<U_k>\cdot<\tau_k+\tau^T_k> + {}^t<U'_k\cdot\tau'_k>) + \nabla\cdot<\rho_k>(D_{uk}+D^T_{uk})\,\nabla^i<u_k>$$

$$+ \nabla\cdot({}^i<E_k>\,D^T_{mk}\,\nabla<\rho_k>) + [<\rho_k>{}^i<U_k> - D^T_{mk}\nabla<\rho_k>]\cdot f_k$$

$$+ <J_{Ek}> + L_{uk} + L'_{uk} + {}^t<\int_{A_k}(-P_k U_k + \tau_k U_k)\cdot n_k\,dA> \qquad (2.93)$$

These, together with the interface balance relations in Eqs. (2.64)
to (2.67) after time averaging constitute the basic equations of
multiphase flow. Equation (2.93) can be replaced by the one based on
internal energy or enthalpy, so also the energy balance relation.

2.6 Closure Relations

Leaving out the effects of turbulence for the present, the basic
equations and balance relations for n phases have the following
number of dependent variables including vectorial quantities:

Averages: ${}^i<\rho_k>$, ${}^i<U_k>$, ${}^i<u_k>$ 3n

$$\alpha_k \qquad\qquad (n-1)$$

with $u_k = E_k - (1/2)U_k\cdot U_k$, $P_k = P_k(\bar\rho_k, u_k)$ and $\sum_k\alpha_k = 1$.

Local variables:

$$\rho_k,\ U_k,\ u_k \qquad\qquad 3n$$

For n(n-1)/2 interfaces arising from n interacting phases:

$$^i\langle U_{sn}\rangle, \ \langle A_k/v\rangle, \ \langle n_k\rangle; \quad U_{sn} = U_s \cdot n_k, \ S(r,t), \ n_k \quad 3n(n-1)$$

S being the local geometry of the interface, and one notes that only the product $U_s \cdot n_k = U_{sn}$ is independent, but not U_s. We have altogether $7n -1 + 3n(n-1)$ dependent variables after taking into account of relations among properties, transport parameters and available constitutive or phenomenological relations. The local variables for evaluating the transfer integrals which themselves can be replaced by proper constitutive relations based on averages and transport parameters.

The number of independent equations are:

Averaged equations:	(2.91)-(2.93)	$3n$
Balance equations:	(after time averaging)	
	(2.64) with generation	$(n-1)$
	(2.64) with balance	$n(n-1)/2$
	(2.65)	$n(n-1)/2$
	(2.66)	$n(n-1)/2$
Local equations:	(2.13)-(2.15)	$3n$
Local balance:	(2.19)-(2.21)	$3n(n-1)/2$

Hence, at least in principle, the basic equations with boundary conditions give a deterministic solution for the averaged dependent variables, including the configuration of the interfaces, via A_k and n_k and surface $S(r,t)$. The closure for the solution is assured. In reality, however, the whole idea of averaging is to avoid solving local equations. In the procedures to follow, the transfer integrals of specific physical systems are reduced (or will be reduced in the future) to constitutive relations in terms of averaged properties and phenomenological or nondissipative coefficients.

Such an approach is illustrated by the mass transfer integral where the evaporation rate is given by, for instance, Eq. (1.30) for n_k bubbles per unit volume:

$$\Gamma_k = v^{-1} \int_{A_k} \rho_k (U_k - U_s) \cdot n_k \, dA = n_k (dm_k/dt) \qquad (2.94)$$

(See, for instance, Sec. 1.4). We also note that at least part of
the transfer integral including shear stress can be expressed as the
drag force on a phase. The momentum integral I_k can be similarly
determined for specific physical systems:

$$I_k = -v^{-1} \int_{A_k} \rho_k (U_k - U_s)(U_k - U_s) \cdot n_k \, dA \qquad (2.95)$$

and similarly for energy transfer integrals. In this way, the local
equations and the local balances will not be part of the independent
equations, nor are local dependent variables.

2.7 Simplified Formulations

Simplified formulations are found in the idealized cases of pure
stratified flow and highly dispersed flow, both consist of partial
differential equations only. The case of dilute suspensions has
broader applications. To be specially pointed out here is the in-
adequacy of pure stratified flow formulation for representing general
multiphase systems.

Pure stratified flow. A pure stratified flow system is defined as one
having phases divided by plane interfaces. The transfer of inertia
force and pressure force across the interface is negligible and the
interfacial momentum transfer is by viscous forces only. The interfacial heat transfer is also diffusional in origin. In this case
turbulence cannot exist because waves at the interface will cause
transfer of inertia force across the interface. Therefore, all
contributions associated with eddy diffusion and Reynolds stress and
eddy conductivity have to be neglected. A set of simplified differential equations is obtained for mass, momentum, and energy
conservations:

$$(\partial/\partial t) \, \alpha_k \, {}^i\!<\rho_k> + \nabla \cdot \alpha_k \, {}^i\!<\rho_k> \, {}^i\!<U_k> = \Gamma_k \qquad (2.96)$$

$$(\partial/\partial t)(\alpha_k \,^i\langle\rho_k\rangle \,^i\langle\mathbf{U}_k\rangle) + \nabla \cdot \alpha_k \,^i\langle\rho_k\rangle \,^i\langle\mathbf{U}_k\rangle \,^i\langle\mathbf{U}_k\rangle = -\alpha_k \nabla \,^i\langle P_k\rangle$$

$$+ \nabla \cdot \alpha_k \,^i\langle\tau_k\rangle + \alpha_k \,^i\langle\rho_k\rangle \cdot \mathbf{f}_k + \mathbf{V}_k \qquad (2.97)$$

$$(\partial/\partial t)(\alpha_k \,^i\langle\rho_k\rangle \,^i\langle u_k\rangle) + \nabla \cdot \alpha_k \,^i\langle\rho_k\rangle \,^i\langle\mathbf{U}_k\rangle \,^i\langle u_k\rangle = \alpha_k \,^i\langle P_k\rangle \nabla\cdot \,^i\langle\mathbf{U}_k\rangle$$

$$+ \nabla \cdot \alpha_k \,^i\langle\mathbf{J}_{qk}\rangle + \alpha_k \,^i\langle\rho_k\rangle \,^i\langle\mathbf{U}_k\rangle \cdot \mathbf{f}_k + \alpha_k \,^i\langle J_{Ek}+\phi_k\rangle + Q_k \qquad (2.98)$$

where V_k is the interfacial viscous drag force per unit volume,

$$V_k = v^{-1} \int_{A_k} \tau_k \cdot n_k \, dA \equiv \alpha_k \,^i\langle\rho_k\rangle F_{kf}(\,^i\langle\mathbf{U}_f\rangle - \,^i\langle\mathbf{U}_k\rangle)$$

$$Q_k = -v^{-1} \int_{A_k} \mathbf{J}_{qk} \cdot \mathbf{n}_k \, dA \equiv \alpha_k \, c_{vk} \,^i\langle\rho_k\rangle G_{kf}(\,^i\langle T_f\rangle - \,^i\langle T_k\rangle) \qquad (2.99)$$

and Q_k is the interacial heat transfer rate per unit volume. F_{kf} and G_{kf} are inverse relaxation times defined according to Eqs. (1.9) and (1.18). All formulations neglecting the transfer of inertia force across the interface are valid only for pure stratified flow (Fig. 2.2). (Prob. 2.9, 2.10)

Dilute suspensions of spheres. Another possible approach for solving the differential-integral equation is by reducing the integrals to derivatives [Soo 1981]. Such a simplification is feasible in the case of a dilute suspension of spheres in a fluid. In the dynamic sense, the latter is defined by the condition of the effect of particle-particle interaction being negligible when compared to particle-fluid interaction. Quantitatively this condition is satisfied when the mean free path of particle-particle interaction λ_k is much larger than the particle-fluid interaction length given by Eq. (1.72). λ_k is given by kinetic analogy to be $[(2)^{\frac{1}{2}} n_k 4\pi a^2]^{-1}$, n_k being the number density of spheres of radius a in suspension. [Soo 1976] (Prob. 2.11, 2.12, 2.13)

It is seen that for the general case of a fluid sphere k in a pressure gradient in a continuum phase f, some of the terms in Eq. (2.59) can be reduced according to, with Eq. (2.56)

$$-\nabla(\alpha_k \, {}^i\!<P_k>) - v^{-1} \int_{A_k} (P_k I - \tau_k) \cdot \mathbf{n}_k \, dA$$

$$= -\alpha_k \, \nabla \, {}^i\!<P_k> + v^{-1} \int_{A_k} \tau_k \cdot \mathbf{n}_k \, dA \equiv F_{kf}\alpha_k \, {}^i\!<\rho_k>({}^i\!<U_f> - {}^i\!<U_k>)$$

$$(2.100)$$

where F_{kf} is the inverse relaxation time for momentum transfer from phase f to phase k; since the pressure gradient in the flow field ∇P_f must be accompanied by a velocity gradient [Soo 1976]. Hence, for a dilute suspension of spheres, the corresponding terms in the momentum equation of phase f is now

$$-\nabla(\alpha_f \, {}^i\!<P_f>) - v^{-1} \int_{A_k} (P_f I - \tau_f) \cdot \mathbf{n}_f \, dA$$

$$\equiv - \nabla \, {}^i\!<P_f> - F_{fk} \, \alpha_f \, {}^i\!<\rho_f> ({}^i\!<U_f> - {}^i\!<U_k>) \qquad (2.101)$$

noting that $\alpha_k + \alpha_f = 1$, when we consider two phases only, k represents a dilute suspension of spheres and equality of action and reaction gives:

$$F_{kf}<\rho_k> = F_{fk}<\rho_f> \qquad\qquad (2.102)$$

To reduce the other interface integrals, the present approximation rewrites Eq. (2.57), for the above phases f and k, as

$$\Gamma_k = \Gamma_{km} + \Gamma_{ko} = -\Gamma_f = -\Gamma_{fm} - \Gamma_{fo} \qquad (2.103)$$

where Γ_{ko} is the generation rate of phase k by mass transfer at nearly constant radius of spheres k, Γ_{km} is the part related to the difference in velocities and remains to be determined. The integral of Γ_k in Eq. (2.57) can be evaluated for the internal motion in k and flow distribution of f around each sphere such as in Fig. 2.7a. These details a simplified by an approximation in Fig. 2.7b which physically represents the case of thin boundary layers around the spheres. For this flow field, we use subscript p for the spheres to distinguish from Fig. 2.7a, with $^i\!<U_k> \sim U_s = U_p$ in a fluid at average velocity $^i\!<U_f>$. The integral in Eq. (2.103) is now given by

$$\Gamma_{fm} + \Gamma_{fo} = -v^{-1} \int_{A_k} {}^i\!<\rho_f> \, (^i\!<U_f> - U_p) \cdot n_f \, dA + \Gamma_{fo} \qquad (2.104)$$

This integral, based on control volume v, can be reduced to that around N_p cavities of radius a_p around each sphere of volume v_p and surface area A_p; n_p is the normal around a sphere, such that

$$\Gamma_{fm} = -v_p^{-1} \int_{A_p} {}^i\!<\rho_f> \, (^i\!<U_f> - U_p) \cdot n_p \, (N_p v_p/v) \, dA$$

$$= \nabla \cdot <\alpha_p \, {}^i\!<\rho_f> \, (^i\!<U_f> - U_p)> = 0 \qquad (2.105)$$

according to Eq. (2.54) and the average of divergence of the integrand is zero because the velocities and density are uniform over all the v_p's. Further, because the velocity difference changes sign around the spherical cavity, $\Gamma_{fm} = 0$.

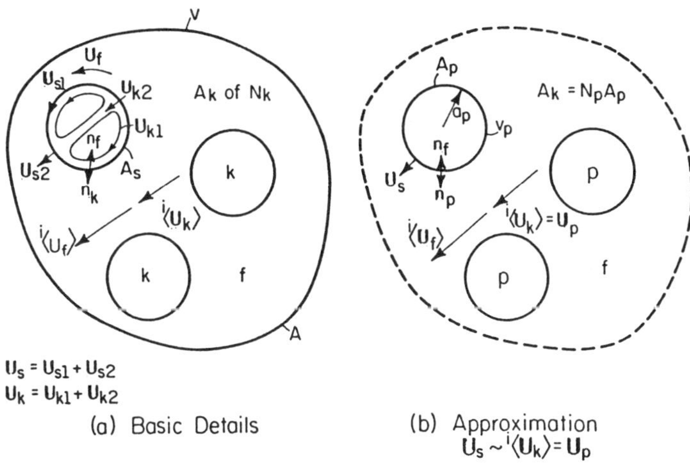

Fig. 2.7 Approximation for Fluid Spheres in Suspension

The momentum transfer integral can be approximated by

$$I_f = -v^{-1} \int_{A_k} \rho_f \, U_f(U_f - U_s) \cdot n_f \, dA$$

$$= v_p^{-1} \int_{A_p} \alpha_p \,^i\!<\rho_f> (^i\!<U_f> - U_p)(^i\!<U_f> - U_p) \cdot n_p \, dA$$

$$+ v_p^{-1} \int_{A_p} \alpha_p \,^i\!<\rho_f> \, U_s \, (^i\!<U_f> - U_p) \cdot n_p \, dA + \Gamma_{fo} \, U_{so} \qquad (2.106)$$

It is seen that (a) the first term after the equality sign does not
vanish because the product is always positive and similar reasoning
as in Eq. (2.105) applies; (b) application of Eq. (2.52) to the
second term of Eq. (2.106) and the fact that the average of the time
derivative because the motion is steady over the duration passing v_p,
which is short. Hence, we have

$$I_f = \nabla \cdot [\alpha_k \,^i\!<\rho_f> (^i\!<U_k> - \,^i\!<U_f>)(^i\!<U_k> - \,^i\!<U_f>)]$$

$$+ (\partial/\partial t)[\alpha_k \,^i\!<\rho_f> (^i\!<U_k> - \,^i\!<U_f>)] + \Gamma_{fo} \, U_{so} \qquad (2.107)$$

The first two terms constitute the virtual mass force acting on the
fluid by the cloud of spheres k of volume fraction α_k. An equal and
opposite force acts on the spheres by the interface momentum balance,
and is the continuum counterpart of the virtual mass force in Eq.
(1.47). The terms $\Gamma_{fo} U_{so}$ are the momentum source due to generation Γ_{fo} and $U_{so} = \,^i\!<U_k>$ for $\Gamma_{fo} > 0$, such as in the case of condensation from a vapor phase k at velocity $^i\!<U_k>$. For the case of solid
spheres, the momentum transfer at its interface and its reaction
gives rise to a distributed stress. The transfer integral I takes a
different form but does not vanish. The first two terms on the
right-hand side of Eq. (2.107) reduces to a form similar to the
virtual mass force term in Eq. (1.47) when $\alpha_k \bar{\rho}_f$ is a constant and Eq.
(2.105) is applied.

Similarly, the interface transfer integral of internal energy
can be reduced to

$$L_f = v^{-1} \int_{A_k} (u_f - u_s)(U_f - U_s) \cdot n_f \, dA + \Gamma_{fo} \, u_{so} \qquad (2.108)$$

For the same approximation in Fig. 2.6b, $(u_f - u_s)$ has only one sign, since $(^i<U_f> - U_p)$ still changes sign around the sphere, the integral over v_p goes to zero. u_{so} here is the latent heat u_{fk} at constant volume; for $\Gamma_{fo} > 0$, $u_{fk} > 0$; that is, for the case where k is the vapor phase, heat is released when condensation occurs. The other interface transfer integrals are reduced such that at the interface

$$-v^{-1} \int_{A_k} J_{qk} \cdot n_k \, dA = Q_{ks} = \alpha_k \, \bar{\rho}_k \, c_{vk} \, G_{kf}(T_f - T_k) \qquad (2.109)$$

G_{kf} is the inverse relaxation time for heat transfer, and in addition to the equality in Eq. (2.102), we have

$$c_{vk}<\rho_k> G_{kf} = c_{vf}<\rho_f> G_{fk} \qquad (2.110)$$

Other integrals in the energy equation are reducible to the rate of work done and dissipation at the interface per unit volume. With the above approximations, the averaged equations are reduced to a set of partial differential equations: (for bubble phase k)

$$(\partial/\partial t) \, \alpha_k \, \bar{\rho}_k + \nabla \cdot \alpha_k \, \bar{\rho}_k \, U_k = \Gamma_k \qquad (2.111)$$

$$(\partial/\partial t) \, \alpha_f \, \bar{\rho}_f + \nabla \cdot \alpha_f \, \bar{\rho}_f \, U_f = \Gamma_f \qquad (2.112)$$

$$(\partial/\partial t)(\alpha_k \bar{\rho}_k U_k) + \nabla \cdot \alpha_k \, \bar{\rho}_k \, U_k \, U_k = \nabla \cdot \tau_{km} + \alpha_k \, \bar{\rho}_k \, f_k$$

$$+ \, \alpha_k \, \bar{\rho}_k \, F_{kf} \, (U_f - U_k) + \Gamma_k \, U_s$$

$$- \, \alpha_k \, \bar{\rho}_f [(\partial/\partial t)(U_k - U_f) + (U_k - U_f) \, \nabla \cdot (U_k - U_f)] \qquad (2.113)$$

$$(\partial/\partial t)(\alpha_f \, \bar{\rho}_f \, U_f) + \nabla \cdot \alpha_f \, \bar{\rho}_f \, U_f U_f = -\nabla P + \nabla \cdot \tau_{fm} + \alpha_f \, \bar{\rho}_f \, f_f$$

$$+ \, \alpha_f \, \bar{\rho}_f \, F_{fk}(U_k - U_f) - \Gamma_k \, U_s$$

$$+ \; \alpha_k \; \bar{\rho}_f [(\partial/\partial t)(U_k - U_f) + (U_k - U_f) \; \nabla \cdot (U_k - U_f)] \quad (2.114)$$

$$(\partial/\partial t)(\alpha_k \; \bar{\rho}_k \; h_k) + \nabla \cdot \alpha_k \; \bar{\rho}_k \; U_k \; h_k = - \nabla \cdot J_{qkm} + J_{EK}$$

$$+ \; \alpha_k \; \bar{\rho}_k \; c_{vk} \; G_{kf}(T_f - T_k) + \Gamma_k \; h_s \qquad\qquad (2.115)$$

$$(\partial/\partial t)(\alpha_f \; \bar{\rho}_f \; h_f) + \nabla \cdot \alpha_f \; \bar{\rho}_f \; U_f \; h_f + (1/2)(\partial/\partial t)(\alpha_k \; \bar{\rho}_k \; U_k^2 + \alpha_f \; \bar{\rho}_f \; U_f^2)$$

$$+ \; (1/2) \; \nabla \cdot (\alpha_k \; \bar{\rho}_k \; U_k \; U_k^2 + \alpha_f \; \bar{\rho}_f \; U_f \; U_f^2) = (\partial P/\partial t) + U_f \cdot \nabla P$$

$$+ \; \nabla \cdot J_{qfm} + J_{Ef} + \alpha_f \; \bar{\rho}_f \; c_{vf} \; G_{fk}(T_k - T_f) - \Gamma_k \; h_s \qquad (2.116)$$

where $U_s = U_k$, $h_s = h_{fg} > 0$ for $\Gamma_k < 0$; h_{fg} is the latent heat of vaporization; internal dissipation has been neglected (also $\Gamma_k > 0$, $U_s = U_f$). In the energy equations, the kinetic energy of the spheres arises from fluid drag and is therefore part of the kinetic energy of the fluid phase f. The $<\;>$ signs are dropped from this point on, although the use of time-volume averaged quantities is understood.

<u>Diffusion under field forces</u>. It is often convenient to solve the dynamic equations by replacing the continuity equation with the diffusion equation, especially when the particulate phase such as in the above is dilute. The continuity equation of the mixture can be expressed in the form:

$$(\partial\rho_m/\partial t) + U_m \; \nabla \cdot \rho_m \equiv d\rho_m/dt_m = -\rho_m \; \nabla \cdot U_m \qquad (2.117)$$

defining d/dt_m. The continuity of phase k can be rewritten as:

$$(\partial\rho_k/\partial t) = -\nabla \cdot (\rho_k \; U_m) + \nabla \cdot [\rho_k(U_m - U_k)] + \Gamma_k \qquad (2.118)$$

Since the flux of k is given by $J_k = \rho_k(U_k - U_m)$, where J_k includes both diffusion and convective fluxes, we have

$$\partial \rho_k / \partial t = -\rho_k \, \nabla \cdot \mathbf{U}_m - \mathbf{U}_m \cdot \nabla \rho_k - \nabla \cdot \mathbf{J}_k + \Gamma_k \qquad (2.119)$$

Substitution of d/dt_m gives

$$d\rho_k / dt_m = -\rho_m \, \nabla \cdot \mathbf{U}_m - \nabla \cdot \mathbf{J}_k + \Gamma_k \qquad (2.120)$$

For mass fraction c_k of phase k, $c_k = \rho_k / \rho_m$, we have

$$\rho_m (dc_k / dt_m) + c_k (d\rho_m / dt_m) = d\rho_k / dt_m \qquad (2.121)$$

with substitution of Eq. (2.117), we get the diffusion equation in the form:

$$\rho_m (dc_k / dt_m) = -\nabla \cdot \mathbf{J}_k + \Gamma_k = (d\rho_k / dt_m) - \rho_k (d\ell n \, \rho_m / dt_m) \qquad (2.122)$$

for any proportion of k to m. For the case of a dilute phase k, $\rho_m \gg \rho_k$, and

$$d\rho_k / dt_m = -\nabla \cdot \mathbf{J}_k + \Gamma_k = (\partial \rho_k / \partial t) + \mathbf{U}_m \cdot \nabla \rho_k \qquad (2.123)$$

is a simplified form and $\mathbf{U}_m$ approaches that of the predominating component. In this way, the equation for the density of k becomes independent of the momentum equation of phase k. The flux $\mathbf{J}_k$ is given by

$$\mathbf{J}_k = -D_{km} \, \nabla \rho_k + (f_k \, \rho_k / F_{kf}) \qquad (2.124)$$

The first term is due to diffusivity k in the mixture which is not necessarily turbulent, and the second term comes from the momentum equation of phase k such as given by Eq. (2.113), when the viscous force is much greater than the inertia force; f_k is the field force per unit mass.

Even for these simple cases, we see that we need transport properties to express terms such as viscous stresses, heat fluxes, drag forces, etc. in pertinent phenomenological relations in order to solve these basic equations or to correlate with experimental results.

While the general equations are applicable to all configurations of interfaces, their solution at this stage of development of multiphase flow is limited to a highly dispersed flow or a pure stratified flow system. In between, extensive empiricism is needed to account for various flow regimes (see, for instance, Govier and Aziz [1972]).

The present volume deals with cases where rigorous mathematical procedure can be applied, leaving ad hoc empirical correlations to handbooks and for future developments from fundamental procedures. Most of the topics therefore deal with particulates (solid particles, droplets, or bubbles).

General treatment of averages of products may also call for introducing in Section 2.4, local deviations from averages [Anderson and Jackson 1967] such as

$$\psi = {}^{i}\!<\psi> + \tilde{\psi} \tag{2.125}$$

so that, from Eq. (2.70)

$${}^{i}\!<U_k U_k> = {}^{i}\!<U_k> \, {}^{i}\!<U_k> + {}^{i}\!<\tilde{U}_k \tilde{U}_k> + {}^{i}\!<\tilde{\rho}_k \tilde{U}_k \tilde{U}_k>/{}^{i}\!<\rho_k> \tag{2.126}$$

and

$${}^{i}\!<U_k u_k> = {}^{i}\!<U_k><u_k> + {}^{i}\!<\tilde{U}_k \tilde{u}_k> + {}^{i}\!<\tilde{\rho}_k \tilde{U}_k \tilde{u}_k>/{}^{i}\!<\rho_k> \tag{2.127}$$

(Prob. 2.14)

These averages of products of local deviations are not necessarily dispersive like the high frequency components (Section 2.5), thus causing closure problems. These local deviations have negligible contributions in the case of dispersed flow systems of particulates. For the system in Fig. 2.7(b), one readily shows that the error in neglecting the deviation of fluid velocity from the averaged value due to velocity distribution around each particle approaches $(|\,{}^{i}\!<U_f> - {}^{i}\!<U_k>|)\ \alpha_k/(1 - \alpha_k)$, assuming potential motion. Furthermore, one can show that

$$|^i\langle \tilde{U}_f \tilde{U}_f \rangle| = O[|^i\langle U_f\rangle - {}^i\langle U_k\rangle|^2 \, \alpha_k (1 - \alpha_k)^{-1}]$$

$$|^i\langle \rho_f U_f U_f\rangle|/{}^i\langle \rho_f\rangle = O[|^i\langle U_f\rangle - {}^i\langle U_k\rangle|^4 \, \alpha_k(1-\alpha_k)^{-1} \, \beta_{sf} \, {}^i\langle \rho_f\rangle]$$

where β_{sf} is the bulk modulus of the fluid f. Hence the local deviations have negligible contributions in the case of either a dilute dispersed system (α_k small) or a dispersed system of fine particles (small $|^i\langle U_f\rangle - {}^i\langle U_k\rangle|$). (Prob. 2.15)

Exercise Problems

2.1 Derive Eq. (2.17) from Eq. (2.15).

2.2 Show that, based on the barycentric frame of reference, and $\sum_k \rho_k E_k = \rho_m E_m$

$$\sum_k \rho_k U_k E_k = \rho_m U_m E_m + \sum_k \rho_k (U_k - U_m)(E_k - E_m) \qquad (2.128)$$

2.3 Show Eq. (2.35) by taking a Taylors series expansion from a reference average of $\langle \psi_0\rangle$. [Whitaker 1969]

2.4 Derive the volume averaging theorems of intrinsic averages. (Hint: v_k is treated as an explicit variable).

2.5 Apply the volume averaging theorem to the continuity equation of an incompressible fluid and interpret the physical meaning of the resulting equation.

2.6 Show that, for a highly dispersed flow ($n = -n_k = -n_f$, and apply mixture conservation), the momentum equation of a phase takes the form [Soo 1981a]:

$$\partial\langle \rho_k U_k\rangle/\partial t + \nabla \cdot \langle \rho_k U_k U_k\rangle = -\alpha_k \nabla \langle P\rangle - \langle P\rangle\nabla\alpha_k + \nabla \cdot \langle \tau_k\rangle$$

$$+ <\rho_k> f_k + V_k + \nabla \cdot <\rho_k (U_k - U_m)(U_k - U_m)> + \Gamma_k U_m \qquad (2.129)$$

Identify the corresponding terms of the transfer integrals.

2.7 For a highly dispersed system of two phases, of which one is dilute, in a one-dimensional closed vessel having an initial concentration distribution, derive a diffusion equation by eliminating the velocity term from the continuity equation. Interpret the meaning of the terms. [Sha and Soo 1979] (Hint: apply the momentum equation in Problem 2.6 for the case of zero net mass flux).

2.8 Show that, when time averaging is carried out with

$$\overline{<\rho_k>}^i = \overline{<\rho_k>}^i_{LF} + \overline{<\rho_k'>}^i \qquad (2.130)$$

with $\overline{<\rho_k'>}^t = 0$, one gets

$$\overline{\overline{<\rho_k'>}^i}^t = - \overline{<\alpha_k'\rho_k'>}^t / \alpha_{kLF} \qquad (2.131)$$

2.9 A fully developed pure stratified flow exists in a duct parallel to the x direction, with the y-axis perpendicular to it. The liquid phase f occupies the thickness between $y = 0$ and $y = -y_{fo}$, the bottom wall, and the gas phase k occupies the thickness $y = 0$ to $y = y_{ko}$, the upper wall. In this two-dimensional system, the pressure gradient in the x-direction is a constant equal to C. Assuming laminar flow in the phases separated by the free surface at $y = 0$, and for given viscosities of each phase, determine the velocity distributions in each phase and the velocity of the interface. From the above results compute the average velocities, viscous stresses, and interface viscous drag force. Compare these to terms in a corresponding averaged momentum equation.

2.10 In the system in Problem 2.9, a surface wave is generated giving deviations from the average velocities of:

$$\tilde{U}_{ky} = \gamma <U_{kx}> \sin (\omega t - kx + \theta)$$

$$\tilde{U}_{sy} = \gamma <U_{kx}> \sin (\omega t - kx)$$

where γ is a proportionality factor, ω is the circular frequency and k is the wave number (2π/wave length), t is the time, θ is a phase angle. The interface is given by a wave of geometry given by

$$\eta = a \sin (\omega t - kx)$$

where a is the amplitude. Show that the y-component of the interface transfer integral I_{ky} averaged over the area 2ax is given by:

$$\rho_k \ \gamma^2 <U_{kx}>^2 a \ k^2 (1 - \cos \theta)/r$$

and that the x-component I_{kx} is zero, and the integral for mass generation is zero.

2.11 Show that the dynamic condition for a dilute suspension is given by the volume fraction of spheres [Soo 1976]:

$$\alpha << [2a/\lambda][<U^2>^{\frac{1}{2}}/<U_k^2>^{\frac{1}{2}}]/3\sqrt{2} \ K \tag{2.132}$$

and

$$\alpha << (3\sqrt{\pi}/2\sqrt{2}) \ (\overline{\nu}/D) \ (\lambda/2a) \ (\overline{\rho}/\overline{\rho}_p) \tag{2.133}$$

for small K. Also apply Eqs. (1.76) and (1.77). For large K

$$\alpha << [(\nu/D) \ (\rho/\rho_p)]^{\frac{1}{2}}/2 \tag{2.134}$$

2.12 Compute, using the results in Prob. 2.11, the condition for a dilute suspension for: (1) $2a = 5\mu m$, $\overline{\nu} = 2 \times 10^{-5}$ m^2/s, $<U^2>^{\frac{1}{2}} = 2$ m/s, $\overline{\rho}_p/\overline{\rho} = 1000$, and $\overline{c}_p/\overline{\rho} = 100$ (case of small K); (2) $2a =$

500 μm, $\bar{\rho}_p/\bar{\rho}_p = 1000$, and $\bar{\rho}_p/\bar{\rho} = 100$ (case of large K).

Ans. (1) 0.005, 0.05; (2) 0.005, 0.15.

2.13 Show that, for one-dimensional two-phase flow through a given area, the portions of area occupied by each phase is equal to the volume fraction of each phase.

2.14 Show that, for a deviation given by $\psi_k = {}^i\langle\psi_k\rangle + \tilde{\psi}_k$, the volume averaging theorems take the forms:

$$\langle\partial\psi_k/\partial t\rangle = \alpha_k(\partial/\partial t)\,{}^i\langle\psi_k\rangle - v^{-1}\int_{A_k} \tilde{\psi}_k U_s \cdot n_k dA \qquad (2.135)$$

$$\langle\nabla\psi_k\rangle = \alpha_k \nabla \cdot {}^i\langle\psi_k\rangle + v^{-1}\int_{A_k} \tilde{\psi}_k n_k dA \qquad (2.136)$$

2.15 Show that the error in neglecting the deviation from the averaged velocity of the fluid phase in Fig. 2.7(b) is as stated in the text at the end of Sec. 2.7. Assume potential motion in a layer (thickness nearly equal to a) around each sphere with velocity (3/2) $({}^i\langle U_f\rangle - {}^i\langle U_k\rangle)$ sin θ, θ being the azimuthal angle from the nose of the sphere. The deviation in density is given by $\tilde{\rho}_f = \beta_{sf}\Delta P_f\,{}^i\langle\rho_f\rangle$; $\Delta P_f/(1/2)\,{}^i\langle\rho_f\rangle|\,{}^i\langle U_f\rangle - {}^i\langle U_k\rangle|^2 = 1 - (9/4)\sin^2\theta$.

Chapter 3

TRANSPORT PROPERTIES AND PROCESSES

3.1 Drag, Heat and Mass Transfer of a Particle Cloud

We note from Eqs. (2.100), (2.101), and (1.11) that momentum transfer from a fluid phase to a dispersed phase can be expressed in terms of a drag coefficient. A simple case is a cloud of solid spheres in a fluid. Where the volume fraction or concentration of solid particles is sufficiently high such that the fluid boundary layer thickness exceeds the interparticle spacing, the drag coefficient of a single sphere is no longer applicable. Determination of the effect of particle concentration has been made from correlating the pressure drop data of fluidized beds. Ergun and Orning [1949], recognizing that the pressure drop (ΔP) over a bed height L in a fluidized bed is caused by simultaneous kinetic and viscous energy losses, proposed the following relation:

$$\frac{\Delta P}{L} = 150 \frac{(1-\varepsilon)^2}{\varepsilon^3} \frac{\bar{\mu}\, U_s}{(2a)^2} + 1.75 \frac{(1-\varepsilon)}{\varepsilon^3} \frac{G_o U_s}{(2a)} \tag{3.1}$$

where U_s is the superficial fluid velocity based on unobstructed flow area for the fluid, ε is the fraction void ($= 1 - \alpha_p$), and the interstitial velocity is $U_\varepsilon = U_s/\varepsilon$; G_o is the mass flow of the fluid based on unobstructed flow area. Equation (3.1) is applicable to flow through a particle bed at velocities below that of minimum fluidization (see Sec. 8.6).

Equation (2.114), with the equality in Eq. (2.102), gives

$$-(dP/dz) \;+\; \alpha_p \, \bar{\rho}_p \, F_{pf} \, U_\epsilon = 0 \tag{3.2}$$

Substitution of Eq. (3.1) gives the inverse relaxation time for momentum transfer from fluid to particles as:

$$F_{pf} = \frac{75}{2} \frac{(1-\epsilon)}{\epsilon^2} \frac{\bar{\mu}}{\bar{\rho}_p \, a^2} + \frac{1.75}{2\epsilon} \frac{\bar{\rho}}{\bar{\rho}_p} \frac{U_\epsilon}{a} \tag{3.3}$$

and the drag coefficient is given by Eq. (1.11):

$$C_D = \frac{8}{3} \frac{a}{U_\epsilon} \frac{\bar{\rho}_p}{\bar{\rho}} F_{pf} = 200 \frac{(1-\epsilon)}{\epsilon^2} \frac{\bar{\mu}}{2a \, U_\epsilon \, \bar{\rho}} + \frac{7}{3\epsilon} \tag{3.4}$$

It is seen that the relation for a single sphere is valid as $\epsilon \to 0.92$ or $\alpha_p = 0.08$ and $U_\epsilon \to U$, the free stream velocity.

The upper limit of fraction solid is given by the case of a packed bed, with pressure drop given by Chilton and Colburn [in Sherwood and Pigford 1952] as

$$\Delta P/L = 2 \, C_f'' \, A_w \, A_p \, A_L \, G_o^2/\bar{\rho} \; (2a) \tag{3.5}$$

where C_f'' is a modified friction factor, A_w is a wall-effect correction factor, A_p is a correction factor for hollow packing, A_L is a correction factor for the wetting of the packing by the solvent circulated, and is equal to unity for a gas-solid system; 2a is the nominal size of packing particle. Similar conversion to particle drag coefficient as in the above gives

$$C_D = (8/3) \, C_f'' \, A_w \, A_p \, A_L \, \epsilon^2/(1-\epsilon) \tag{3.6}$$

which fits the relation in Eq. (3.4) for $A_w = A_p = A_L = 1$ for $\epsilon \sim 0.42$. The above relations of C_D are compared in Fig. 3.1. Note that for spherical particles, $\epsilon \sim 0.42$ corresponds to an inter-particle spacing of 1 diameter, while $\epsilon \sim 0.92$ corresponds to that of

2 diameters.   The Reynolds number is based on a single particle. Cluster formation leading to a lower drag coefficient of a group of particles, may be regarded as a source of instability in a fluidized bed [Kaye & Boardman 1962].

<u>Heat and mass transfer</u>.   In conjunction with the drag or pressure drop characteristics of a cloud of particles and the modification of relations based on a single sphere, corresponding changes in the heat and mass transfer relations are expected.   A comprehensive summary of dimensional correlations is given by Chu et al. [1953] for mass transfer factor j_d in both fixed and fluidized beds, based on molar concentration

$$j_d = N_{Sh}/N_{Sc} \, N_{Re} = (k_c \, M/G_o) \, (\overline{\mu/\rho} \, D)^{2/3} \tag{3.7}$$

where k_c is the rate constant for mass transfer, M is the molecular weight of the fluidizing medium, $(\overline{\mu/\rho} \, D)$ is the Schmidt number.   They gave, for

$$2a \, G_o/\overline{\mu} \, \alpha > 30, \; j_d = 1.77 \, (2a \, G_o/\overline{\mu} \, \alpha)^{-0.44}$$

and $\hspace{10cm}$ (3.8)

$$2a \, G_o/\overline{\mu} \, \alpha < 30, \; j_d = 5.7(2a \, G_o/\overline{\mu} \, \alpha)^{-0.78}$$

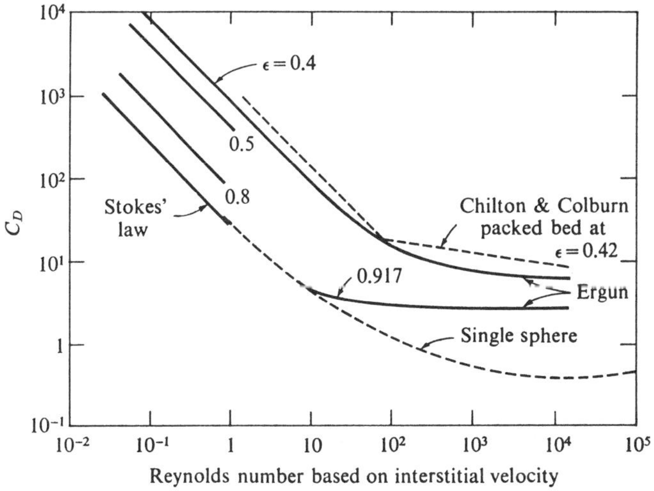

Figure 3.1 Drag coefficient of spheres in a cloud.

They further suggested the j-factor for heat transfer by analogy

$$j = N_{Nu}/N_{Pr} \, N_{Re} = (h/c \, G_o)(N_{Pr})^{2/3} \sim j_d \qquad (3.9)$$

where h is the heat transfer coefficient and c is the specific heat
of the fluid.

Mullin and Treleaven [1962] measured the rate of mass transfer
from spheres at fraction voids down to $\varepsilon = 0.52$. Their results are
shown in Fig. 3.2, for both fluidized beds and fixed beds. In Fig.
3.2, $N_{Re} = U \, (2a)/\bar{v}$, where U is the average velocity, N_{Sh}
$= k'_c \, (2a)/D$, k'_c is the mass transfer coefficient, and D is the
diffusivity, the Schmidt number $N_{Sc} = \bar{v}/D$. General agreement with
Chu et al. [1952] is seen with $k'_c = k_c M/\rho$, $N_{Sh} N_{Sc}^{-1/3} = j_d \, N_{Re}$,
$j_d \propto N_{Re}^{-0.44}$.

Ferron and Watson [1962] studied the heat transfer from solid
particles in a fluidized bed, but did not correlate with fraction
void.   The effect of fraction void was recognized, however. Their
results (bed height/diameter = L/2R = 1) were in agreement with those
of Wamsley and Johanson [1954], and Kettenring et al. [1950] (L/2R =
0.6 to 1.25) as shown in Fig. 3.3, h_∞ is the heat transfer

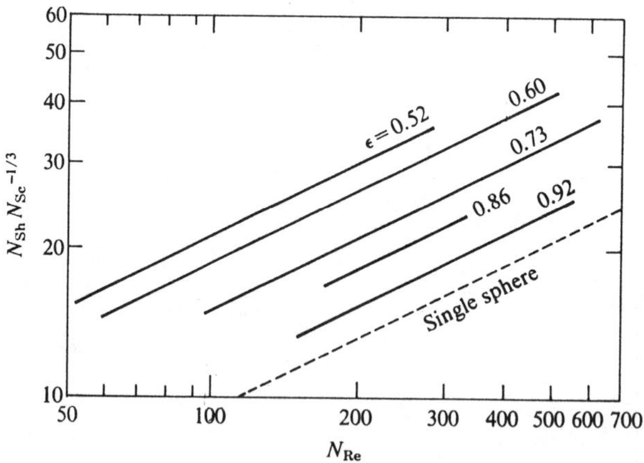

Figure 3.2   Influence of particle concentration on rate of
mass transfer [Mullin et al. 1962]

coefficient for complete mixing.

<u>Bubbles and droplets</u>.   Corresponding to momentum and heat transfer, the rate of generation or decrease of bubble phase k, Γ_{ek} by evaporation (or condensation) can be expressed in terms of, from Eq. (1.30):

$$\Gamma_{ek} = n_k(dm_{ek}/dt) = -M_{kf} \, (\bar{\rho}_s - \bar{\rho}_k) \qquad (3.10)$$

where n_k is the number density of phase k in the fluid f and d m_{ek}/dt is given by Eq. (1.32) for bubble growth by evaporation from superheated liquid and by Eq. (1.34) for evaporation produced by superheated vapor and M_{kf} is the inverse relaxation time for mass transfer.   In the latter case we get for small $T_s - T_k$ and $T = (T_s + T_k)/2$,

$$M_{kf} \sim 3(\alpha_k/a_k) \, C \, (\hat{R}T/2\pi)^{1/2} \qquad (3.11)$$

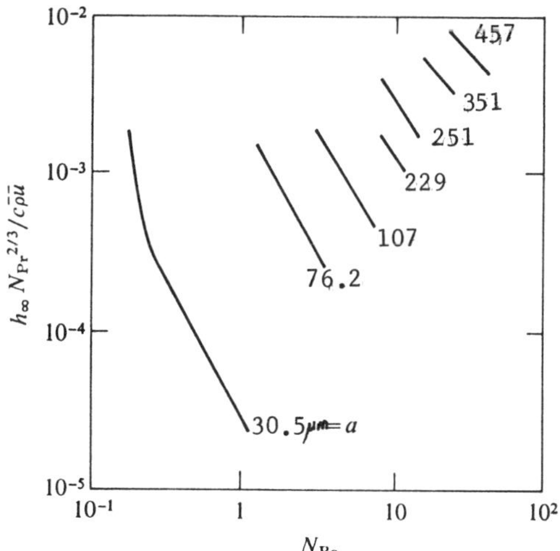

Figure 3.3   Heat-transfer coefficient for complete mixing in dense bed of solid spheres (L/D from 0.6 to 1.25) [Ferron et al. 1962]

For discrete size range of particles chosen for multiphase calculation, the generation rate of species (k+1) by evaporation into k is given by;

$$\Gamma_{e(k+1)} = -\Gamma_{ek}(m_{k+1}/m_k) \tag{3.12}$$

In the case of droplets, the generation rate is given by:

$$\Gamma_{ep} = n_p(dm_{ep}/dt) = -M_{pg}(\bar{\rho}_s - \bar{\rho}_g) \tag{3.13}$$

and, corresponding to Eq. (1.30),

$$M_{pg} = (\rho_p/\bar{\rho}_p) \; (3/2) \; N_{sh} \; D_{pg}/a_p^2 \tag{3.14}$$

<u>Irreversibility of momentum transfer from particle to fluid</u>. In the absence of shear motion, body force, and chemical reaction, the equation for the conservation of momentum for the mixture of one species of particle (subscript p) and fluid (without subscript) is given by ($d/dt = \partial/\partial t + U_i \, \partial/\partial x_i$, $d/dt_p = \partial/\partial t + U_{pi} \, \partial/\partial x_i$):

$$\rho(dU_i/dt) + \rho_p \; (dU_{pi}/dt_p) = -(\partial P/\partial x_i) \tag{3.15}$$

which is true when momentum is transferred reversibly between particle and fluid, that is, treating the particle phase as a true continuum. This would lead to the conclusion that in a diffuser, as solid particles are slowed down, they contribute to the pressure rise also. This is not always the case: even in the laminar range of relative motion, Froessling (see Schlichting [1979]) showed that before separation of boundary layer occurs, the boundary layer thickness δ for flow a sphere is given by:

$$(\sqrt{2} \; \delta/2a) \; [2a(\Delta U)/\bar{\nu}] \; \lesssim \; 3 \tag{3.16}$$

where ΔU is the relative speed. Therefore if the cloud of particles is to exert a sufficient drag on the gas, the interparticle spacing must be smaller than $(2\delta + 2a)$ given by Eq. (3.16), or the number density of solid particles n_p must give

$$n_p^{-1/3} \gtrsim \delta \qquad (3.17)$$

as an approximation to account for possible fluctuation in the spacing. This relation leads to the condition

$$N_e \equiv [2a(U - U_p)/\bar{\nu}] \, (2a)^{-2} \, n_p^{-2/3} \gtrsim - 9/2 \sim -5 \qquad (3.18)$$

for successful transfer of momentum of mass motion with the particles accelerating the gas in the same manner as the gas may accelerate the particles. Otherwise, the gas simply acts as a free stream with the kinetic energy of the particles dissipated completely in their wakes. This physical condition is depicted in Fig. 3.4a, which also leads to the conclusion as given in Fig. 3.1. Eq. (3.15) therefore can be modified as

$$\rho(dU_i/dt) + K_m \, \rho_p \, (dU_{pi}/dt_p) = -\partial P/\partial x_i \qquad (3.19)$$

with the "effectiveness" $K_m = 1$ for the case when solid particles are accelerated by the gas, and $K_m < 1$ for the case when solid particles are decelerated by the gas. The magnitudes of K_m have been determined by and experimental study of Tomita et al. [1980]. They correlated the pressure recovery in a sudden expansion with particles of glass and polyvinylchloride of 133 to 900 μm and introduced a parameter ζ as a measure of recovery of particle momentum to static pressure. This is represented by, for inlet condition 1 at diameter $2R_1$ and expansion to state 2 at diameter $2R_2$ such that:

$$P_1 + (\rho_1 \, U_1^2/2) = P_2 + (\rho_2 \, U_2^2/2) + (\zeta \, \rho_1 \, U_1^2/2) \qquad (3.20)$$

They correlated with a Richardson number:

$$N_{Ri} = (\rho_p/\rho) \, 2R_2 F/[1 + (R_1/R_2)]U_1 \sim -3\pi c_1 \, N_e^{-1} \qquad (3.21)$$

where c_1 is an empirical constant for length $\Delta z = c_1\, n_p^{-1/3}$, over which deceleration of particles occurs. When correlated to K_m and N_e, one gets

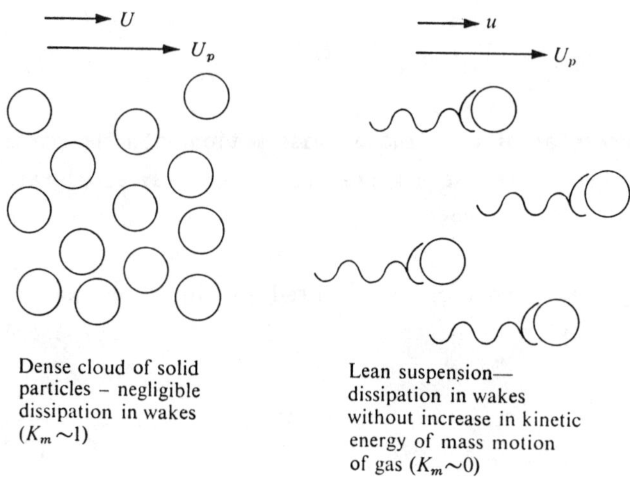

(a) Transfer of momentum of mass motion from cloud of solid particles to gas

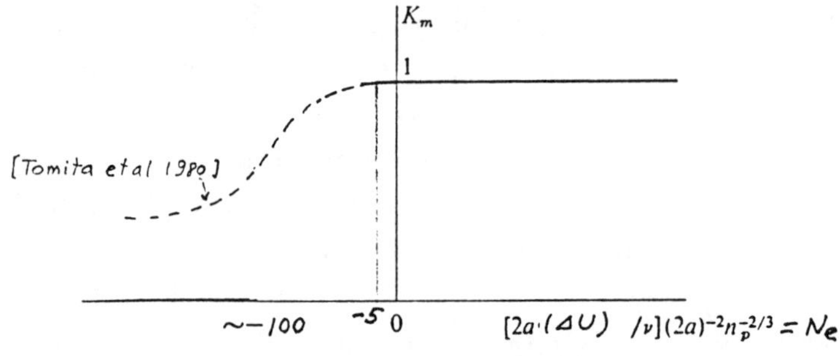

(b) "Discontinuous" representation of effectiveness based on Froessling

$$\Delta U = |\underline{U} - \underline{U}_p| \quad \text{for} \quad |\underline{U}| > |\underline{U}_p|$$
$$\Delta U = -|\underline{U} - \underline{U}_p| \quad \text{for} \quad |\underline{U}| < |\underline{U}_p|$$

Figure 3.4   Effectiveness of momentum transfer from solid particles to the gas [Soo 1965]

ζ	N_{Ri}	K_m	N_e
0	10	0	-10
-0.1	5x10	0.2	-2x10
-0.15	3x10	0.4	-3x10
-1	3	1	-5

These results are plotted as shown by the dotted line in Fig. 3.4b, along with the boundary layer approximation of Eq. (3.18).

3.2   Interaction of Particles with Surfaces and Momentum Transfer.

As a preliminary to treating interactions among phases of a multiphase system, we consider the nature of impaction of particles in a suspension with a surface. Applications such as spray painting, sand blasting, particle collection, and spray cooling, involve motion of a suspension toward a surface. Langmuir and Blodgett [1948] treated the problem as one of potential motion of the fluid but with viscous drag on the particles in the cloud. The equation of motion of a particle can be represented by, for $\bar{\rho}_p \gg \bar{\rho}$,

$$d\,U_p/dt = F(U - U_p) \tag{3.22}$$

where U, U_p are the velocity vectors of the fluid and the particles.   For the coordinate system in Fig. 3.5, the boundary condition is given by the x-component: $x = -\infty$, $U_x = U_{px} = U_0$;

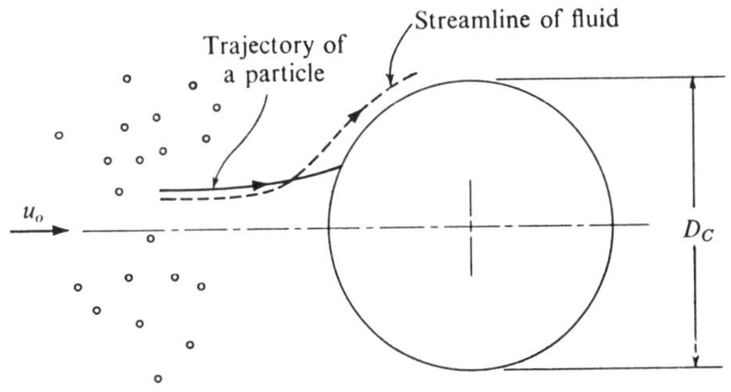

Figure 3.5   Impaction of particles with a body.

$U(x,y,z)$ is given and it is assumed that it is unaffected by the presence of the particles.

The nature of impaction is given by the fraction impacted, η. η is defined as the ratio of the cross-sectional area of the original stream from which particles of a given size are impacted because the trajectories intersect with the collector surface, to the projected area of the target in the direction of flow or the cross-sectional rea of the original stream. For the sake of generality of the results, Eq. (3.22) is non-dimensionalized by introducing a reference Reynolds number $(N_{Re})_o = 2a\bar{\rho}\,U_o/\bar{\mu}$, so that the local particle Reynolds number is given by

$$N_{Re} = (N_{Re})_o\,|U - U_p|/U_o \tag{3.23}$$

and to account for large relative motion, F is denoted as

$$F = F\star\,9\bar{\mu}/2a^2\,\bar{\rho}_p \tag{3.24}$$

where

$$F\star = C_D(N_{Re})/24 = F\star(N_{Re}) \tag{3.25}$$

which is equal to 1 for the Stokes law regime. Equation (3.22) now becomes, neglecting gravity effect,

$$\Psi(dU_p\star/dt\star) = F\star(U\star - U_p\star) \tag{3.26}$$

where $U\star = U/U_o$, $U_p\star = U_p/U_o$, $t\star = tU_o/D_c$, D_c is the characteristic dimension of the target, and

$$\Psi = \bar{\rho}_p\,U_o(2a)^2/18\bar{\mu}\,D_c = U_o/F_1\,D_c = N_{m1} \tag{3.27}$$

is a ratio of relaxation time for fluid to particle momentum transfer to the flow time and is, in general, designated as a momentum transfer number. When it is based on Stokes drag ($N_{Re} < 1$, or $F\star = 1$), it is, by precedence, called the Langmuir parameter. The non-Stokesian behavior is accounted for by a parameter:

$$\Phi = (N_{Re})_o^2/2\Psi = 9(\overline{\rho}/\overline{\rho}_p)(U_o D_c/\overline{\nu}) \tag{3.28}$$

and $(N_{Re}) = (2\Phi\Psi)^{1/2}$. Calculations of fraction impacted for potential flow about cylinders, spheres, and ribbons were performed by Langmuir and Blodgett [1948], as well as for viscous flow. The relation for impaction on a sphere is given in Fig. 3.6, together with the experimental results of Ranz and Wong [1952] and Walton and Woolcock [1960]. Fraction impacted of ellipsoids of a fineness ratio of 2 was given by Morr and Soo [1973] and a fineness ratio of 5 was given by Dorsch et al. [1955]. (Prob. 3.1)

The disagreement of the theoretical results of Langmuir and Blodgett and the experimental results bears notice. Ranz and Wong performed experiments with isodispersed sulfuric acid mists with mean droplet radii of 0.18 to 0.65 μm in a wind tunnel with air speeds of 10 to 100 m/s. Targets included a wire of 77 μm diameter and sphere of 0.9 mm. Walton and Woolcock used a target formed by a water droplet of 0.25-1.00 mm radius suspended by a glass capillary; an isodispersed aerosol of methylene blue of spherical particles of 1.25 to 2.5 μm radii flowed over the water droplet. Even greater discrepancy

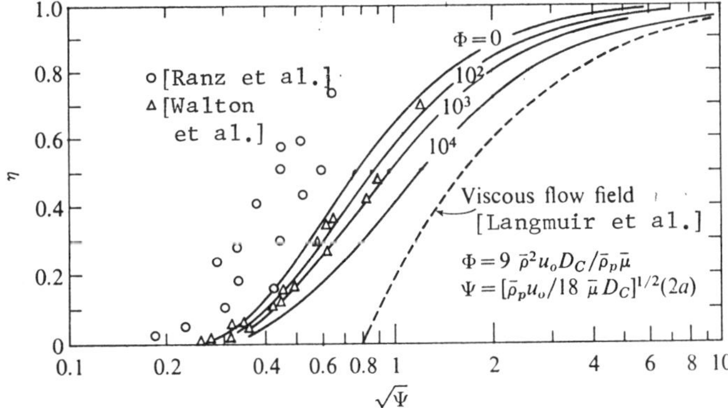

Figure 3.6 Impaction on a sphere [Ranz & Wong 1952]

from the theoretical results based on inertia effect was reported by Yao et al. [1971]. The latter treated flow of suspended particles of 0.01, .1, and 1 μm particles in water at 5 C and 25 C over a collector sphere of diameter D_c = 0.5 mm in a filter bed for water filtration with U_o = 0.001 to 0.003 m/s. An important source of discrepancy is seen to be the effect of diffusivity of particles [Soo 1973].

In a general system, the effect of inertia of particles, gravity effect, and the effect of diffusion are shown in Fig. 3.7; the effect of diffusivity is such that a particle may hit the back of a sphere. For this system, the density distribution of particles at steady flow is given by a diffusion equation in the form:

$$U \cdot \nabla \rho_p = D_p \nabla^2 \rho_p + [1 - (\overline{\rho}/\overline{\rho_p})] \, (g/F) \, \partial\rho_p/\partial z \qquad (3.29)$$

and the fraction impacted is given by integrating the mass flux of particles over the sphere and dividing by the free stream mass flux into the projected target area. In the limit, D_p is the Brownian diffusivity (Eq. 1.46).

Yao et al. [1971] treated the overall fraction impacted as a linear combination of:

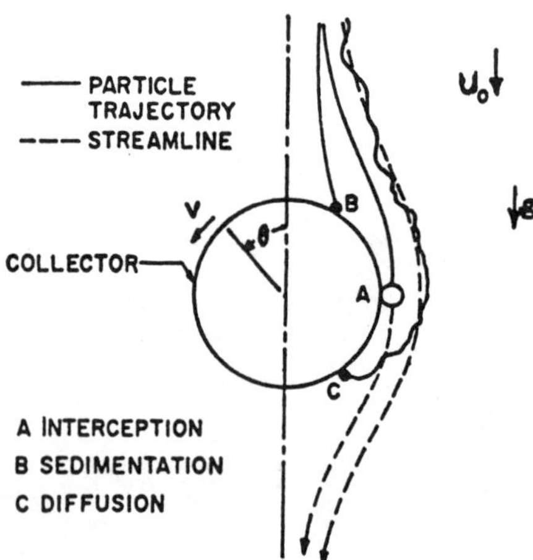

Figure 3.7   Basic transport mechanisms in water filtration
            [Yao et al. 1971]

$$\eta = \eta_D + \eta_I + \eta_G \qquad (3.30)$$

for diffusion (subscript D), interception (I), and sedimentation (G) and gave approximate relations which are now expressed as:

$$\eta_D = 4.04 \, (U_0 D_c/D_p)^{-2/3} = 4.04 \, (D_p/D_c^2 \, F)^{2/3} \, \psi^{-2/3} \qquad (3.31)$$

$$\eta_I = (3/2) \, (2a/D_c)^2 = 27 \, (D_c U_0/\bar{\nu})^{-1} \, \psi \qquad (3.32)$$

$$\eta_G = (\bar{\rho}_p - \bar{\rho}) \, g/F \, U_0 \bar{\rho}_p = (\bar{\rho}_p - \bar{\rho}) \, \bar{\rho}_p^{-1} (g/F^2 D_c) \, \psi^{-1} \qquad (3.33)$$

It is seen that diffusion increases η. The parameter $N_{DF} = D_p/D_c^2 F$ is a ratio of relaxation time to diffusion time and is denoted as the diffusion response number. The results of Yao et al. for small diffusivities (10^{-15} to 10^{-11} m^2/s) are shown in Fig. 3.8 together with the results of a survey by Ranz [1956]. Large values of N_{DF} around 10^{-3} accounts for the effect of turbulent diffusion which are present in some of the experimental points in Fig. 3.8.

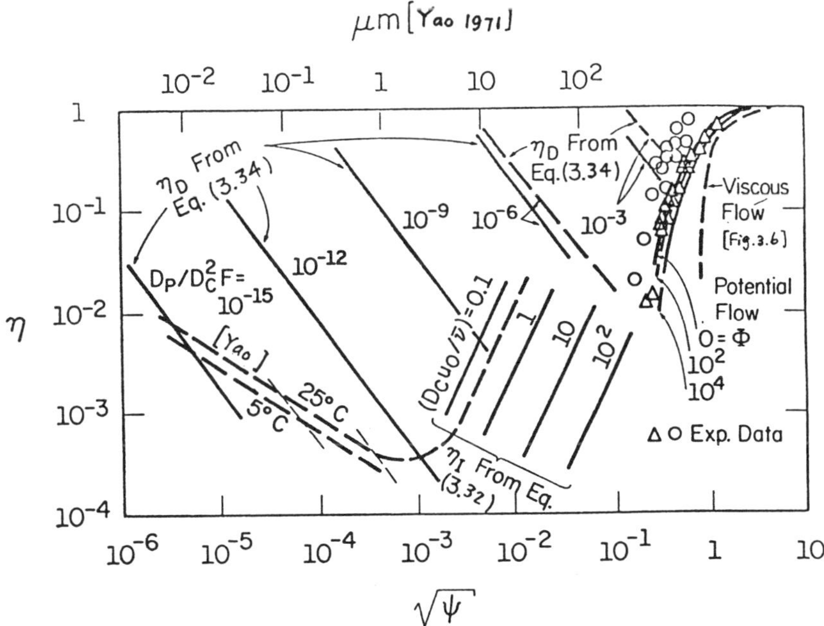

Figure 3.8 Fraction impacted on a sphere and influence of diffusivity of particles. [Soo 1973]

The limitation of Eq. (3.31) is seen in the approximation due to Levich [1962] which includes the existence of a boundary layer for the diffusion phenomenon and the existence of a completely absorbing surface of D_c. The former calls for a Peclet number $N_{Pe} = U_oD_c/D_p \gtrsim 1000\ N_{Re}$, for $N_{Re} = D_cU_o/\bar{v}$; and a completely absorbing surface call for $\rho_p = 0$ at the surface. He gave, for small $N_{Pe} < 1$, that is, for the diffusion rate greater than the convection rate,

$$\eta_D = 8\ N_{DF}\ \psi^{-1}\ [1 + 0.64\ N_{DF}^{-1/3}\ \psi^{1/3}] \tag{3.34}$$

and $N_{DF}\ \psi^{-1} = N_{Pe}^{-1}$. Figure 3.8 shows parametrically the effect of diffusion in terms of the diffusion response number N_{DF}, and Eq. (3.34) for large N_{DF} are plotted in dash-dot lines for N_{DF} of 10^{-6}, and 10^{-3}. We also see that $\eta > 1$ due to diffusion is a possibility. Moreover, the linearity represented by Eq. (3.30) does not hold for large ψ. One notes that fraction impacted is identical to collection efficiency for a completely absorbing surface. Collection efficiency including the effect of sticking probability at the surface needs to be accounted for. An approximate relation for collection efficiency based on finite sticking was obtained from solving Eq. (3.29) [Soo 1973].

<u>Surface forces and sticking probability.</u> The force of adhesion of particles on a clean surface or a surface with a layer of deposit of particles influence deposition of particles. In a review, Corn [1961] showed that adhesive forces are either electrical or liquid (viscosity and surface tension) in origin. The electrical forces include those due to contact potential difference and dipole effect, space charges, and electronic structure.

The adhesive force between a plane solid and spherical solid particle is given by Krupp [1967]:

$$F_v = (\bar{h\omega}/8\pi\ z_o^2)a \tag{3.35}$$

where $\bar{h\omega}$ is a Lifschitz-van der Waals constant, h is the quantum constant (Planck constant/2π), $\bar{\omega}$ is a frequency which is not clearly
related to the dielectric constants of the adherents, z_0 is the
distance between the adherants at the instant of maximum attraction
during separation, $z_0 \sim 4 \times 10^{-10}$ m.   For quartz particles of
2.8 μm radius, the sticking probability on polyamide and glass fibers
of 19 μm diameter was first reported by Loeffler and Muhr [1972].
Data of sticking probabilities σ_w for quartz and glass particles of 5
and 10 μm diameter on 20 μm polyamide and glass fibers are shown in
Fig. 3.9; it was noted that rebounding starts at about 5-15 cm/s and
decreases with increase in speed [Loeffler 1977].

In the adhesion of an individual particle to particle multilayers adhering to a substrate, a space charge effect exists between
individual particles giving a force equal to qE; q is the charge of a
particle and E is the electric field.   In particle multilayers the
interparticle distance is large compared with the particle diameter,
and a force originates from the repulsion of charges of an equal
sign.   The force of adhesion between polymer particles (5 and 30
microm in diameter) and a layer of amorphous selenium was measured by
centrifuging after a contact time of 1 hr.   Charges were obtained
triboelectrically to 3×10^4 electrons per particle to 4×10^5

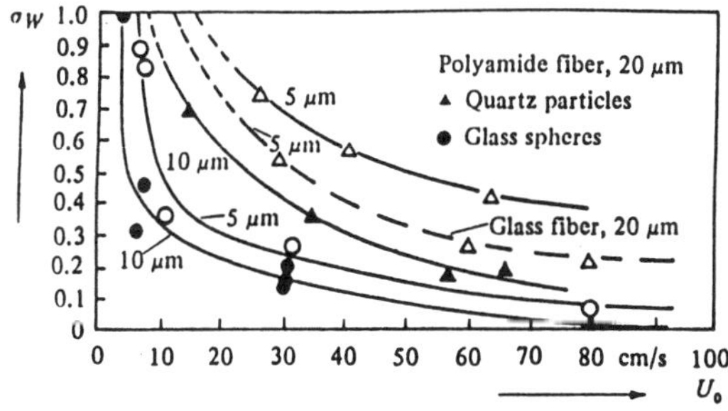

Figure 3.9   Sticking probability of quartz particles and glass
            spheres on 20 μm fibers of polyamide and glass as a
            function of velocity [Löffler 1977].

e/particle.   Adhesive forces of both multilayers and selenium surfaces were measured.   An effective particle charge, that is, the portion of the total charge q that is seen by the partner of adhesion, $\overline{h\omega} \sim 0.2$ ev to 12 ev [Krupp 1967] (0.2 with water layer [Rumpf 1972]).   Mazzone et al. [1987] determined the force exerted by a liquid bridge between two spheres.

<u>Effect of electrostatic forces.</u> The magnitude of fraction impacted is strongly influenced by the electrostatic charges and potentials of the particles and the collector.   Based on fundamental principles of electrostatics, Kraemer and Johnstone [1955] formulated and determined the mechanics of deposition of charged aerosols on collecting surfaces.   Neglecting the inertia force, the equation of motion (3.22) is modified to

$$0 = F(U - U_p) - (F_e/m_p) \tag{3.36}$$

where F_e is the electrical force acting on each of the particles (subscript 2); quantities pertaining to the collector is denoted with a subscript 1.   The principles of electrostatics [Maxwell 1904] give for the case of spherical collector with a constant charge Q:

$$F_{2e} = -\left(\frac{\varepsilon_r-1}{\varepsilon_r+2}\right) \frac{a^3 Q_1^2}{2\pi\,\varepsilon_o r_{12}^5} + \frac{Q_1\,q}{4\pi\,\varepsilon_o r_{12}^2} + q^2\left[\frac{R_c}{4\pi\,\varepsilon_o r_{12}^3} - \frac{R_c\,r_{12}}{4\pi\,\varepsilon_o(r_{12}^2 - R_c^2)^2}\right] - \frac{q^2\,R_c^3 n_p}{3\,\varepsilon_o\,r_{12}^2} \tag{3.37}$$

and for a spherical collector at constant potential V:

$$F_{2e} = -\left(\frac{\varepsilon_r-1}{\varepsilon_r+2}\right) \frac{8\,V_1^2\,R_c^2\,\pi\,\varepsilon_o\,a^3}{r_{12}^5} + \frac{V\,q\,R_c}{r_{12}^5} - \frac{q^2\,R_c^3\,n_p}{3\varepsilon_o\,r_{12}^3}$$

$$- \frac{q^2\,R_c^2\,\pi\,n_p\,R^2}{4\pi\,\varepsilon_o\,r_{12}^2} + q^2\left[\frac{R_c}{4\pi\,\varepsilon_o\,r_{12}^3} - \frac{R_c\,r_{12}}{4\pi\,\varepsilon_o\,(r_{12}^2-R_c^2)^2}\right] \tag{3.38}$$

where ε_r is the dielectric constant of the particles, ε_o is the permittivity of free space, r_{12} is the distance between the centers of the collector and a particle, R_c is the radius of the spherical collector, n_p is the number density of the particle cloud, R is the

radius of the cloud of particles surrounding the collector, and F is
the inverse relaxation time for momentum transfer from fluid to
particle in suspension and $F = 6\pi\mu a/m_p C$, from Eq. (1.8), to account
for small aerosols.   Expressing in terms of spherical coordinates
with axial symmetry, we get

$$U_{pr} = dr/dt = U_r + (C\,F_{er}/6\pi\bar{\mu}a) \tag{3.39}$$

$$U_{p\theta} = rd\theta/dt = U_\theta + (C\,F_{e\theta}/6\pi\bar{\mu}a) \tag{3.40}$$

where r is the radial coordinate and θ is the azimuthal angle.

Introducing dimensionless variables $r^* = r/R_c$, $t^* = tU_0/R_c$,
where U_0 is the free-stream velocity of the aerosols.   On
substituting Eqs. (3.37) and (3.38) into Eqs. (3.39) and (3.40), one
gets, for velocity field of fluid U given by potential theory:

$$dr^*/dt^* = (1-r^{*-3})\cos\theta - (N_{ev})_I\, r^{*-5} + [(N_{ev})_E - (N_{ev})_S - (N_{ev})_G]r^{*-2}$$

$$+ (N_{ev})_M\, [r^{*-3} - r^*(r^{*-2} - 1)^{-2}] \tag{3.41}$$

$$d\theta/dt^* = -(r^{*-1} + \tfrac{1}{2}r^{*-4})\sin\theta \tag{3.42}$$

and for viscous flow,

$$dr^*/dt^* = (1 - \tfrac{3}{2}r^{*-1} + \tfrac{1}{2}r^{*-3})\cos\theta - (N_{ev})_I\, r^{*-5}$$

$$+ [(N_{ev})_E - (N_{ev})_S - (N_{ev})_G]\, r^{*-2} + (N_{ev})_M\, [r^{*-3} - r^*(r^{*2}-1)^{-2}] \tag{3.43}$$

$$d\theta/dt^* = -(r^{*-2} - \tfrac{3}{4}r^{*-2} - \tfrac{1}{4}r^{*-4})\sin\theta \tag{3.44}$$

The dimensionless parameters N_{ev} are called electroviscous numbers, correlating various electrostatic forces to viscous forces;
they are defined according to:

The ratio of image force to viscous force is

$$(N_{ev})_M = Cq^2/24\pi^2 \, \varepsilon_o \bar{\mu} \, U_o \, aR_c^2.$$

The ratio of space-charge force to viscous force is

$$(N_{ev})_S = Cq^2 \, R_c \, n_p/18\pi \, \varepsilon_o \, \mu \, aU_o.$$

The ratio of force due to induced charge for collector at constant voltage to viscous force is

$$(N_{ev})_G = Cq^2 \, n_p \, R_c^2/24\pi \, \varepsilon_o \, \bar{\mu} \, U_o \, a.$$

The ratio of image force to viscous force is

$$(N_{ev})_I = Ca^2 \, Q_1^2 \, (\varepsilon_r-1)/(\varepsilon_r+2) \, 12\pi^2 \, \varepsilon_o \, \bar{\mu} \, U_o \, R_c, \quad \text{for collector at constant charge,}$$

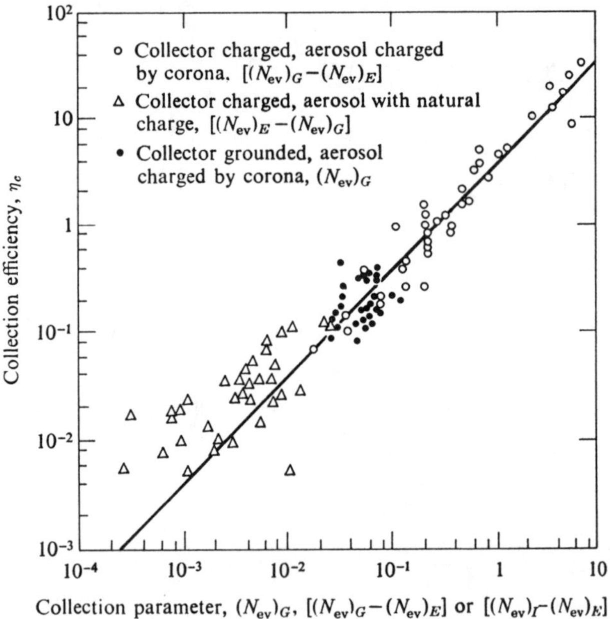

Figure 3.10   Collection of dioctyl phthalate aerosol particles on spherical collector [Kramer et al. 1955]

$$= 4C \ \varepsilon_o \ V_1^2 \ a^2 (\varepsilon_r - 1)/(\varepsilon_r + 2) \ 3\bar{\mu} \ U_o R_c^3, \quad \text{for collector at constant voltage.}$$

The ratio of coulombic force to viscous force is:

$$(N_{ev})_E = Cq \ Q_1/24\pi^2 \ \varepsilon_o \ \bar{\mu} \ U_o \ aR_c^2 \quad \text{for collector at constant Q,}$$

$$= Cq \ V_1/6\pi \ \bar{\mu} \ U_o \ aR_c \quad \text{for collector at constant V.}$$

All these parameters serve an analogous purpose as the momentum transfer number in Eq. (3.27).

Kraemer and Johnston [1955] performed extensive computations for a number of combinations of (N_{ev})'s and carried out experiments with impaction of dioctyl phthalate aerosols. Their results are summarized in Fig. 3.10, giving the collection efficiency or fraction impacted as defined after Eq. (3.22). In Fig. 3.10, (N_{ev}) applies to the case of aerosols charged and the collector grounded; $[(N_{ev})_G - (N_{ev})_E]$ applies to the case with the aerosols charged, but the collector is held at constant charge or potential; $[(N_{ev})_I - (N_{ev})_E]$ applies to the case where only the collector is charged or held at a constant potential.

Transfer of momentum. Take the case of collision of a cloud of particles of radius a, mass m, number density n_p, and a velocity U with a target sphere of radius R, with single scattering. With specular reflection, the rate of change of axial component of momentum of a particle approaching a point of contact represented by the geometry in Fig. 3.11 is given by

$$m \ U_o - [m \ U_o \ r^* \cos(\pi - \theta - \phi)] \qquad (3.45)$$

$$= m \ U_o \ [1 + r^* \cos(\theta + \phi)]$$

where r^* is the ratio of reflected speed to the incoming speed. For limiting contact angle θ for the tangential particle path ($\theta = \pi/2$ for $\eta = 1$) the force acting on the target sphere is given by

$$F = n_p \, m \, U_o \int_o^{\theta_m} [1 + r^* \cos(\theta+\phi)] \; (R d\theta \cos\theta) \; (2\pi \, R \sin\theta) \quad (3.46)$$

One notes that for $\eta < 1$, $\theta_m < \pi/2$, $\phi \geq \theta$, Eq. (3.46) integrates to

$$F \leq \eta \, \pi \, n_p \, m \, U_o^2 \, R^2 = \eta \, \pi \, R^2 \, \rho_p \, U_o^2 \quad (3.47)$$

regardless of the value of r*, including the case where all particles are stuck on impact. It is readily shown that this is not the case for a target consisting of a circular cylinder with axis normal to the direction of the flow stream. The drag coefficient of a sphere due to particle impaction is given by:

$$C_{Dp} = F/(\tfrac{1}{2} \, \rho_p \, U_o^2 \, \pi \, R^2) \leq 2 \, \eta \, (\Psi, \, \Phi) \quad (3.48)$$

analogous to that of free molecular flow. Validation of the basic formulation such as Eq. (3.46) was made by measuring the drag force exerted on a cylinder by a particle cloud [Soo 1967]. (Prob. 3.2, 3.3)

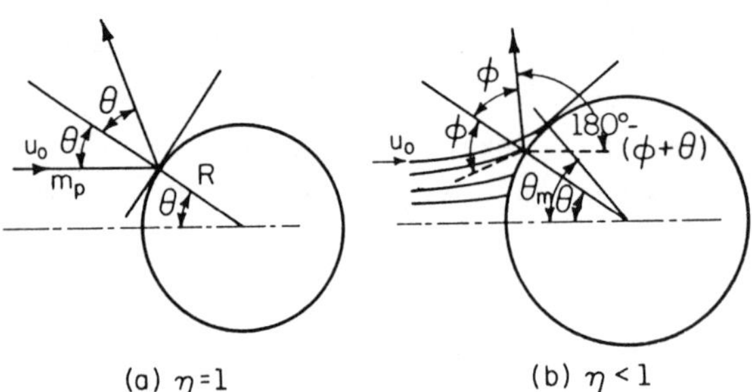

Figure 3.11   Collision and specular reflection of
particles from a sphere

3.3 Transport Processes among Particle Clouds - Single Scattering.

Mutual interactions including collisions among solid particles in a fluid suspension occur by their difference in response to a field force: gravity, fluid flow field or electromagnetic field. When the suspension is sufficiently dilute, single scattering of each collision may be assumed.   Using the idealization of spherical particles, transport processes due to particle-particle collisions can be treated on a common basis. They include momentum, heat, mass, and charge transfers. These transport processes can be characterized by their relaxation times or their inverse.

Fraction impacted.   When one considers a stationary sphere 1 or radius a_1 bombarded by a cloud of spheres 2 of radius a_2 and material density $\bar{\rho}_{p2}$ and conveyed by a fluid of material density $\bar{\rho}$, viscosity $\bar{\mu}$, and free stream velocity U_o, only a fraction of the particles 2 whose centers lie within the cylinder formed by the velocity vector U_o and the projected area $\pi(a_1 + a_2)^2$ will hit sphere 1 because the drag exerted by the fluid as balanced by the inertia of the particles. For relative motion of particles 2 to particle 1 at a relative speed $\Delta U_p = |U_{p1} - U_{p2}|$, with inertia force due to reduced mass, Eq. (3.26) becomes:

$$(m_1^{-1}+m_2^{-1})^{-1} \frac{(\Delta U_p)^2}{2(a_1+a_2)} \frac{d\ U_2^*}{dt^*} = F_{p12} = F_{p2f}F^* (U^*-U_{p2}^*)m_2\ \Delta U_p \quad (3.49)$$

with $U^* = U/\Delta U_p$, $U_{p2}^* = U_{p2}/\Delta U_p$, $t^* = t\Delta U_p/2(a_1 + a_2)$, and the parameters Ψ_{12} and Φ_{12} now become:

$$\Psi_{12} = \frac{\bar{\rho}_{p2}\ U_o(2a_2)}{18\bar{\mu}\ 2(a_1 + a_2)} = \frac{U_o}{F_{p2f}\ 2(a_1 + a_2)} = N_{m2} \quad (3.50)$$

and

$$\Phi_{12} = \frac{(N_{Re})_2^2}{2\Psi_{12}} = 9\ (\frac{\bar{\rho}}{\rho_{p2}}) [\frac{U_o\ 2(a_1 + a_2)\ \bar{\rho}}{\bar{\mu}}] \quad (3.51)$$

where F_{p2f} is the inverse relaxation time for momentum transfer from the fluid to the particle 2, $(a_1 + a_2)$ is the radius of exclusion, N_{m2} is the momentum transfer number from the fluid to the particle 2, $(N_{Re})_o$ and F_{p2f} are as defined by Eqs. (3.23) and (3.24). The fraction impacted as a function of Ψ_{12} and Φ_{12} is as given by Fig. 3.6. (Prob. 3.4)

<u>Momentum transfer</u>. Transfer of momentum from the fluid and from species 2 particles to species 1 particles are given by Eq. (1.9) and the drag force in Eq. (3.47) and can be expressed as:

$$d\ U_{p1}/dt = F_{1f}\ (U - U_{p1}) + F_{12}\ (U_{p2} - U_{p1}) \qquad (3.52)$$

Based on the above determination of fraction impacted and including the effect of reduced mass, the inverse relaxation time for momentum transfer from particles 2 to particles 1 is given by:

$$F_{12} = \frac{F_{p12}}{m_{p1}(U_{p2}-U_{p1})} = n_{12}\ \rho_{p2}\ \pi(a_1+a_2)^2\ (m_1 + m_2)^{-1}\Delta U_{12} \qquad (3.53)$$

For two clouds of interacting particles in a given volume, equality of action and reaction leads to:

$$\rho_{p1}\ F_{12} = \rho_{p2}\ F_{21} \qquad (3.54)$$

for $n_{12} = n_{21}$.

Equation (3.53) has been validated experimentally by Arastoopour et al. [1982] using large particles (1) of 5 to 8 mm diameter of glass, ceramics, and steel and small particles (2) of 30 μm silica sand at relative velocities of 2 to 10 m/s. (Problem 3.5)

<u>Collision of elastic spheres</u>. Another elementary interaction is that of the phenomenon of impacting elastic spheres. The basic formulation was given by Herz (1881) and Rayleigh (1900) [Timoshenko 1934]. For two spheres 1 and 2 of elastic materials of moduli E_1, E_2 and Poisson ratios ν_1 and ν_2, impacting with relative motion represented in Fig. 3.12, the two centers approach each other by a distance α_1' smaller than the sum of radii a_1 and a_2:

$$\alpha_1' = [5 \, (\Delta U \cos \theta)^2/4 \, n_1 n]^{2/5} \tag{3.55}$$

where $n = (16/9\pi^2)^{1/2} \, (k_1 + k_2)^{-1} \, (a_1^{-1} + a_2^{-1})^{-1/2}$, $k = (1 - \nu^2)/\pi E$, and $n_1 = m_1^{-1} + m_2^{-1}$, m being the mass of a particle. The relative motion is denoted by and angle θ and relative speed $\Delta U = |U_{p1} - U_{p2}|$. The maximum compressive force at the area of contact P is given by:

$$P = n \, \alpha_1'^{3/2} \tag{3.56}$$

and the duration of contact is given by:

$$\Delta t = 2.94 \, (\alpha_1'/\Delta U \cos \theta) \tag{3.57}$$

The radius of maximum area of contact is given by:

$$a_{12} = (3\pi/4)^{1/3} \, P^{1/3} \, (k_1 + k_2)^{1/3} \, (a_1^{-1} + a_2^{-1})^{-1/3} \tag{3.58}$$

The area of contact over the duration Δt therefore have the characteristic (maximum for spheres) value:

$$A_{12}\Delta t = \pi \, a_{12}^2 \, \Delta t = 21.91(a_1^{-1}+a_2^{-1})^{-3} \, (\Delta U)^{-1} \, (N_{Im})^{4/5} \, (\cos \theta)^{-1/5}$$

$$\equiv (A_{12}\Delta t)_o \, (\cos \theta)^{-1/5} \tag{3.59}$$

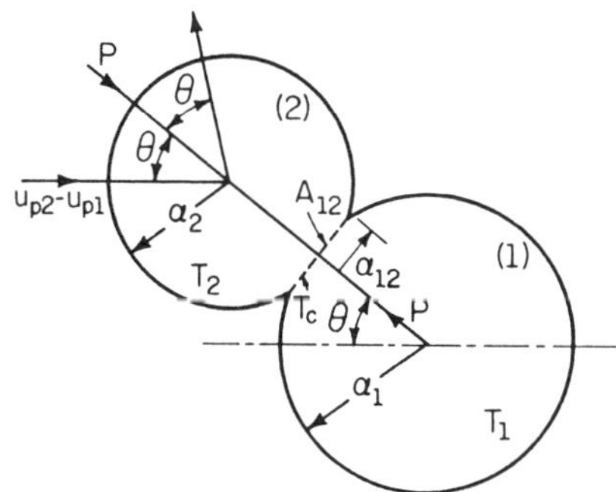

Figure 3.12 Collision and contact of two elastic spheres

relating the geometry to the dynamics of interaction, where N_{Im} is denoted as an impaction number:

$$N_{Im} = (\Delta U)^2 \ (1+r^*) \ 2^{-1} \ (m_1^{-1}+m_2^{-1})^{-1} \ (a_1^{-1}+a_2^{-1})^3 \ (k_1+k_2) \qquad (3.60)$$

relating the dynamic force of impact to the elastic force exerted by the materials. Note that N_{Im} has been redefined and differs from that in earlier publications.

<u>Heat transfer</u>. For a particle 1 collided by a cloud of particles 2 in a suspension, the temperature of particle 1 can be assumed to change uniformly. The time constant for transient heat conduction is to the order of α_1/a_1^2 (s^{-1}), where α_1 is the thermal diffusivity of material constituting particle 1. The energy equation of particle 1 can be written as:

$$dT_{p1}/dt = G_{1f}(T - T_{p1}) + G_{12}(T_{p2} - T_{p1}) \qquad (3.61)$$

where G_{1f} is the time constant for heat transfer from the fluid to particle 1 (Eq. 1.20), and G_{12} is that for heat transfer from particle 2 to 1 by collision. Heat transfer by collision occurs according to the mechanism depicted in Fig. 3.12 with area and duration of contact given by Eq. (3.59). For characteristics thicknesses d_1' and d_2' (which may be of the order of a_1 and a_2) over which the temperature gradients for heat conduction occur, an energy balance gives

$$(\bar{\kappa}_{p1}/d_1') \ (T_1 - T_c) = (\bar{\kappa}_{p2}/d_2') \ (T_c - T_2) \qquad (3.62)$$

where T_c is the temperature at the contacting surface and $\bar{\kappa}_{p1}$, $\bar{\kappa}_{p2}$ are thermal conductivities of materials constituting particles 1 and 2. The time rate of change of temperature of each particle is given by:

$$c_{p1} \, m_1 \, d \, T_{p1}/dt = -(\bar{\kappa}_{p1}/d_1') \, A_{12} \, (T_1 - T_c)$$

$$= -(\kappa_{p2}/d_2') \, A_{12}(T_c - T_2) = -c_{p2} \, m_2 \, d \, T_{p2}/dt \qquad (3.63)$$

where c_{p1} and c_{p2} are specific heats of particles 1 and 2 respectively. By eliminating T_c, Eq. (3.63) can be integrated to give

$$T_{p1} \sim T_{pi} + (<T_{p2}> - <T_{p1}>) \left\{ 1 - \exp \left[- \frac{(\bar{\kappa}_{p1} \, \bar{\kappa}_{p2})^{\frac{1}{2}} \, A_{12} t}{m_1 \, c_{p1} \, (d_1' d_2')^{\frac{1}{2}}} \right] \right\} \quad (3.64)$$

where $<T_p>$'s are the mean particle temperatures over time t. Over duration Δt of contact and since the exponential term in Eq. (3.64) is usually small, the amount of heat transferred is given by:

$$q \sim (<T_{p2}> - <T_{p1}>) \, \frac{(\bar{\kappa}_{p1} \, \bar{\kappa}_{p2})^{\frac{1}{2}} \, A_{12} \, \Delta t}{(d_1' \, d_2')^{\frac{1}{2}}}$$

$$\equiv h_{12} \, (T_{p2} - T_{p1}) \, A_{12} \, \Delta t \qquad (3.65)$$

The resultant time constant for heat transfer is thus:

$$G_{12} = h_{12} \, 4\pi \, a_1^2/m_1 \, c_{p1}$$

$$\sim \left[\eta_{12} \, \frac{\Delta U (\bar{\kappa}_{p1} \, \bar{\kappa}_{p2})^{\frac{1}{2}}}{m_1 \, c_{p1} \, 2(a_1 a_2)^{\frac{1}{2}}} \right] \int_0^{\pi/2} A_{12}(\Delta t) \, (a_1 + a_2)^2 \, 2\pi \, \sin\theta \, d\theta \qquad (3.66)$$

or

$$G_{12} = C_{12} \, (\rho_2/c_{p1}) \, [(\bar{\kappa}_{p1} \, \bar{\kappa}_{p2})^{\frac{1}{2}}/\bar{\rho}_{p1} \, \bar{\rho}_{p2}]/a_1 a_2 \qquad (3.67)$$

with the collision coefficient given by

$$C_{12} = (45/64) \, \pi \, 21.91 \, \eta_{12} \, \frac{(a_1 a_2)^{\frac{1}{2}}}{(a_1 + a_2)} \, (N_{Im})^{4/5} \qquad (3.68)$$

and G_{12} and G_{21} are related according to:

$$c_{p1}\, \rho_{p1}\, G_{12} = c_{p2}\, \rho_{p2}\, G_{21} \tag{3.69}$$

Note that in Eq. (3.67) $\overline{\kappa}_{p1}/c_{p2}\, \overline{\rho}_{p1}$ is the thermal diffusivity of the material of particle 1. The quantity $[\kappa_{p1}/c_{p1}\, \rho_{p1}\, a_1 a_2]\, (\overline{\kappa}_{p2}/\overline{\kappa}_{p1})^{-\frac{1}{2}}$ correlates thermal diffusion in particle 1 with characteristic diffusion length $(a_1 a_2)^{\frac{1}{2}}$. By taking the ratio of G_{12}/F_{12} one readily shows that, in general, the thermal response is far slower than momentum response.

When $a_1 \gg a_2$, $m_1 \gg m_2$, the above relations can be applied to the case of collision with a wall. The heat transfer coefficient is given by, for particles impacting with an intensity $\langle U_p^2 \rangle^{\frac{1}{2}}$: $(d_1' \sim a_2)$

$$h_w = n_{21}\, n_{p2}(\overline{\kappa}_{p2}/a_2)\, \langle U_p^2 \rangle^{\frac{1}{2}}\, (A_{12}\, \Delta t) \int_0^{\pi/2} (\cos\theta)^{-1/5}\, \sin\theta\, d\theta$$

$$= n_{12}\, (\rho_{p2}/\overline{\rho}_{p2})\, (\overline{\kappa}_{p2}/a_2)\, (3/4\pi)(5/4)\, 21.91\, \langle N_{Im} \rangle^{4/5} \tag{3.70}$$

Problems 3.6 and 3.7 show that impaction of particles is not so effective as a means of heat transfer.

It is noted that since the temperature of the particle phase is that of the body temperature of the particles, heat conduction by surface contact can be expressed in terms of an accommodation coefficient α_c defined by:

$$\alpha_c = (T_{pr} - T_{pi})/(T_w - T_{pi}) \tag{3.71}$$

where T_{pi} is the temperature of the incoming particle, and T_{pr} is that of a reflected particle. Since $T_{pw} \sim (T_{pi} + T_{pr})/2$, the heat flux J_{qp} due to impaction of particles at the wall is given by:

$$J_{qp}/c_p\, \dot{m}_{pw} = (T_{pr} - T_{pi}) \sim (-2\alpha_c/2 - \alpha_c)(T_{pw} - T_w) \tag{3.72}$$

where $\dot{m}_{pw}$ is the mass flux of impact at the wall. The accommodation coefficient α_c is given by, from Eq. (3.65):

$$c_p m_p (T_{pr} - T_{pi}) \sim \langle A \, \Delta t \rangle \, \frac{(\bar{\kappa}_p \, \bar{\kappa}_w)^{\frac{1}{2}}}{a_p} (T_w - T_{pi}) \tag{3.73}$$

and Eq. (3.71); giving small α_c in single scattering.

Erosion and attrition. A mode of mass transfer by particle-wall interaction is erosion of the wall or attrition of the particles depending on their properties and the dynamic of impact. The properties include hardness, yield strength, and surface roughness of each. The dynamics of impact are related to the kinetic energy of the particles, angle of impact, and friction coefficient of the surfaces. Attrition is usually a short term concern of the condition of the particles after handling while erosion is a cumulative phenomenon relating to the maintenance and replacement cost of equipment. Erosion is a major concern of, for instance, a coal handling system. Attrition, however, is a major concern in grain handling.

The energies to remove a given volume of material are denoted as [Neilson and Gilchrist 1968]: ϵ_d, related to the ultimate yield stress for ductile wear σ_d; and ϵ_b, related to the ultimate impact resistance for brittle wear σ_b.

To relate the mechanisms of erosion to basic material properties, we need to identify the normal compressive force P and the tangential machining force P_d, area of contact A and duration of contact Δt given by Eq. (3.59) and the geometry in Fig. 3.12 with particle 1 reduced to a plane wall (subscript w) impacted by particles 2 (subscript p) at velocity U and angle θ made with the normal to the wall. We further assume that the deviation from the elastic state of the material is small. With P given by Eq. (3.56) and Δt given by Eq. (3.57), the normal impulse of impact is given by:

$$P_b \, \Delta t = (2.94) \, (5/16) \, m \, U(1 + r^*) \tag{3.74}$$

and the tangential impulse of impact is:

$$P_d \, \Delta t = f \, P_b \, \Delta t \tag{3.75}$$

where f is the coefficient of friction, with (A Δt) given by:

$$(A \, \Delta t) = 21.91 \, a^3 \, U^{-1} \, N_{Im}^{4/5} \, (\cos\theta)^{-1/5} = (A \, \Delta t)_o \, (\cos\theta)^{-1/5} \tag{3.76}$$

and the impact number is now:

$$N_{Im} = U^2 (1 + r^*) \, 2^{-1} \, m \, a^{-3} \, (k_1 + k_2) \tag{3.77}$$

where $0 \le r^* \le 1$, $r^* \to 1$ as $\theta \to \pi/2$. These basic relations are applicable to two different modes of erosion: the ductile mode (subscript
d) and the brittle mode (subscript b).   Erosion via tangential
machining tends to be more prominent for ductile materials (metals,
for instance) than for brittle materials (ceramics, for instance).
Erosion of a real materials is a result of the combined effect of
these modes.

The erosion in volume loss w_d per impact is given by Soo [1977]:

$$w_d = \eta_d \, [C_d \, (P_d/A) - \sigma_d] \, U \, \sin\theta \, (A \, \Delta t)/\epsilon_d \tag{3.78}$$

where C_d corrects the non-sphericity of the actual particles, stress
concentration at sharp edges, roughness of the surface impacted, and
non specular reflection from an actual surface, $C_d > 1$; η_d is a
mechanical efficiency of impact, including the effects of gliding,
scattering, lifting of the particle by the gas stream, $\eta_d < 1$
usually.   In terms of a volumetric erosion rate per unit area, W_d
produced by a dust stream of concentration ρ_p, we can express this
rate as a dimensionless erosion energy parameter:

$$E_d^* \equiv W_d \, \epsilon_d / \rho_p \, U^3 \, C_d f(1 + r^*) \, (2.94)(5/16) \, \eta_d$$

$$= \sin\theta \, [1 - K_d^* \, (\cos\theta)^{-1/5}] \equiv F_d(\theta, K_d^*) \tag{3.79}$$

and the resistance parameter K_d^* is given by

$$K_d^* = \sigma_d/C_d \ (P_d/A)$$

$$= 23.85 \ (\sigma_d/C_d) \ a^3 \ (mU^2)^{-1} \ (1 + r^*)^{-1} \ N_{Im}^{4/5} \qquad (3.80)$$

Since $(1 + r^*)^{1/5}$ is nearly 1, K^* can be treated as a parametric constant and the function $F_d \ (\theta, K_d^*)$ characterizes the ductile wear at various angles θ. The maximum value of F_{dm} occurs at angle θ_m given by:

$$K_d^* = 5(\cos \theta_m)^{11/5} \ (4\cos^2 \theta_m + 1)^{-1} \qquad (3.81)$$

Experimental results show that the ductile mode which is typical of metal targets has a maximum erosion occurring at an angle θ between 60 and 70 degrees [Smeltzer et al. 1970].

The brittle mode which is typical of glass and ceramics is characterized by an erosion rate decreasing with angle θ to near but greater than 0 degree [Finnie 1967]. The mechanism is one of cracking of the target surface by fatigue and brittle failure for stress above an impact yield stress σ_b. The erosion in terms of volume loss per impact w_b is given by:

$$w_b = \eta_b \ [C_b(P_b/A) - \sigma_b] \ U \cos \theta \ (A \ \Delta t)/\varepsilon_b \qquad (3.82)$$

where C_b has an analogous meaning as C_d given before, although it may differ in magnitude even for similar particles and surfaces. The volumetric erosion rate per unit area W_b produced by a dust stream of concentration ρ_p can be expressed in the form of a dimensionless erosion parameter E_b^*:

$$E_b^* \equiv W_b \ \varepsilon_b/\rho_p \ U^3 \ C_b(1 + r^*) \ (2.94)(5/16) \ \eta_b$$

$$= \cos \theta \ [1 - K_b^* \ (\cos \theta)^{-1/5}] \equiv F_b \ (\theta_1 K_b^*) \qquad (3.83)$$

where K_b^* has the same form as K_d^* with quantities of subscript d substituted by subscript b. The combined effect of ductile and brittle wear is verified experimentally [Goldberger and Nack 1964,

Fraas 1975] although uncertainty exists in the basic parameters of material properties. An example is given in Fig. 3.13 [Goldberger and Nack 1964]. Levy et al. (1986) further found scale formation (corrosion) and thickness loss from 9Cr-1Mo steel at temperatures of 900°C of air suspending 130 μm alumina particles. A coating of alumina reduced flame corrosion but was susceptible to brittle wear (Ritter et al. 1986).

When applied to wear by random motion of particles in a fluidized bed with an intensity $<U_p^2>^{\frac{1}{2}}$, averaging over all directions from 0 to $\pi/2$ for θ, and magnitudes of speeds gives, for ductile wear:

$$E_d^* = W_d\ \varepsilon_d/f\rho_p\ <U^2>^{3/2}\ C_b(1 + r^*)\ 2.94\ (5/16)\ (2/3\ \sqrt{\pi})\ \eta_b$$

$$= 1 - 0.9586\ <K_d^*> \tag{3.84}$$

where $<K_d^*>$ is given by replacing U by $<U^2>^{\frac{1}{2}}$ in both K_d^* and N_{Im}. For

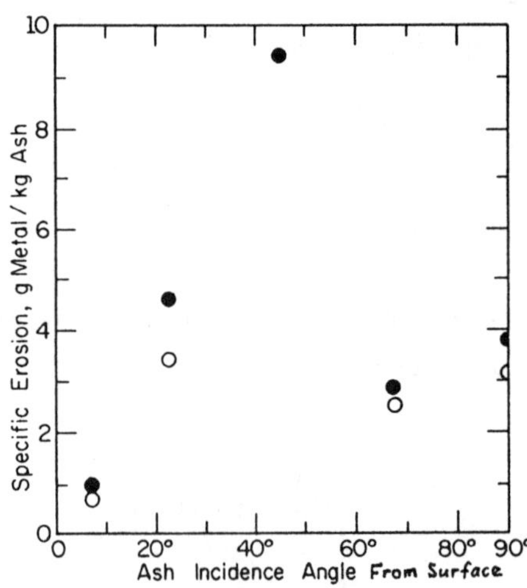

Figure 3.13   Specific erosion ($W\ \bar\rho_2/\rho_1\ V$) versus ash incidence angle [Goldberger and Nack, 1964]; stagnation temperature 760°C, U = 285 m/s, material: S-816, dust loading 25 g/hr ● - maeasured, O - calculated, $K_D^* \sim 0.80$

brittle wear, we get:

$$E_b^* = W_b \; \varepsilon_b/\rho_p \; <U^2>^{3/2} \; C_b(1 + r^*) \; 2.94 \; (5/16)(2/3\sqrt{\pi}) \; \eta_b$$

$$= 1 - 0.8981 \; <K_b^*> \tag{3.85}$$

with the averages obtained in a similar way as in the case of ductile wear. Experimental trend is seen in Fig. 3.14 for metallic surfaces imbedded in a fluidized bed combustor in comparison to the result of direct impingement by a stream of particles [National Research Development Corp. 1971]. Wear by a sliding bed of bulk materials can be similarly treated [Soo 1977]. Erosion of turbine blades by water droplets in wet steam corresponds to brittle wear [Huang et al. 1973]. (Prob. 3.8)

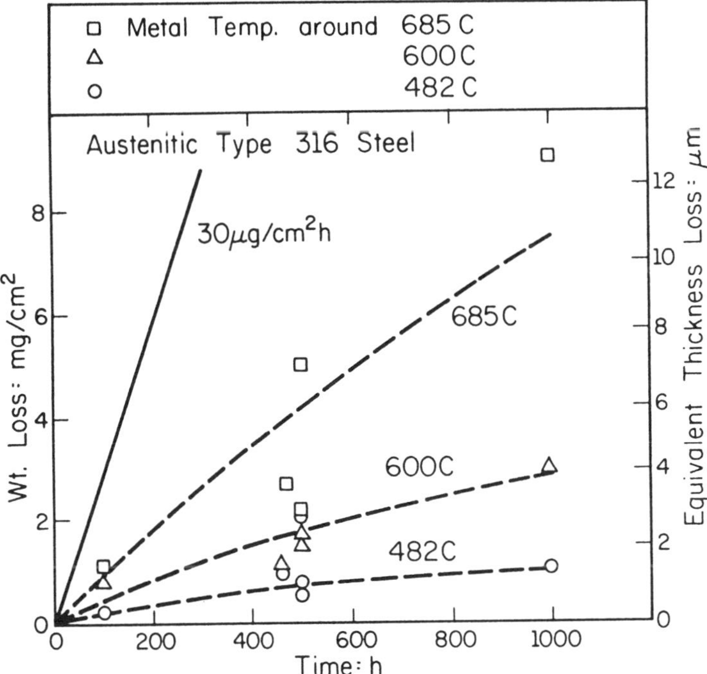

Figure 3.14   Weight loss of metallic surfaces embedded
              in fluidized bed combustor [NRDC 1971]

Attrition of materials by impact has been reported by Mills and Mason [1978]. Degradation of quartz particles conveyed in various pipe materials (steel, nylon, and fiberglass) was noted, particularly at the bends. In the case of quartz particles on steel, for an initial mean size of 180 μm, the product of fragmentation distributes around 40 μm, that is, a bimodal distribution in size was the result. The pressure at the point of contact during a collision may crush the particle locally. The maximum local compressive stress at the point of impact is given by Eq. (3.56) and can be reduced to the form:

$$P = 0.17 \, E_p (\bar{\rho}_p U_p^2 \cos \theta / E_p)^{1/5} (1 - \nu_p^2)^{-4/5} \tag{3.86}$$

for $E_w \gg E_p$. The above numerical coefficient is for spherical particles. The amount of dust generated is seen to be proportional to the difference between P and the fracture strength of the particle materials. An asymptotic size is reached when the particle size is such that the impact velocity is greatly reduced in the fluid boundary layer.

   Charge transfer. A direct analogy exists between transfer of electric charges by surface contacts and the heat transfer mechanism. The current density through the contact area A_{21} is given by (Fig. 3.12):

$$J_{21} = h_{e21}(V_2 - V_1)$$

$$= (\sigma_1/d_1')(V_c - V_1) = (\sigma_2/d_2')(V_2 - V_c) \tag{3.87}$$

where h_{e21} is the charge transfer coefficient through the contact surface, V_1, V_2, and V_c are the potentials of spheres 1 and 2 and the contact area, and σ_1, σ_2 are the electrical conductivities of the materials. The time rate of change of potential of each particle is

$$C_1(dV_1/ct) = -C_2(dV_2/dt)$$

$$= J_{21}A_{21} = A_{21}h_{e21}(V_2 - V_1) \tag{3.88}$$

where C_1, C_2 are the capacitances of the particles respectively, $C = 4\pi\epsilon a$, ϵ is the permittivity, and

$$h_{e21} = \left(\frac{\sigma_1\sigma_2}{d_1'd_2'}\right)^{1/2} \left[\left(\frac{\sigma_2 d_1'}{\sigma_1 d_2'}\right)^{1/2} + \left(\frac{\sigma_1 d_2'}{\sigma_2 d_1'}\right)^{1/2}\right]^{-1} \tag{3.89}$$

after eliminating V_c.

For suspended particles, having charges q_1 and q_2, and for work functions ϕ_1 and ϕ_2 and of the materials and surfaces of each, the potential difference is:

$$V_2 - V_1 = \left(\phi_2 + \frac{q_2}{4\pi\epsilon_2 a_2}\right) - \left(\phi_1 + \frac{q_1}{4\pi\epsilon_1 a_1}\right) \tag{3.90}$$

Similar derivation as that of Eqs. (3.64) and (3.65) gives the charge transfer per impact:

$$Q_{21} = - Q_{12} \sim h_{e21} A_{21} \Delta t \, (V_2 - V_1) \tag{3.91}$$

The charge transfer coefficient h_{e21} given by Eq. (3.89) is based on clean surfaces. In reality, both h_{e21} and ϕ's are influenced by the absorbed surface layer and humidity. (Prob. 3.11) Moreover, each particle has a saturation charge q given by

$$V = \phi + q/4\pi\epsilon a = \text{constant} \tag{3.92}$$

for its contact with another surface. The average charge transfer by impact is seen to be a function of the impact number.

The rate of charge transfer is measured by an inverse relaxation time H_{21}, with

$$H_{21} = \eta_{21}\pi(a_2 + a_1)^2 n_1 \Delta U \, (5/2) \, h_{e21}(A_{21}\Delta t)_o/C_2 \tag{3.93}$$

by analogy to the heat transfer process. Note that η_{21} here is affected by the electric field of interacting particles and may exceed 1 via mutual attraction (Fig. 3.10). A reciprocal relation also exists:

$$C_2 n_2 H_{21} = C_1 n_1 H_{12} \qquad\qquad (3.94)$$

where the n's are number densities of particle clouds 1 and 2. It is further seen that for particles of similar materials in a cloud, Eq. (3.92) gives q/a = constant, via mutual collision.

The above results on charge transfer is also the basis of electrostatic probes for a particle suspension. For the case of sphere 2 impacted by a cloud of spheres 1, Eq. (3.85) gives the change of the potential of sphere 2 according to:

$$C_2 dV_2/dt = n_{21} n_1 \Delta U_{21} \int_{o}^{\pi/2} 2\pi \, Q_{21}(a_1 + a_2)^2 \sin\theta \, d\theta \qquad (3.95)$$

(Fig. 3.12). Since C = dQ/dV, the total electric current due to successive charge transfer is:

$$i_2 = (5/4) \, 2\pi \, (a_2 + a_1)^2 \, n_{21} n_1 (\Delta U_{21})(V_2 - V_1) h_{e21} (A_{21}\Delta t)_o \qquad (3.96)$$

with mean value of $(V_2 - V_1)$ over Δt. This relation is the basis of an electrostatic probe theory for the measurement of mass flow of a suspension; with V_1 given by Eq. (3.89) and an adjustable probe voltage V_2 [Cheng and Soo 1970]. The average charge transfer per impact on a 12.7 mm steel ball by 16 μm coal dust at 18.3 to 36.6 m/s is nearly $Q = 4.268 \times 10^{-18} \, (N_{Im})^{4/5}$ C. Modification by surface conditions is shown in Prob. 3.9. Extension to the case of charge transfer in turbulent pipe flow shows a significant bearing on the electrostatic safety in pneumatic transport [Soo 1981]. (Prob. 3.10)

3.4 Viscosity and Thermal Conductivity of Particle Clouds.

To account for the shear resistance and heat transfer in a multiphase system, one must have a knowledge of the transport properties of phase interactions. Take the simple case of a suspension of identical spherical particles in a gas, interactions between these two phases give rise to, for instance, viscosities due to fluid-

fluid, fluid-particle, and particle-particle interactions. That due to fluid-fluid interactions is obviously the viscosity of the gas itself; those due to fluid-particle and particle-particle inter- actions remain to be dealt with. When there is a distribution in particle sizes such as giving rise to 1-2 interaction depicted in Fig. 3.12, additional particle-particle and particle-fluid inter- actions must be accounted for. Moreover, since there are 3 viscosi- ties due to phase interactions in a two-phase system, while there are only two tensors for the shear rates, we must obtain the viscosity of each phase in the mixture [Soo 1970]. Similar considerations apply to heat fluxes.

<u>Dilute suspensions</u>. A simple case is that of a dilute suspension defined in Sec. 2.7. In this case, the shear stress and heat flux in the fluid is unaffected by the presence of the particles and those of the particle cloud arise from particle-fluid interactions only, since the effect of particle-particle interaction is negligible. The shear resistance of the particle cloud arises from that of diffusion of momentum via bodily transport of particles of a given momentum per unit volume via diffusivity D_{pm}, the diffusivity of particle phase p in the mixture [Soo 1969]. Take the example of a "shear stress" in terms of gradient of x- component of momentum in the y-direction,

$$\tau_{xy} = D_{pm}(\partial \rho_p U_{px}/\partial y) \sim D_{pm}\rho_p(\partial U_{px}/\partial y) \tag{3.97}$$

Similarly the "heat flux" in the x-direction is due to bodily trans- port of heat by the particles via diffusion:

$$J_x = - D_{pm}(\partial c_p\rho_p T_p/\partial x) \sim - D_{pm} c_p\rho_p(\partial T_p/\partial x) \tag{3.98}$$

In the case of a dilute suspension, $D_{pm} \sim D_{pf}$, the diffusivity of particles in the fluid (Sec. 1.7), the predominating component of the mixture. The viscosities of components in the mixture are now:

$$\mu_{fm} = \bar{\mu}_f, \; \mu_{pm} = \mu_{pf} \sim D_{pf}\rho_p \tag{3.99}$$

and the thermal conductivities of components in the mixture are now:

$$\kappa_{fm} = \bar{\kappa}_f, \quad \kappa_{pm} = \kappa_{pf} \sim D_{pf}c_p\rho_p \tag{3.100}$$

$\bar{\mu}_f$ and $\bar{\kappa}_f$ are the viscosity and thermal conductivity of the pure fluid material.   Note that μ_{pf} and κ_{pf} also conform to the kinetic theory of molecules by analogy.

<u>Viscosity of particle-particle interaction</u>.   When the particle concentration is high enough, shear motion gives rise to viscosity of particle-particle interaction.   Consider a sphere of radius r in a shear stream of gradienty $\partial U/\partial y$ of a particle cloud of number density n and mass m, and denote the velocity U as zero in the center plane of the sphere, the relative velocity (Fig.   3.15) is given by:

$$\Delta U = (\partial U/\partial y)y = (\partial U/\partial y) \ r \ \sin \ \theta \ \cos \ \phi \tag{3.101}$$

where ϕ is the azimuthal angle measured from the plane normal to that of $\Delta U = 0$.   The projected area of impact is $r \ \sin \ \theta d\phi \ d\theta \ \cos \ \theta$ and the change of momentum is $m\Delta U(1 + r^*\cos 2\theta)$, assuming single scattering and specular reflection as in Sec.   3.3.   The force acting on the sphere from $\theta = 0$ to $\theta = \pi/2$ is, using fraction impacted η as

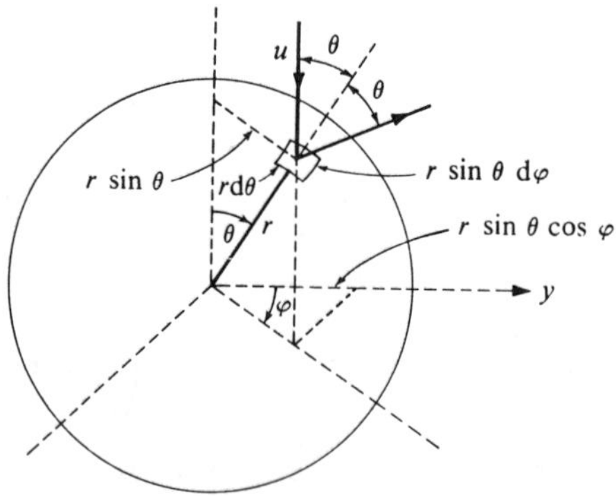

Figure 3.15   Impaction due to shear motion.

an empirical parameter,

$$F \sim 2\eta \int_0^{\pi/2} \int_0^{\pi/2} n(\Delta U)^2 m \ (1 + r^* \cos 2\theta) \ r^2 \sin \theta \cos \theta \ d\theta \ d\phi$$

$$= \eta(\pi/8) \ nm(\partial U/\partial y)^2 r^4 [1 - (r^*/3)] \qquad (3.102)$$

Note that $\eta \leq 1$ here is different from the η given in Sec. 3.3. It is seen that over a thickness $a_1 + a_2$ as denoted in Fig. 3.12, the total shear stress on the cloud 1 due to velocity gradient tensor Δ_2 of cloud or particles 2 is given by:

$$\tau_{12} \simeq -\eta_{12} n_1 (a_1 + a_2)^5 (\pi/8) \ n_2 (m_1^{-1} + m_2^{-1})^{-1} [1 - (r^*/3)] |(\Delta_{p2} : \Delta_{p2})^{\frac{1}{2}}| \Delta_{p2}$$

$$\equiv - \mu_{p12} \ \Delta_{p2} \qquad (3.103)$$

or

$$\mu_{p12} = (\pi/8) \ \eta_{12} \rho_{p1} \rho_{p2} (m_1 + m_2)^{-1} (a_1 + a_2)^5 [1 - (r^*/3)] |(\Delta_{p2} : \Delta_{p2})^{\frac{1}{2}}|$$

$$(3.104)$$

where Δ_p has the Cartesian components

$$\Delta_{p2ij} = (\partial U_{p2i}/\partial x_j) + (\partial U_{p2j}/\partial x_i) \qquad (3.105)$$

and

$$\Delta_p : \Delta_p = \sum_i \sum_j \Delta_{pij} \Delta_{pji} \qquad (3.106)$$

It is seen that the viscosity of the particle cloud due to the simple model of interaction falls into the class of Ostwald-de Waele model (see, for instance, Bird et al. [1960]) of non-Newtonian fluid, with $\tau = |m|\sqrt{(\Delta : \Delta)/2}|^{n-1}| \ \Delta$; m and n are empirical constants. Note that for $r^* = 1$, $a_1 = a_2$, $\bar{\rho}_{p1} = \bar{\rho}_{p2}$, $\mu_{p11} = \eta \ \alpha_1^2 \ \bar{\rho}_p a^2 (\Delta : \Delta)^{\frac{1}{2}}$ (Prob. 3.11). The proportionality was confirmed by the experimental results of Savage [1983].

<u>Viscosity and thermal conductivity of particle clouds</u>. For a case of particle-particle interaction such as represented in Fig. 3.12, but with a large enough and the number density of particles 2 large enough, multiple scattering may occur. This is, when a particle 2 reflected by particle 1 after a collision meets another particle 2 and is reflected back to have another collision with particle 1, in the way of transition from free molecular flow to viscous flow of particles of species 2. Specifically, when the mean free path of species 2, Λ_{p2} is much smaller than a_1, or

$$\Lambda_{p2} = [\sqrt{2}\, n_{p2} 4\pi a_2^2\, \eta_{22}]^{-1} < a_1 \tag{3.107}$$

That is, when the volume fraction particles of species 2 is:

$$\alpha_2 = \rho_{p2}/\bar{\rho}_{p2} > (a_2/a_1)\, \eta_{22}/3\sqrt{2} \tag{3.108}$$

We have an apparent viscosity μ_{p2} of particles 2 which would be set in random motion by multiple collision with an intensity of motion $<U_{p2}^2>^{\frac{1}{2}}$, or [Sodha et al. 1963]

$$\mu_{p2} \sim (1/3)\, \rho_{p2} <U_{p2}^2>^{\frac{1}{2}} \Lambda_{p2} = a_2\, \bar{\rho}_{p2}\, <U_{p2}^2>^{\frac{1}{2}}/9\sqrt{2}\eta_{22} \tag{3.109}$$

in which case the time constant of momentum transfer F_{p12} will take the form of Eq.   (3.24) with $\bar{\mu}$ replaced by μ_{p2}. Similarly, the apparent thermal conductivity κ_{p2} of cloud of particles 2 takes the form:

$$\kappa_{p2} \sim \mu_{p2} c_{p2} = c_{p2} a_2 \bar{\rho}_{p2}\, <U_{p2}^2>^{\frac{1}{2}}/9\sqrt{2} \tag{3.110}$$

by kinetic analogy. The inverse relaxation time for heat transfer G_{p12} is now, following Eq. (1.20):

$$G_{p12} = 3G_1^* \,\kappa_{p2}\alpha c_{12}/c_{p1}\bar{\rho}_{p1} a_1^2 \tag{3.111}$$

where G_1^* is a similar correction for convection in the way of F_1^*, α_{c12} is the accommodation coefficient for the 21 collisions (Eq. 3.71).   Cases of strong particle-particle interactions may overshadow the presence of a fluid. (Prob. 3.12).

<u>Transport properties of a particle-fluid mixture.</u>   When applied to the situation where there is no lag of particle motion from fluid motion, Einstein [1906]   showed that the viscosity μ_m of an incompressible fluid containing solid spheres is given by:

$$\mu_m = \bar{\mu}(1 + 2.5\alpha) \tag{3.112}$$

for small α, where α is the proportion of the total volume occupied by the particles.   Further details are given in Happel and Brenner [1965].   Taylor [1932] modified this relation to apply to liquid droplets of viscosity $\bar{\mu}_p$ of the liquid material to give:

$$\mu_m = \bar{\mu}[1 + 2.5\alpha \ (\bar{\mu}_p + \tfrac{2}{5} \bar{\mu})(\bar{\mu}_p + \bar{\mu})^{-1}] \tag{3.113}$$

When applied to a liquid containing small volumes of air bubbles, Taylor [1954] accounted for the effect of their compressibility by introducing a second coefficient of viscosity ζ:

$$\zeta = (4/3) \ \bar{\mu}/\alpha \tag{3.114}$$

When a suspension is dense enough, particle-particle interaction tends to overshadow the effect of the presence of the fluid [Savage 1983].   Motion of the particle phase is influenced by the drag force exerted by the fluid and shear force of the particles (Eq. 3.104). Savage included α_p^2 in an empirical function; Schuegerl [1971] followed a similar correlation of experimental data for the viscosity of fluidized beds but retaining the Newtonian behavior in the functional relation.

Not much is known about the thermal conductivity of particle cloud. For equal velocity of phases, Goring and Churchill [1961] gave:

$$\kappa_m = \bar{\kappa} \, \{1 - (1 - \alpha \, [1 - (\bar{\kappa}_p/\bar{\kappa})])\} \qquad\qquad (3.115)$$

for a dilute system.

3.5 Flow regimes of a cloud of particles in a turbulent fluid.

For very low density of particles such that the number of collisions among the particles is negligible when compared to that with the wall, we have an analogous situation to that of flow of rarefied gases. The shear stress at a solid wall τ_{pw}, following that analogy, can be represented by:

$$\tau_{pw} \sim \rho_{pw} \, U_{pw} <U_{pw}^2>^{\frac{1}{2}}/2\sqrt{\pi} = (1/2) \, C_{pf} \rho_{po} U_{po}^2 \qquad (3.116)$$

where $<U_{pw}^2>^{\frac{1}{2}}$ is the intensity of particle motion [Soo et al. 1966] near the wall and U_{pw}, U_{po} are the mean speed of particles at the wall and in the free stream. C_{pf} is the friction factor, and

$$(U_{po}/U_{pw}^2>^{\frac{1}{2}}) \, C_{pf} \sim (\rho_{pw}/\rho_{po})(U_{pw}/U_{po})/\sqrt{\pi} \qquad (3.117)$$

The next situation is that of significant collision among the particles, but the scale of motion of particles is small. The motion of the particle phase is analogous to viscous motion with slip. The "viscosity" of the particle phase is, in a sense, a microscopic representation of particle-particle interactions. In a region where the particle density is ρ_p, the shear stress τ_{pw} exerted by the particles on a flat plate and friction coefficient in this range, by analogy to laminar boundary layer motion of a fluid [see, for instance, Schlichting 1979], is given by

$$\tau_{pw}/(1/2) \, \rho_{po} U_{po}^2 = C_{pf} = 0.664/(N_{Re,x})^{\frac{1}{2}} \qquad (3.118)$$

where

$$N_{Re,x} = U_{po}x\,\rho_{pw}/\mu_{pw}$$

$$= 9\sqrt{2}\,(U_{po}/<U_{pw}^{2}>^{\frac{1}{2}})\,(\rho_{pw}/\bar{\rho}_p)\,(x/a) \tag{3.119}$$

and x is measured from the leading edge along the flat plate, with μ_{pw} given by the hard-sphere model by kinetic analogy (Eq. 3.109). Equation (3.118) can be expressed in a similar form as Eq. (3.117) as:

$$\left(\frac{U_{po}}{<U_{pw}^{2}>^{\frac{1}{2}}}\right) C_{pf} = 0.664\,[9\sqrt{2}\,(\frac{\rho_{pw}}{\bar{\rho}_p})\,(\frac{x}{a})]^{-\frac{1}{2}}[\frac{U_{po}}{<U_{pw}^{2}>^{\frac{1}{2}}}]^{\frac{1}{2}} \tag{3.120}$$

Equations (3.117) and (3.120) are plotted in Fig. 3.16 for $\rho_{po}U_{po}/\rho_{pw}U_{pw} = 3.67$. The transition between the two ranges occurs at

$$[9\sqrt{2}\,(\rho_{pw}/\bar{\rho}_p)\,(x/a)]\,[U_{po}/<U_{pw}^{2}>^{\frac{1}{2}}] \sim 20 \tag{3.121}$$

Below the value of 20, we may have a range of free-particle motion due to lack of particle-particle interaction. For the case of transport of 0.15 mass ratio of 200 μm glass particles in air at room

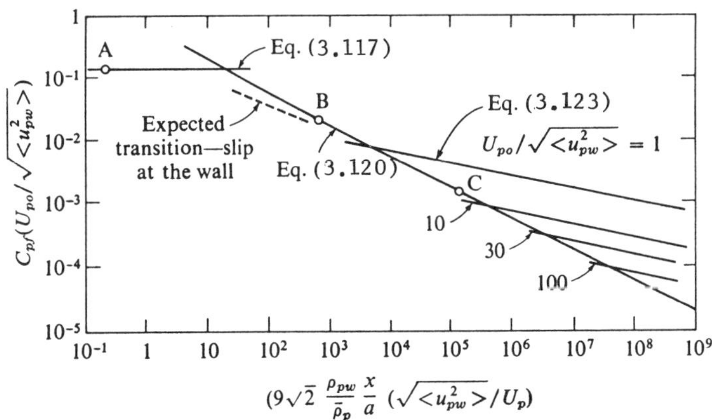

Figure 3.16 Friction factor of flat plate due to solid particles [Soo 1962]

condition ($\bar{\rho}$ = 1.14 kg/m^3) and at a velocity of 30.5 m/s, giving $<U_{pw}^2>^{1/2} \sim$ 1.8 m/s [Soo et al. 1960], the product in Eq. (3.121) is 0.2 at x = 0.3 m. The particles in this case have negligible mutual collision (point A in Fig. 3.16). However, slip motion may still occur due to particle-fluid interaction [Soo 1969] at a velocity given by:

$$U_{pw} = L_p (\partial U_p / \partial y)|_w \tag{3.122}$$

with y coordinate normal to the flat plate, and L_p is the particle-fluid interaction length (Eq. (1.72).

The third range is such that large-scale turbulent motion of particles occurs. The density of particles is high enough for correlated random motion of particles exists by mutual collision. In this case the particles and the fluid act as a mixture of heavy and light gases. The shear stress and friction factor due to solid particles are:

$$\tau_{pw}/(1/2)\ \rho_{po} U_{po}^2 = C_{pf} = 0.0592\ (N_{Re,x})^{-1/5} \tag{3.123}$$

which is plotted in Fig. 3.16. Extending the above example, if we change the mass ratio to 5 and a to 1 μm, at x = 0.3 m, the product in Eq. (3.121) is 666 (point B in Fig. 3.16). However, if we raise the pressure to 2 MPa and change a to 0.1 μm, the same product becomes 1.33×10^5 (point C), close to the condition of turbulence. For a ratio of $U_{po}/<U_{pw}^2>^{1/2}$ = 16.6, turbulence of the particle phase occurs when the product in Eq. (3.121) is greater than 10^6. The experiments of Wen and Simons [1959] include largely the range of slip motion and cases of slug motion of the dense phase.

Therefore, even when particles are suspended in a turbulent fluid, a dilute particle phase may execute what corresponds to laminar slip motion as evidenced by experiments (Ch. 5). Motions in a dense suspension will be treated in Ch. 8.

3.6 Brownian Motion, Coagulation, and Agglomeration.

We have so far dealt with situations in which the motion of solid particles is due to fluid motion only. This is true for relatively large particles, say, greater than the order of 1 μm in size. Due to thermal excitation, these particles may exhibit significant Brownian motion, with mean squared velocity $<U_p^2>$ given by:

$$\tfrac{1}{2}\, m_p <U_p^2> \sim \tfrac{3}{2}\, kT \qquad (3.124)$$

where k is the Boltzmann constant.

The random nature of Brownian motion frequently brings individual particles into close proximity where electrostatic or polarization forces (van der Waals forces) attract them to each other, causing coagulations. In the simple model due to Smoluchowski [1917], N spherical particles of equal size in a dilute dispersion were considered so that only binary encounter is significant. Take one particle as stationary, imagine a sphere of radius R surrounding the particle such that any other particle entering this sphere will be attracted to the first and become bonded to it. The diffusional flux across that surface depends on the average rate at which particles cross as a result of Brownian motion. For particles of radius a, crossing occurs with R < 2a. The concentration n of these particles satisfies the diffusion equation:

$$\partial n/\partial t = D\, r^{-2}(\partial/\partial r)\, (r^2 \partial n/\partial r) \qquad (3.125)$$

Where D is Brownian diffusivity of the particles given by:

$$D = kT/3\pi\mu a \qquad (3.126)$$

and the boundary conditions are:

$$n = n_0 \text{ for } r > R, \; t = 0$$
$$n = 0 \quad \text{at} \quad r = R, \; t > 0 \qquad (3.127)$$
$$n = n_0 \text{ as } \quad r \to \infty$$

The solution of Eq. (3.125), satisfying Eq. (3.127) is

$$n = n_0 [1 - \frac{R}{r} + \frac{2}{\sqrt{\pi}} \frac{R}{r} \int_0^{(r-R)/2\sqrt{Dt}} \exp(-z^2) dz] \qquad (3.128)$$

The flux of particles crossing the surface at R is

$$J = D(\partial n / \partial r)_{r=R} = (Dn_0/R)[1 + (R/\sqrt{\pi Dt})] \qquad (3.129)$$

The total number of particles that cross the surface $r = R$ or total number of collisions per unit time is

$$\dot{N} = \int J dS = D 4\pi n_0 R \, [1 + \frac{R}{\sqrt{\pi Dt}}] \qquad (3.130)$$

The rate of rapid coagulation is given by that at $t \gg R^2/D$, the diffusion time, as

$$\dot{N} = 4\pi DRn_0 \qquad (3.131)$$

For like particles, the total number of contacts per unit volume of dispersion per second is

$$N = 8\pi DRn_0^2 = 16 \; \pi Dan_0^2 \qquad (3.132)$$

The rate of change is therefore

$$dn/dt = -8\pi DRn^2 \qquad (3.133)$$

Integration gives, for constant value of DR:

$$n = n_0/[1 + (t/\tau)] \qquad (3.134)$$

where $\tau = 1/8\pi DRn_o$ is the characteristic coagulation time, over which coagulation would have halved the original number of particles.

Another aspect of agglomeration is by adhesion of solid particles to solid surfaces. Factors influencing adhesion are: London-Van der Waals forces, humidity, surfaces, contact area, time of contact, static electricity, viscous surface coating, temperature, to name a few (see, for instance, Corn [1961]).

3.7 Condensation, Evaporation, Droplet Coalescence and Break-up.

The multiphase formulation accounts for, as phase change, any change from species 1 to species 2 particles by condensation, evaporation, coalescence, or break-up.

Condensation of water vapor in the atmosphere may begin on small ions but principally with aerosols of 0.1 to 10 μm as nuclei. Knowledge on basic nucleation mechanism is still incomplete although once a droplet is formed subsequent condensation process is quite well understood. (see, for instance, Mason [1957])

The dynamics of coalescing or shattering of particles are not accounted for by the relaxation mechanism of particle-particle interaction treated in Sec. 3.3. It is seen that when a particle 1 and a particle 2 collide and combine, the momentum of the particle 3 so generated is given by the conservation of momentum as

$$m_3 U_3 = m_1 U_1 + m_2 U_2 \tag{3.135}$$

The energy dissipated is given by

$$\frac{1}{2} m_1 U_1^2 + \frac{1}{2} m_2 U_2^2 - \frac{1}{2} m_3 U_3^3 = \frac{1}{2} \left(\frac{m_1 m_2}{m_1 + m_2}\right) (U_1 - U_2)^2 \tag{3.136}$$

(Prob. 3.13) which might produce a rise in temperature or cause the resulting drop to oscillate which may lead to shattering of the

drop.   Liquid droplets were also observed to collide without
coalescing.

A simple case of growth of a large droplet (acretion) in a fog
was treated by Mason [1957]. A water drop larger than neighboring
droplets acquires also a larger terminal velocity than the the
rest. This results in its collision with the smaller neighbors lying
in its path. If a fraction η_{co} of the droplets 2 colliding (with
fraction impacted η_{12}) with the larger droplet coalesce with it, the
larger droplet of radius a grows at a rate given by:

$$\frac{dm_1}{dt} = 4\pi a_1^2 \, \bar{\rho}_p \frac{da_1}{dt} = \eta_{12}\eta_{co}\pi a_1^2 \, \bar{\rho}_{p2} \, (U_{p1} - U_{p2}) \qquad (3.137)$$

Note that droplet 1 is slowed down by impaction with droplets 2
according to Eq. (3.52) with the gravity effect added. (Prob. 3.14)

Droplet shattering. When a liquid droplet or jet is subjected to a
gas velocity in excess of some critical value, it will disintegrate
or shatter.   Two general types of breakup were observed: one is by
distortion of the droplets forming irregular ligaments; another is by
blowing out of liquid in the form of bubbles or lenticular shapes.
For non-viscous liquid suddenly exposed to a gas stream at constant
velocity U, Hinze [1949] gave the critical condition as

$$(\delta/a)/N_{We} = (\delta/a)/(\bar{\rho}U^2 a/\sigma) = -0.17 \qquad (3.138)$$

where δ is the displacement in the radial direction, $\bar{\rho}$ is the density
of the gas.   The case $\delta = $ a agrees with the critical Weber
number $N_{We} \sim 6$ given by Lane [1951].   Morrell [1961] gave -0.2 for
the above ratio when applied to a jet.   Morrell showed that for an
action time t_a, the natural period τ of a jet is

$$\tau = 2\pi(\bar{\rho}_p a^3/6\sigma)^{\frac{1}{2}} \qquad (3.139)$$

where $\bar{\rho}_p$ is the density of the liquid, and the decay of dynamic pressure is given by

$$\bar{\rho}U^2 = \bar{\rho}_o\bar{U}_o \exp(-t/t_a) \qquad (3.140)$$

where subscript o refers to initial conditions. Fig. 3.17 gives $\tau/2\pi t_a$ vs. $(\delta/a)/N_{We}$. For a sphere the quantity 6 in Eq. (3.139) is replaced by 8.

The breakup by slipping action of a fluid was given by Taylor [1956] correlating the liquid boundary layer thickness and the slipping rate:

$$\bar{\rho}_o U^2/(\zeta T\alpha/\beta) = \text{constant} \qquad (3.141)$$

as a first approximation, where ζ is the ratio of actual tensile strength to the ideal value, T is the absolute temperature, α is the coefficient of thermal expansion, β is the coefficient of

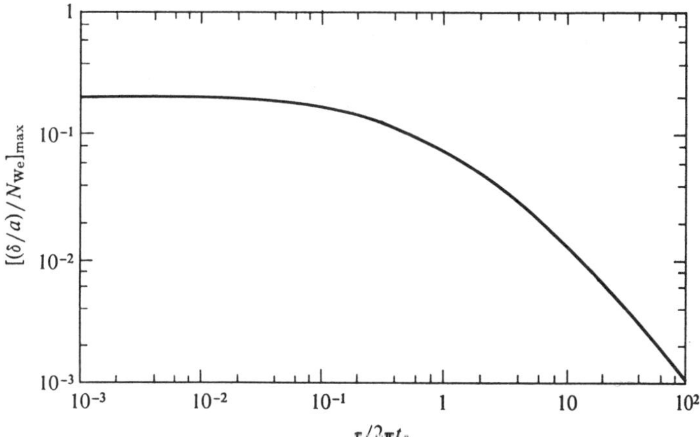

Figure 3.17 Maximum nondimensional displacement of N_{We} as a function of $\tau/2\pi t_a$ [Morell 1961].

compressibility. Morrell suggested the model as

$$f \, \bar{\rho}_0 U_0^2 \, a/\sigma = K; \text{ for } t_a \geq \tau, \; a \leq a_1$$

$$\rho_0 U_0^{\,2}/(\zeta T \alpha/\beta) \sim H; \text{ for } t_a \leq \tau, \; a \geq a_1$$

$$(3.142)$$

where $f = f(\tau/2\pi t_a)$, K and H are constants, and a is the value of radius a at $t_a = \tau$. These conditions were confirmed by experimental data. Analogous correlations have been made on splitting of bubbles by large relative motion of a liquid [Sevik and Park 1972].

<u>Agglomeration</u>. The nature of agglomeration of particles of similar or different material by their relative motion or scavenging of dust particles by liquid droplets can be seen via a one-dimensional consideration. The relative motion may be produced by acceleration of a suspension of particles of different sizes, or by a countercurrent liquid spray to a dusty gas. The basic interaction can be illustrated by a one-dimensional model with finite and constant relative velocity (ΔU) between species 1 and species 2 particles. We take $a_2 \gg a_1$ and treat species 2 particles as the collector with a mean sticking probability of σ. The collection rate of a_1 on a_2 for cloud density ρ_{p1} of a_1 particles, is given by $\sigma n_{12}(\Delta U)\rho_{p1}\pi \, a_2^2$ per a_2 particle per unit time. For each volume of particle cloud of number density n_2 of a_2 particles, we have $n_2\sigma n_{12}(\Delta U)\rho_{p1}\pi a_2^2$ as the rate of agglomeration per unit volume. As the collection proceeds particles 2 will have less particles 1 to collect, it is readily shown (Prob. 3.18) that the collection efficiency over a distance L traversed by each particle of species 2 is given by:

$$\eta_c = 1 - \exp\left[-\frac{3}{4}\frac{\dot{m}_2}{\dot{m}_1}\frac{\rho_{p1}}{\rho_{p2}}\frac{\Delta U}{U_{p2}}\frac{L}{a_2}\sigma n_{12}\right]$$

$$(3.143)$$

where m's denote the flow rates in mass per unit time, U_{p2} is the velocity of species 2 particles. For liquid spray into a bulk of dusty gas, $\Delta U = U_{p2}$, L is the stopping distance, and the ratio of volume flow rates can be replaced by volume ratio in a batch process. Various spray geometry can also be accounted for [Cheng 1977]. (Prob. 3.15)

Agglomeration as phase change. For the cases of agglomeration or disintegration involving a distribution of particle sizes, Twomey [1966] gave a relation for the rate of change of number density of $n(v_1)$ droplets of volume v_1 as:

$$dn(v_1)/dt = - n(v_1) \int_0^\infty K(v_1,v) \, n(v)dv$$

$$+ \frac{1}{2} \int_0^{v_1} K(v,v_1 - v) \, n(v) \, n(v_1 - v) \, dv \qquad (3.144)$$

where $K(v_1, v)$ is the coagulation coefficient for number density of droplets of volume v_1 coagulating with that of droplets of volume v. The first term on the right-hand side gives the number of v_1 lost by increase in size while the second term gives the number of drops of v_1 generated by coagulation. He showed that clouds with 50 droplets/cm^3 produce rain droplets of 100-400 μm in 20-30 minutes.

When bubbles (or droplets) with a varying size distribution is approximated by bubbles of discrete bubble masses of varying population such that mass m_ℓ may have $\ell = \ldots, (k-1), k, (k+1), \ldots$ For the case of evaporating bubbles of species k, its rate of generation consists of [Sha & Soo 1979]:

$$\Gamma_k = - \Gamma_{ek} + \Gamma_{e(k-1)} \, (m_k/m_{k-1})$$

$$- \sum_\ell K_{k\ell} \, \rho_k^2 \rho_\ell (\rho_k + \rho_\ell)^{-1} + \sum_\ell K_{(k-\ell)\ell} \, \rho_{k\ell}\rho_\ell \qquad (3.145)$$

The first two terms are the net generation rate including loss of species k by growing into size (k+1) and gain of species k by growth of species (k-1) as given by Eq. (3.12). The third term is the loss by coalescence of species (k-ℓ) to ℓ to give k, $K_{k\ell}$ is the binary coalescence coefficient for species k and ℓ. This coefficient is not

well known, but some experimental results are available [Kirkpatrick
& Lockett 1978, Narayanan et al. 1974] for the purpose of
approximation.

Exercise Problems

3.1 Compute the fraction impacted of a sphere of 10 microm diameter
    by particles of 1 μm diameter and 1000 kg/m material density, and
    at a relative velocity of 10 m/s in air at 25 C and 1 bar.

                                                          Ans. 0.82

3.2 Assuming specular reflection, show that the force per unit length
    exerted by a particulate cloud of density ρ_p and free stream
    velocity U_0 on an infinitely long cylinder of radius R is given
    by, for $\eta = 1$,

$$F/L = 2\rho_p U_0^2 R \left[1 + \frac{r^*}{3}\right] \tag{3.146}$$

    Note that this relation was validated by experiments of Chilton
    et al. [1963]. Compute the drag coefficient.

3.3 Compute the net force exerted on a sphere of 25 mm diameter when
    it is placed in a dilute suspension of solid particles in air at
    room condition.   The mass flow ratio of particles to air is
    0.8.   The particles are 100 microm in diameter with a material
    density of 1000 kg/m^3 and at a velocity of 3 m/s. The density of
    air is 1.15 kg/m^3.   Ans. 0.00533 N

3.4 Compute the fraction impacted of a particle cloud of two dif-
    ferent size of particles in relative motion with $\Delta U = 0.1$ m/s.
    The particles have similar material density of 1000 kg/m and the
    viscosity of the gas is 10^{-5} kg/m,s , for
    (a) $2a_1 = 10$ μm, $2a_2 = 20$ μm;
    (b) $2a_1 = 1$ μm, $2a_2 = 2$ μm.   Ans. 0.81, 0.3

3.5 With coarse particles (subscript 1) and fine particles (2) in relative motion, Arastoopour et al. [1982] denoted the drag force on particles 1 of volume v_1 by particles 2 as:

$$F_{12} = \beta v_1 (U_2 - U_1)^2 \qquad (3.147)$$

show that β corresponds to:

$$\beta = (3/4)\, C_{D12}\, \bar{\rho}_{p2} \alpha_2 / 2a_1 \qquad (3.148)$$

For $2a_1 = 0.5$ cm and $\bar{\rho}_{p1} = 7600$ kg/m^3, $2a_2 = 0.3$ mm and $\bar{\rho}_{p2} = 2640$ kg/m^3, $\alpha_2 = 0.01$, an experimental value of $(3/4)\, C_{D_{12}}\, \bar{\rho}_{p2} = 2342$ was determined. Compare this value to that calculated according to Eq. (3.53). Ans. $\eta_{12} = 0.53$

3.6 Determine the heat-transfer coefficient due to collision of a cloud of particles (p) with a wall (w) for cloud density ρ_p and intensity of random motion of particles $\langle U_p^2 \rangle^{\frac{1}{2}}$. Take both particle and wall as made of steel ($\nu = 0.3$, $E = 2.07 \times 10^5$ MPa, $\bar{\rho}_p = 7000$ kg/m^3, $\bar{\kappa}_p = 28.72$ W/m C, $c_p = 0.5$ kJ/kg C), $r^* = 1$. Compute for $\langle U_p^2 \rangle^{\frac{1}{2}} = 0.3$ m/s, $\alpha_p = 0.1$, $a = 10$ μm and $a = 100$ μm. Ans. 0.469, 0.0469 W/m^2 C

3.7 Compute the ratio of inverse relaxation times of heat transfer to that of momentum transfer for the same data as in the previous problem. (Assume $a_1 = a_2$). Ans. 1.38×10^{-5}, 1.38×10^{-6}

3.8 Derive Eqs. (3.84) and (3.85).

3.9 A charge transfer coefficient of 5×10^{-8} mho/cm for collision of 16 microm coal dust on a 12.7 mm steel sphere at 36.6 m/s was measured [Cheng and Soo 1970]. Compare it to the theoretical result based on material properties. Ans. Very large contact resistance.

3.10 When an electrostatic probe is shaped from an infinite cylinder placed normal to the flow direction of a suspension, compute the probe current per unit length of this probe.

3.11 For a mixture of coal and water of a = 1 mm, $\bar{\rho}_p$ =1300 kg/m^3, α_p = 0.6, compute the viscosity μ_{pp} due to particle-particle interaction in laminar flow with mean velocity of 1.52 m/s, pipe radius 0.23 m, r* ~ 1, η = 1.   Ans. 0.026 kg/m-s

3.12 A suspension of 60% volume fraction of solid consists of slag spheres (p) of c_p = 752 J/kg K, $\bar{\kappa}_p$ = 0.59 W/m K, $\bar{\rho}_p$ = 2720 kg/m^3, 2a = 0.78 mm. E_p = 5640 MPa, ν_p = 0.25. A copper surface (w) moves past at 0.6 m/s (also the assumed random velocity of p), E_w = 11,000 MPa, ν_w = 0.35, $\bar{\kappa}_w$ = 386 W/mK . Compute the heat transfer coefficient to the particles.   (1) Assume single scattering of particles , (2) assume an accommodation coefficient of 0.01 due to multiple scattering.   [Dunsky et al. 1966]
Ans. 0.209, 184 W/m^{2}K

3.13 Show that energy is dissipated when two droplets of masses m_1 and m_2, each of velocities U_1 and U_2 coalesce.

3.14 A rain drop of 1 mm diameter initially increases its size as it falls through a fog of very fine droplets of cloud density 0.001 kg/m^3. Assume the drag coefficient of the drop is 0.44, falling in air of density 1 kg/m^3, and fraction impacted and coalescence of 1, compute its increase in size after falling 30 m and its terminal velocity.   Ans. 0.0075 mm

3.15 In a counter flow spray chamber of 0.58 cm diameter, water is sprayed at the rate of 40 cc/s, mean size 500 μm into a dusty gas containing coal particles 1.65 μm diameter at an air flow rate of 0.083 m^3/s counter current. The spray velocity is initially 3.34 m/s with mean distance of traverse of 1.91 m before being collected. Compute the collector efficiency. Assume the spray occupying a length equal to the mean distance of traverse.
Ans. 91.8%

Chapter 4

EFFECTS OF WAVES AND ELECTRICITY
AND SURFACE BOUNDARY CONDITIONS

4.1 Interaction with Radiation

Interaction of radiation, or more precisely, electromagnetic waves, with a cloud of particles is significant in many applications such as dust clouds and combustion flames at high temperatures, as well as in thermal insulations using particulate materials at very low temperatures. It is readily seen that a detailed treatment is beyond the scope of this book. Rather, we shall deal with a few simple cases to identify the significance of the phenomena.

When dealing with transmission of radiant energy through a cloud of particles of sufficient optical depth, the effects of absorption, emission, and scattering must be taken into account. The Milne equation (see, for instance, Chandrasekar [1960]) gives the transfer of energy along a ray as:

$$d\ I(s,\theta,\phi)/ds = -\rho_p\ \beta_m\ I(s,\theta,\phi) + \rho_p\ \alpha_m\ I_{bb}(s)$$

$$+ (\rho_p\ \sigma_m/4\pi) \int_o^{2\pi} \int_o^{\pi} I(s,\theta',\phi')\ S(\theta,\phi,\theta',\phi')\ \sin\theta'\ d\theta'\ d\phi' \tag{4.1}$$

where I is the monochromatic intensity of radiation, I_{bb} is that of the blackbody in J/m^2, stearad, and is given by the Planck function:

$$I_{bb}(T) = (2h\nu^3/c^2) \exp(-h\nu/kT) \; [1 - \exp(-h\nu/kT)]^{-1} \qquad (4.2)$$

for temperature T, h is the Planck constant, k is the Boltzmann constant, c is the speed of light, ν is the frequency; s is the distance
along a ray, ρ_p is the mass density, β_m is the monochromatic mass-
extinction coefficient in m^2/kg, α_m is the monochromatic mass-
absorption coefficient, σ_m is the monochromatic mass-scattering
coefficient, ϕ is the polar angle in radians, θ is the azimuthal
angle, and $X(\theta,\phi,\theta',\phi')$ is a scattering function.   Eq. (4.1)
expresses the change in monochromatic intensity with respect to distance along the ray at s,θ,ϕ with respect to the reference direction
in terms of the extinction of the ray, the increase in intensity due
to thermal-energy emission from the particles in the elemental
volume, and the energy scattered into θ,ϕ from rays traversing the
elemental volume from all directions.   The scattering function S is
defined such that

$$\sigma_m \; I(s,\theta',\phi') \; S(\theta,\phi,\theta',\phi') \; d\,\Omega' \; d\nu \; dm \; d\Omega/4\pi$$

is the rate at which radiant energy is scattered by the differential
element of mass dm from the pencil of rays enclosed in the solid

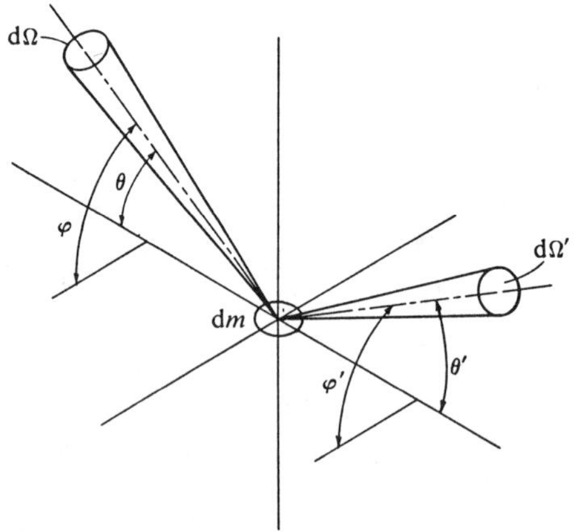

Figure 4.1 Coordinate scheme for scattering function

angle $d\Omega'$ having intensity $I(s,\theta',\phi')$ within the frequency range ν and $\nu + d\nu$ into solid angle $d\Omega$ characterized by θ, ϕ, as shown in Fig. 4.1. The integral of scattering function has the range

$$\Sigma = (4\pi)^{-1} \int_{0}^{4\pi} S(\theta,\phi,\theta',\phi') \, d\Omega' \qquad (4.3)$$

and $0 < \Sigma < 1$, depending on absorption. When it is equal to 1, the scattering is conserved and absorption is neglected. The effect of absorption is accounted for by the relation:

$$\beta_m = \sigma_m + \alpha_m \qquad (4.4)$$

Eq. (4.1) can be reduced by imposing certain geometrical and optical restrictions. A system consisting of a uniform plane cloud of particles suspended in a transparent medium and bounded by infinite surfaces which emit and reflect radiation in a diffuse manner is taken. The particles are specified as homogeneous spheres of uniform diameter and known refractive index. As shown in Fig. 4.2, x is the geometrical distance from surface 1 measured along a normal to the surface. Due to symmetry, the intensity will vary only with respect to polar angle θ measured from the normal to plane 1 and is independent of the azimuthal angle, ϕ. Introducing optical depth τ according to Fig. 4.2:

$$d\tau = \rho_p \, \beta_m \, dx = \rho_p \, \beta_m \, ds \, \cos\theta \equiv \rho_p \, \beta_m \, ds\mu \qquad (4.5)$$

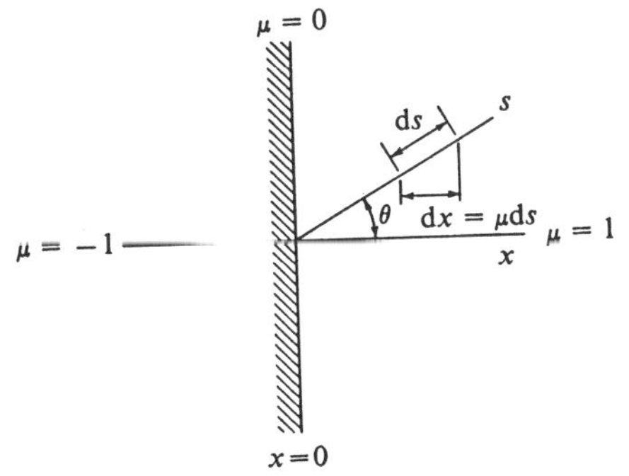

$\mu = 0$

ds s

θ $dx = \mu ds$ $\mu = 1$

$\mu = -1$ x

$x = 0$

Figure 4.2 Coordinate scheme for axially symmetric case

Eq. (4.1) can be expressed in a form with τ and $\cos\theta$ or μ as independent variables. With the assumption of isotropy, the equation can
be solved in relatively simple form for cases of isothermal scattering media [Love and Grosh 1964] and diffuse nonisothermal media
[Viskanta and Grosh 1961] for our illustrations.

Isothermal scattering media. With substitution of Gaussian quadrature formula [Chandrasekhar 1960] for the integral, Love and Grosh
treated a one-dimenional system of an isothermal scattering medium
consisting of a dilute cloud of particles of sufficient optical depth
at temperature T_a between two parallel diffuse walls of reflectivities ρ_1 and ρ_2 and at temperatures T_1 and T_2. Other given quantities are the density of the particle cloud of spheres of radius a,
complex refractive index $m^* = n^* - i\,k^*$. The extinction cross
section K_e and the mass extinction coefficient and mass scattering
coefficient are given by Chromey [1960] according to

$$\beta_m = 3\,K_e/4a\,\bar{\rho}_p \tag{4.6}$$

and

$$K_e = 24\,n^*k^*\alpha\,[(n^{*2} + k^{*2}) + 4(n^{*2} - k^{*2}) + 4]^{-1} \tag{4.7}$$

where α is the particle size parameter

$$\alpha_j = 2\pi a/\lambda_j = \pi x_j\,2a\,kT/ch \tag{4.8}$$

where λ_j is the wave length; $\lambda_j = c/\lambda_j$, c being the speed of light
and ν_j is the frequency $\nu_j = x_j kT/h$ with x_j as a variable for the
quadrature formulation. Knowing K_e, the optical depth τ_0 is given
by:

$$\tau_0 = \int_0^{x_0} \rho_p\,\beta_m\,dx \simeq (3/4)\,(\rho_p/\bar{\rho}_p)\,K_e(x_0/a) \tag{4.9}$$

Values of β_m, σ_m, and K_e computed from tables of Chromey are shown in
Figs. 4.3 and 4.4 for $m^* = 1.25 - 1.25i$ for iron and $m^* = 2.00 - 0.60i$

for carbon in the visible range of wavelengths. Derivations of Love and Grosh gave a solution based on the quadrature formula of the net radiant heat flux at wall 1, $\dot{q}_{net,1}$, in the form

$$\dot{q}_{net,1} = T_1^4 \sum_{j=1}^{n} A_j M_j - T_2^4 \sum_{j=1}^{n} A_j N_j - T_a^4 \sum_{j=1}^{n} A_j Q_j \qquad (4.10)$$

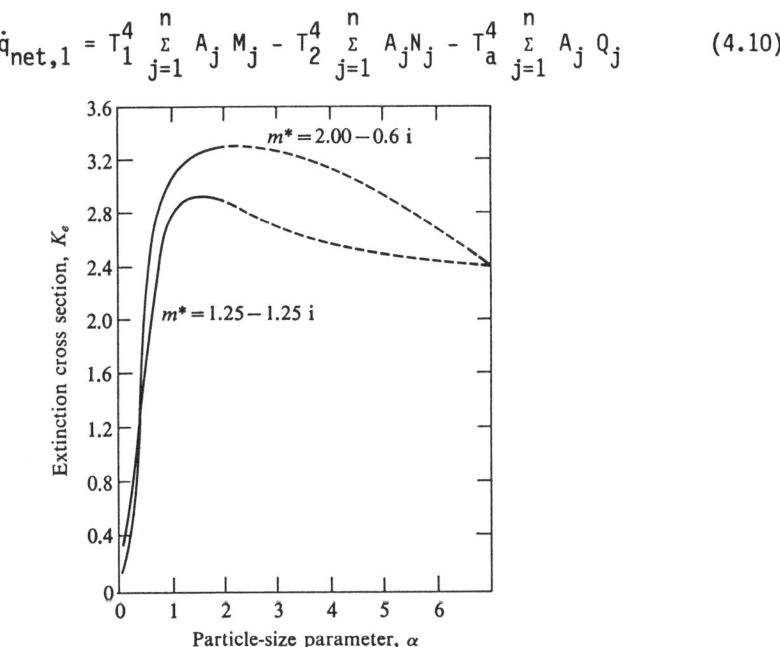

Figure 4.3   Extinction cross section K_e plotted as a function of particle size for the complex refractive indixes 2.00-0.60i and 1.25-1.25i

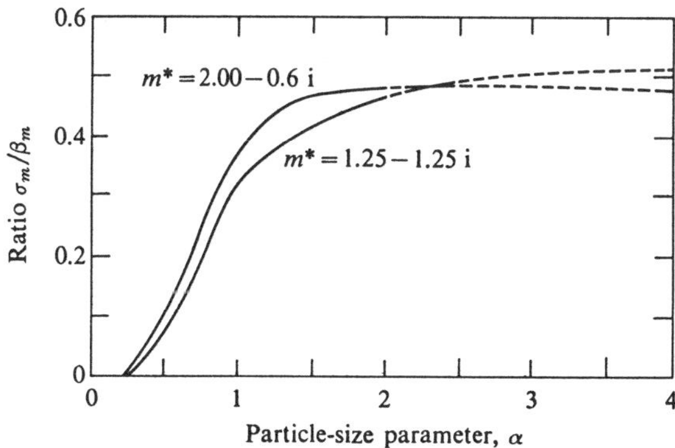

Figure 4.4   The ratio of scattering to extinction σ/β plotted as a function of particle size for the complex refractive indices 2.00-0.60i and 1.25-1.25i

where A_j's are universal and have the following values for a fifth-order approximation (n from 1 to 5), in 10^{-10} W/m^2K^4:

$$A_1 = 1.149, \; A_2 = 41.26, \; A_3 = 100.74, \; A_4 = 35.70, \; A_5 = 1.311.$$

The frequencies at which M_j, N_j, Q_j, must be determined correspond to T_1, T_2, and T_a and x_j or the particle size parameter α_j. Corresponding to the fifth-degree polynomial, for $\alpha_j = \alpha_j^* 2aT \times 10^{-4}$ and a in μm,

$$\alpha_1^* = 0.178, \; \alpha_2^* = 0.952, \; \alpha_3^* = 2.423, \; \alpha_4^* = 4.775, \; \alpha_5^* = 8.519.$$

The parameters M, N, and Q are given in Figs. 4.5 and 4.6 for given index of refractions and reflectivities of the bounding walls. For small τ_0 and α, the effect of anisotropic scattering is closely approximated by the isotropic approximation. Note that even for $\tau_0 = \infty$, a cloud of carbon particles is still not a blackbody. This simple case of an isothermal scattering medium is included to show the significance of the effect of scattering of radiation for heat transfer between two walls (Prob. 4.1) and its effect on a heat transfer to a flowing suspension (Prob. 4.2).

<u>Diffuse radiation in nonisothermal conducting media</u>. In general, at a large particle concentration, the above cloud temperature will not remain uniform. Some simplification is afforded when scattering can be neglected and when the particle concentration is large such that they can be treated as opaque layers of plates [Viskanta and Grosh 1961]. In this case, the medium may be represented as isotropic and homogeneous, and these plates are isothermal and diffuse. The equation of transfer of a monochromatic ray may be expressed as, since $\beta_m \sim \alpha_m$:

$$\mu d\, I(x,\mu)/dx = -\rho_m \, \alpha_m(x) \, \{I(x,\mu) - [n_m(x)]^2 \, I_{bb}(x)\} \qquad (4.11)$$

where n_m is the index of refraction. The net output of radiation from an element of unit volume per unit time is, for an absorption

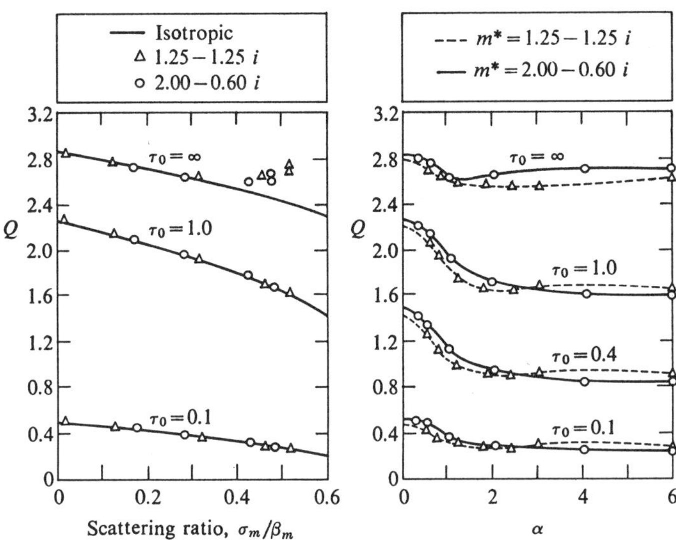

Figure 4.5 A comparison of the computed parameter Q for scattering by spheres having refractive indices of 2.00-0.60i and 1.25-1.25i. The reflectivity of both walls equal 0.1

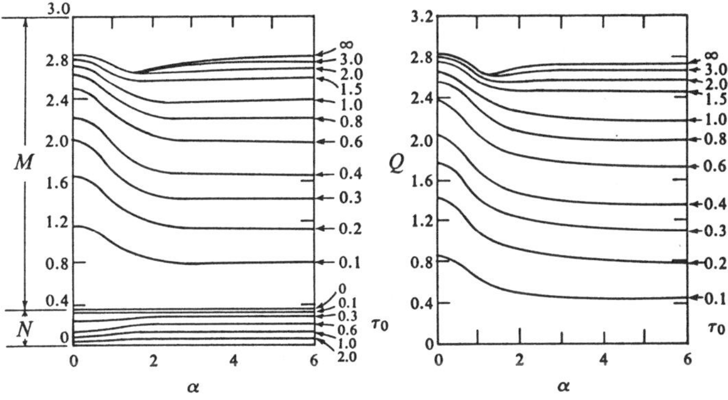

Figure 4.6 A graphical representation of the computed parameters for spheres having a refractive index of 1.25-1.25 i. The reflectivity of wall 1 equals 0.1 and of wall 2 equals 0.9

coefficient $\alpha_m' = \alpha_m \rho_m$ in m^{-1},

$$\varepsilon_n(x) = \alpha_m'(x) \ \{4[n_m(x)]^2 \ E_{bb}(x) - \varepsilon'(x)\} \tag{4.12}$$

The first term accounts for emission and the second term for the amount of energy which incident on the element and the fraction of this energy that is absorbed, and

$$\varepsilon'(x) = \int_{\Omega=4\pi} I(x,\mu) \ d\Omega \tag{4.13}$$

where Ω is the solid angle.

The energy equation for simultaneous conduction and radiation in this absorbing medium of thermal conductivity is

$$(d/dx) \ [\kappa(dT/dx)] = \int_0^\infty \varepsilon_n(x) \ d\nu = \varepsilon_{no} \tag{4.14}$$

Eqs. (4.11) and (4.14) suffice to determine the temperature distribution and heat transfer in the medium for given boundary conditions. They are, as before,

$$I(x,\mu) = (o) \ \text{at} \ x = 0 \ \text{and} \ \mu < 0$$

$$\tag{4.15}$$

$$I(x,\mu) = I(x_o) \ \text{at} \ x = x_o \ \text{and} \ \mu > 0$$

Simplification of the above relations is made possible by taking κ and α_m' as constants and introducing a dimensionless quantity, radiative conductivity number N_{RC} for reference temperature T^*,

$$N_{RC} = \kappa \ \alpha_m'^2 \ T^*/4 \ \alpha_m' \ \sigma_r \ T^{*4} = \kappa \ \alpha_m'/4\sigma_r T^{*3} \tag{4.16}$$

where σ_r is the Stefan-Boltzmann constant $(2\pi^5 k^4/15c^2 h^3 = 5.67 \times 10^{-8}$ $W/K^4 m^2)$ and introducing $\theta = T/T^*$, Eqs. (4.11) and (4.14) are now reduced to:

$$N_{RC} \, d^2 \Theta/d\tau^2 = n_m^2(\tau) \, \Theta^4(\tau) - (1/2)\{[R(\tau) \, E_2(\tau)/\sigma_r T*^4]$$

$$+ [R(\tau_0)E_2(\tau_0-\tau)/\sigma_r \, T*^4] + \int_0^{\tau_0} n_m^2(\tau') \, E_1(|\tau-\tau'|) \, \Theta^4(\tau')d\tau'\}$$

$$(4.17)$$

where R is the radiosity, πI, E_n is the exponential integral:

$$E_n(\tau) = \int_0^1 \mu^{n-2} \exp(-\tau/\mu)d\mu = \int_1^\infty e^{-\tau\mu} \, \mu^{-n} \, d\mu \qquad (4.18)$$

and integration over μ, ϕ, and ν is accounted for with $d\Omega = -d\mu d\phi$. The boundary conditions are now

$$\Theta(\tau) = \Theta(0), \qquad \tau = 0 \quad (x = 0)$$
$$\Theta(\tau) = \Theta(\tau_0), \qquad \tau = \tau_0 \quad (x = x_0)$$

$$(4.19)$$

Note that large N_{RC} means small radiation effect.   The result of calculation by Viskanta is shown in Fig. 4.7, for $\Theta(0) = 0.1$ and $\tau_0 = 1.0$; and $\Theta = 0.5$ and $\tau_0 = 0.1$, showing that for small N_{RC}, the temperature of the medium is nearly uniform, especially for small optical depth τ_0.

One of the approximate solutions was due to Rosseland [1931]. For optically thick media "near" thermodynamic equilibrium, the

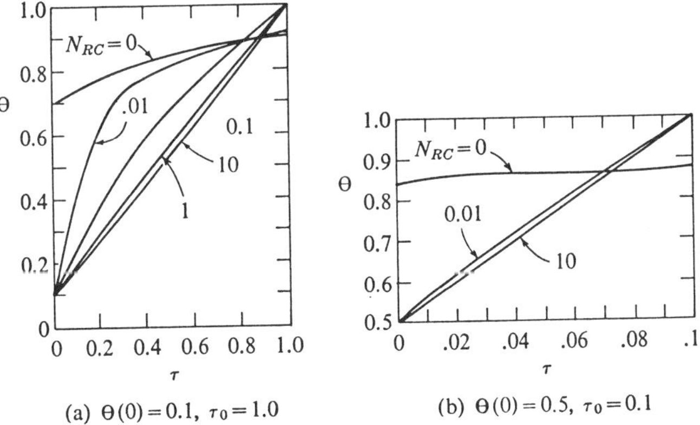

(a) $\Theta(0)=0.1$, $\tau_0=1.0$ (b) $\Theta(0)=0.5$, $\tau_0=0.1$

Figure 4.7   Variation of Θ with $\Theta(0)$ and optical thickness, $\Theta = T/T*$ [Viskanta & Grosh 1960]

radiant flux is given by:

$$E = -(1/3\alpha_m') \text{ grad } \varepsilon_0' = -(16n_m^2 \sigma_r T^3/3\alpha_m')\nabla T \qquad (4.20)$$

that is, an approximation based on diffusion of radiant flux vector. Eq. (4.14) becomes

$$(d/dx) [\kappa_{eff} \, d(T/dx)] = 0 \qquad (4.21)$$

with

$$\kappa_{eff} = \kappa + \kappa_{rad} = \kappa + (16 \, n_m^2 \, \sigma_r \, T^3/3\alpha_m') \qquad (4.22)$$

where κ_{rad} is the radiative conductivity. The nonlinearity in temperature as shown in Fig. 4.7 is seen from solving Eq. (4.21). (Prob. 4.3). For large particle concentration, the absorption coefficient α_m' is given by

$$\alpha_m' = \alpha\pi \, a^2 n_p \qquad (4.23)$$

where α is the absorptivity of the surface of the particle; n_p is the number density of particles. For a gas-solid system consisting of opaque solid particles, since reflection is, in general, diffuse, the index of refraction of a mixture is that of the gas phase alone. For transparent or partially transparent solid particles, the index of refraction depends on their surface condition. For rough surfaces and low transmissivity, the index of refraction will be that of the gas phase as one limiting condition, and the Lorentz [1915] relation provides

$$\frac{n_m^2 - 1}{n_m^2 + 2} \sim \frac{n_{gm}^2 - 1}{n_{gm}^2 + 2} + \frac{\rho_p}{\overline{\rho}_p} \frac{n_{pm}^2 - 1}{n_{pm}^2 + 2} \qquad (4.24)$$

as the other limit, especially for very small transparent solid spheres. In the above relation n_{gm} is the index of refraction of the gas and n_{pm} is the index of refraction of the solid material.

4.2 Interaction with Sound Waves

Propagation of sound waves through a suspension is basically a momentum transfer phenomenon. Applications include absorption of noise by a dispersion of solid particles and liquid droplets in a gas, determination of mean particle size, and sonic agglomeration and dispersion [see, for instance, Hueter and Bolt 1955]. Another case of interest is the propagation of sound through a liquid containing gas bubbles [Blitz 1963].

<u>Suspension of solid particles</u>. Propagation of sonic and ultrasonic waves in aerosol suspensions has been studied by many [Richardson 1962]. The dispersion (modification of speed of sound) and attenuation (absorption or dissipation of energy) by particles in a suspension can be accounted for by solving Eq. (1.47) for particle oscillation induced by wave propagation through the fluid [Soo 1967]:

$$(4\pi/3)\ a^3\ \bar{\rho}_p\ \dot{U}_p = X + (4\pi/3)\ a^3\ \bar{\rho}\ \dot{U} \tag{4.25}$$

where the dotted quantities are time derivatives, X denotes the sum of the drag force, the virtual mass force, and the Basset force in the form

$$X = \frac{4\pi}{3}\ a^3\ \bar{\rho}\left[\frac{1}{2} + \frac{9}{4\sqrt{N_{Re}}}\right] \frac{d(U-U_p)}{dt} + \frac{4\pi}{3}\ a\bar{\mu}\left[\frac{2a}{\sqrt{\pi\nu}} \int_{-\infty}^{t} \frac{\dot{U}_p-\dot{U}}{(t-\tau)^{\frac{1}{2}}}\ dt\right]$$

$$+ 3\pi a^3\ \bar{\rho}\ \frac{\omega}{\sqrt{N_{Re}}}\left[1 + \frac{1}{\sqrt{N_{Re}}}\right]\ (U - U_p) \tag{4.26}$$

where $(N_{Re}) = \omega(2a)^2 8\ \bar{\nu}$, ω is the circular frequency of the wave.

For mass ratio of solid to gas m_p^*, the equation of plane wave in the mixture can be expressed as

$$\rho\ \frac{\partial^2 U}{\partial t^2} = \gamma\ P\ \frac{\partial^2 U}{\partial x^2} + \frac{3}{4}\ \frac{m_p'''^*}{\pi a^2\ \bar{\rho}_p}\ \left(\frac{\partial x}{\partial t}\right) \tag{4.27}$$

where t and x are time and space coordinates, $\bar{\rho}$ and P the mean density and pressure of the gas, and γ the apparent ratio of specific heats of the mixture. In Eq. (4.27), the volume occupied by the solid particles and the effect of scattering (10^{-6} times that due to friction [Sewell 1910]) are neglected. Taking $\bar{\mu}$ and $\bar{\rho}$ as constants, and for $U = U_0 \exp[i(\omega t - kx)] \exp(-\alpha'x)$, k being a wave number, and α' the attenuation coefficient, Eq. (4.25) gives:

$$\frac{U_p}{U} = \frac{\beta^2 + (1+\alpha) \, [(\bar{\rho}/\bar{\rho}_p) + \alpha] - i[1 - (\bar{\rho}/\bar{\rho}_p)] \, \beta}{(1+\alpha)^2 + \beta^2} \tag{4.28}$$

with

$$\alpha = G^* + H^*, \quad \beta = F^* + H^* \tag{4.29}$$

and

$$F^* = (N_{Re})^{-1/2} \, (9/4) \, (\bar{\rho}/\bar{\rho}_p) \, [1 + (N_{Re})^{-1/2}] \tag{4.30}$$

the friction force parameter;

$$G^* = (\bar{\rho}/\bar{\rho}_p) \, [(1/2) + (9/4) \, N_{Re}^{-1/2}] \tag{4.31}$$

the virtual mass force parameter; and

$$H^* = 2\sqrt{2} \, (\bar{\rho}/\bar{\rho}_p)/(N_{Re})^{1/2} \tag{4.32}$$

the resistance parameter due to the Basset force.

Neglecting heat transfer, substitution of U and U_p into Eq. (4.27) gives:

$$\frac{a_m^2}{a_g^2} = [1 + m_p^* \, (1 - \bar{\rho}/\bar{\rho}_p) \, \frac{(1+\alpha)\alpha + \beta^2}{(1+\alpha)^2 + \beta^2}]^{-1} \tag{4.33}$$

from the real part, where $a_g = (\gamma P/\rho)^{\frac{1}{2}}$, the speed of sound in the pure gas, $a_m = \omega/k$ is that in the mixture, for $\alpha'a_g/\omega \ll 1$. The imaginary part gives the attenuation coefficient α':

$$\frac{\alpha' a_g}{\omega} = \frac{m_p^*}{2} \frac{a_m}{a_g} (1 - \overline{\rho}/\overline{\rho}_p) \frac{\beta}{(1+\alpha)^2 + \beta^2} \tag{4.34}$$

which is similar to the result of Urick [1948]. For $G^* \sim H^* \sim 0$, $\overline{\rho}_p/\overline{\rho} >> 1$, and $F^* << 1$, Eq. (4.34) reduces to Sewell's result excluding the effect of scattering:

$$\alpha' \, a_g/\omega = m_p^* \, F^*/2 \tag{4.35}$$

for negligible effect of virtual mass force and Basset force. (Prob. 4.4)

Fig. 4.8 shows the dispersion for $\overline{\rho}_p/\overline{\rho}$ of 1000 and 100 with 0.3 kg of magnesia per kg of air, for various values of $(2a)^2 \omega/\overline{\nu}$. The dispersion due to heat transfer is not usually felt when $(2a)^2 \omega/\overline{\nu}$ is large, but is felt when that quantity is low (slow motion) [Epstein and Carhart 1953]. The lower limit is that of a gaseous mixture. This theoretical trend of dispersion was demonstrated by the experimental results of Temkin and Dobbins [1966] using oleic acid particles below 6 μm diameter in nitrogen.

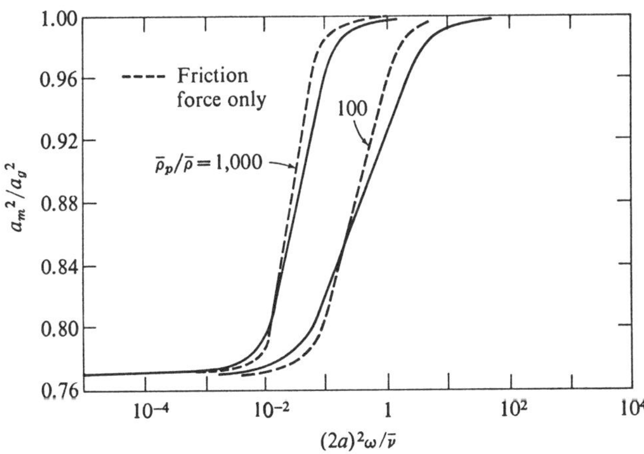

Figure 4.8   Dispersion of sound by 0.3 kg of magnesia per kg of air [Soo 1967]

Fig. 4.9 shows the attenuation parameter $\alpha' a_g/\omega$ versus $(2a)^2 \omega/\bar{\nu}$ according to Eq. (4.34) with comparisons to experimental results. At large values of $(2a)^2 \omega/\bar{\nu}$, resistance due to the Basset force term becomes significant, suggesting larger actual attenuation than Sewell's relation due to modification of the flow field. This effect is more prominent in the experimental results of Hartman and Focke [1940] because of a small material density ratio than in those of Laidler and Richardson [1938]. Also shown is the comparison to analytical results of Epstein and Carhart [1953] which approaches Sewell's at high frequencies, and the experimental results of Knudsen et al. [1948]. The spread of particle size tends to obscure the trend as given by the solid lines in Fig. 4.9. Validation of reduced

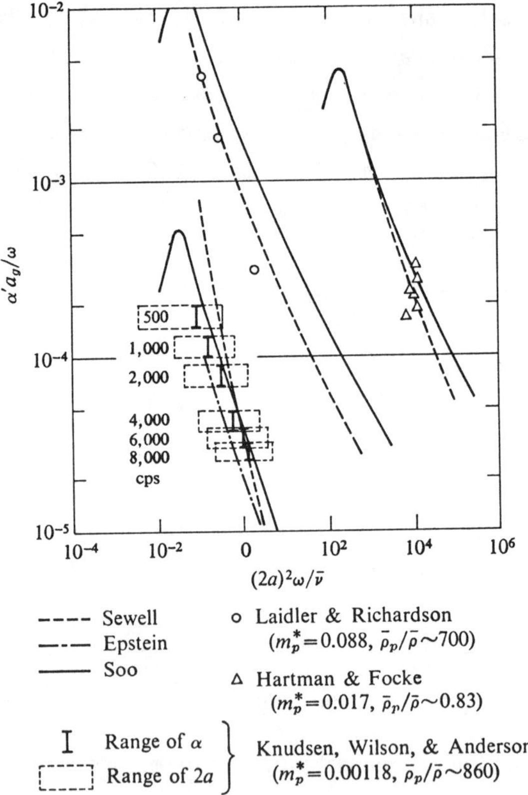

Figure 4.9 Attenuation per wavelength compared to experimental data [Soo 1967]

attenuation at low values of $(2a)^2 \, \omega/\bar{\nu}$ was shown by the experimental results of Temkin and Dobbins. The asymptotic value is readily shown (Prob. 4.4). The limit is the attenuation coefficient of a gaseous mixture.

The effect of scattering may become significant for large particles, for a unit depth of a suspension, Urich [1948] gave:

$$\alpha'_s \; a_g/\omega = (1/48) \; m^\star_p \; (\bar{\rho}/\bar{\rho}_p) \; (\omega \; 2a/a_g)^3 \; (a_g/a_m)^3 \tag{4.36}$$

<u>Liquid-vapor mixture</u>. The simplest cases of a liquid-vapor mixture are liquid water containing steam bubbles and steam containing water droplets. When considering wet steam or bubbly water simply as a molecular mixture, its speed of sound is readily determined. Using the definition in the Steam Tables [Keenan and Keyes 1950]: v_f, s_f for the specific volume and entropy of saturated liquid and v_g, s_g for those of saturated vapor (all these properties are functions of temperature T or pressure P only), quality x for the fraction by mass of vapor, $s_{fg} = s_g - s_f$, $v_{fg} = v_g - v_f$, and $s = s_f + xs_{fg}$, $v = v_f + xv_{fg}$ for the mixture, the speed of sound of a molecular mixture of a given quality x is given by:

$$a^2_m = (\partial P/\partial \rho)_s = -v^2 (\partial P/\partial v)_s \tag{4.37}$$

Derivation based on thermodynamic relations at equilibrium (Prob. 4.5) gives:

$$- \left(\frac{\partial P}{\partial v}\right)_s = \left(\frac{dP}{dT}\right)^2 \left[\left(\frac{ds_f}{dT} + x \frac{ds_{fg}}{dT}\right) - \left(\frac{dP}{dT}\right) \left(\frac{dv_f}{dT} + x \frac{dv_{fg}}{dT}\right) \right]^{-1} \tag{4.38}$$

with derivatives along the saturation curves. Eq. (4.37) with Eq. (4.38) can be used to compute the speed of sound in a molecular mixture of wet steam. The fraction by mass of vapor can be converted to volume fraction of vapor α via

$$\alpha = [1 + (1-x/x) \; (v_f/v_g)]^{-1} \tag{4.39}$$

The result of computation is shown in Fig. 4.10. We note that, due to the inertia of the liquid phase, the speed of sound in wet steam is lower than that in a pure liquid or a pure vapor (see, for instance, Weast [1975]). The effect of finite bubble size at small x or droplet size at large x [England 1966, Hanna et al. 1979] and experimental results of speed of sound as low as 150 m/s [Dryndrozhik

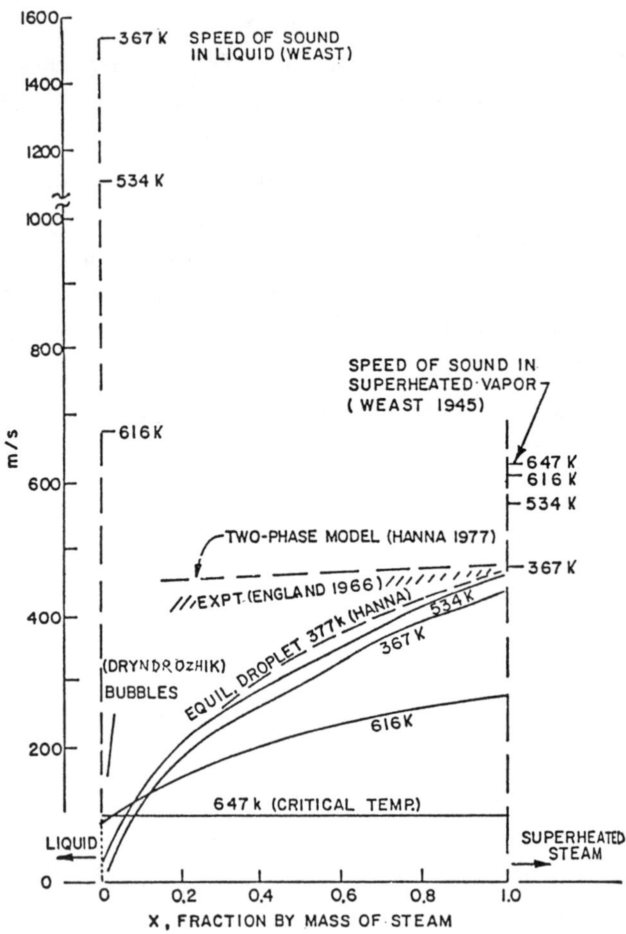

Figure 4.10   Speed of sound in wet steam

1975] are shown also in Fig. 4.10. These results show that the speed of sound in a liquid-vapor mixture is still not known in detail.

Acoustic properties of air bubbles in water were studied by Carstensen and Foldy [1947] to determine the effect of bubbles in the wakes of ships and submarines in the propagation of sound. Measurements were made on the attenuation of sound through a screen (431 x 76 mm to 152 mm and various vertical length) of bubbles and the amount of reflection of the sound from such a screen, with various number densities of bubbles over a range of sizes. Bubbles were produced by a microdisperser. Bubble radii ranging from 0.5 to 1 mm were measured optically and acoustically, the latter method according to the resonant circular frequency ω_0 of a bubble, and

$$\omega_0 = (3\gamma\, P_0/\bar{\rho}_p\, a^2)^{\frac{1}{2}} \tag{4.40}$$

where $\bar{\rho}_p$ is the density of the vapor.

Theoretical study in this case accounted for the distribution in bubble size and number distribution in space, and change in bubble configuration; losses due to viscosity and thermal conductivity were included. A comparison of theoretical and measured damping constant

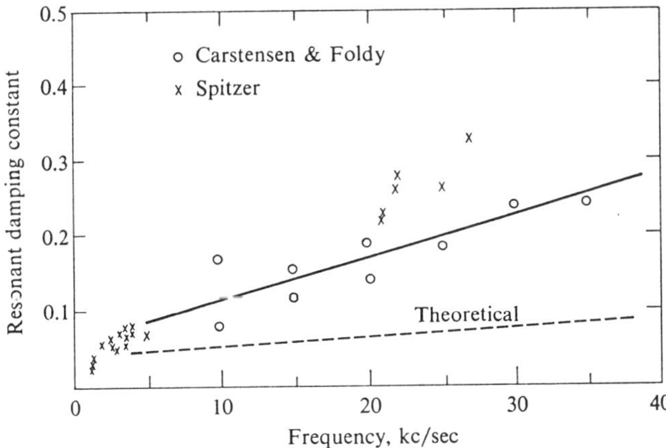

Figure 4.11 Comparison of theoretical and experimental results [Carstensen and Foldy 1947]

is shown in Fig. 4.11. Typical trends of attenuation and reflection are shown in Fig. 4.12.

The sonic velocity of a vapor-liquid mixture is strongly related to the critical flow of such a mixture to be dealt with in the next chapter.

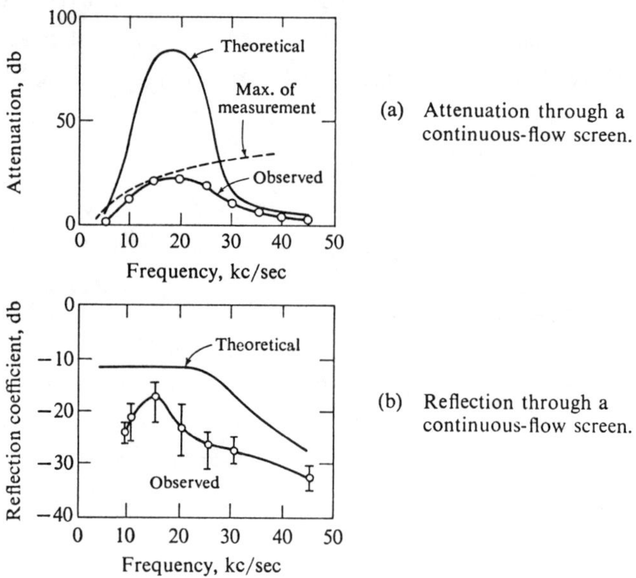

Figure 4.12   Typical results of measurements and computation [Carstensen & Foldy 1947]

4.3 Interaction with an Ionized Gas

The problem of attenuation of radio waves by combustion product of a metallized propellant rocket and the concentration of ash particles in a combustion magnetohydrodynamic generator led to studies of interactions of solid particles in a suspension in an ionized gas [Soo 1967, Soo & Dimick 1965].

In a gas-solid suspension at high temperatures, a sufficient amount of emitted electrons becomes free due to scattering. The field around a solid particle has to be treated according to an average free-electron density n_e, given by (Eq. 1.89):

$$n_e = n_{cs} \exp[-\phi_e/k\,T] \tag{4.41}$$

where ϕ_e is the equivalent thermionic potential energy. Summing over a field which is much larger than each solid particle-electron cloud system (see, for instance, Landau & Lifschitz [1958]),

$$\phi_e = \phi + (Z_p\, e^2/4\pi\, \epsilon_0 r) + \int_V \Sigma\, (n_s c^2/4\pi\, \epsilon_0\, r_s)\, dv$$

$$\sim \phi + (Z_p\, e^2/4\pi\, \epsilon_0\, r)\, \exp(-r/R_0)$$

$$\sim \phi + (Z_p\, e^2/4\pi\, \epsilon_0\, a) \tag{4.42}$$

where n_s is the space-charge density at r_s from the solid particle under consideration, and R_0 is the Debye-Huekel length given by summing over location i, or

$$R_0^2 = (\epsilon_0\, kT/e^2)\, (\Sigma_i\, n_{pi}\, Z_{pi})^{-1} \gtrsim (n_p)^{-2/3} \tag{4.43}$$

In the study by Soo and Dimick [1964], the case in which substantial extent of ionization may occur among the gaseous atoms of a gas-solid suspension was considered. At equilibrium, the condition of charge neutrality gives, for non-reactive solid particles in an inert gas,

$$n_e = n_z + Z\, n_p \tag{4.44}$$

where n_e and n_z are electron and ion concentration in the gaseous phase respectively.   Ionization equilibrium in the gaseous phase gives

$$n_e\, n_z \simeq n^2/P\, K_p \tag{4.45}$$

for low extent of ionization (say, less than 1%), where n is the concentration of neutral atoms, P the pressure, and K_p the equilibrium constant.   Taking first-degree ionization for the present,

$$n_e n_z \simeq n(2g_z/g)\ (2\pi m_e kT/h^2)^{3/2}\ \exp(-I/kT) \tag{4.46}$$

where g, g_z are the statistical weights of the atom and ion respectively, h is the Planck constant, m_e is the electronic mass, I is the ionization potential of first-degree ionization.

Equilibrium between the phases is reached when there is no net gain or loss of charge due to random motion.   At the surface of the solid particle, the number of ions diffusing toward the solid particle per unit time, $\dot{N}_z$, is equal to that of electrons diffusing toward the particle, $\dot{N}_e$, minus that of electrons leaving by thermionic emission, or from Eqs. (1.91), (4.41), and (4.42),

$$\dot{N}_e - (4\pi a^2\, A_o\, T^2/e)\ \exp(-\phi/kT) = \dot{N}_z \tag{4.47}$$

where, with consistent degree of approximation, $\alpha_e \equiv N_{et}$ in Eq. (1.85):

$$\dot{N}_e = 2a^2\, n_e(2\pi kT/m_e)^{1/2}\ \exp\,\alpha_e \tag{4.48}$$

$$\dot{N}_z = 2a^2\, n_z(2\pi kT/m_z)^{1/2}\ \exp(-\alpha_e) \tag{4.49}$$

m_z being the mass of an ion; $\alpha < 0$ for negatively charged solid particles, according to Eq. (1.86). Eqs. (4.48) and (4.49) apply to all values of α_e. Substitution of these relations into Eq. (4.47) gives

$$n_e \exp \alpha_e - (2\pi m_e A_0 T^2/e)(2\pi m_e kT)^{-1/2} \exp(-\phi/kT)$$

$$= (m_e/m_z)^{1/2} n_z \exp(-\alpha_e) \tag{4.50}$$

The equilibrium value of α_e is given, by eliminating n_e and n_z from Eq. (4.44), (4.46), and (4.50) to be

$$\frac{[B - (m_e/m_z)^{\frac{1}{2}} \alpha_e \exp(-\alpha_e)][B - \alpha_e \exp \alpha_e]}{4(m_e/m_z)^{\frac{1}{2}} \sinh^2 [\alpha_e + (1/4) \ln (m_z/m_e)]} = K \tag{4.51}$$

where the parameters B and K are given by (with Eq. (1.93))

$$B = 2\pi \, m_e \, e \, A_0 \, T^2 \exp(-\phi/kT)/(2\pi m_e kT)^{\frac{1}{2}} \, 4\pi \, \epsilon_0 \, a \, kT \, n_p$$

$$= (1-\bar{r})(n_{cs} \, e^2/4\pi \, \epsilon_0 \, akT \, n_p) \exp(-\phi/kT)$$

$$= (m_e \, e^2/2\epsilon_0 \, a \, n_p \, h^2)[2n_b(2\pi m_e kT/h^2)^{-\frac{1}{2}}]^{\frac{1}{2}} \exp[-(\phi+\Delta\epsilon/2)/kT]$$

$$K = ne^4(2g_z/g)(2\pi m_e kT/h^2)^{3/2} \exp(-I/kT)(4\pi \, \epsilon_0 \, a \, kT \, n_p)^{-2}$$

$$= (n_{cs}e^2/4\pi\epsilon_0 a \, kTn_p)^2 (n/n_{cs})^2 (2g_z/g)[(2\pi \, m_e kT/h^2)^{3/2}/n]\exp(-I/kT)$$

Eq. (4.51) reduces for K = 0 to the case of negligible ionization in the gaseous phase. The parameter K relates the competition between thermionic emission and collection of electrons from thermal ionization of the gas by the electrostatic capacitance of the solid particles in the nature of a floating probe [Barger 1961]. (Prob. 4.6, 4.7)

Eq. (4.51), in general, gives α_e as a multivalued function for given B and K. However, the only values of α_e which are physically significant are those giving positive (or zero) values of n_e and n_z; that is, α_e's are restricted to those giving

$$n_e = \frac{(n_p \, 4\pi\varepsilon_o \, akT/e^2) \, [B - (m_e/m_z)^{\frac{1}{2}} \, \alpha_e \, \exp(-\alpha_e)]}{[\exp \alpha_e - (m_e/m_z)^{1/2} \, \exp(-\alpha_e)]} > 0 \qquad (4.52)$$

and

$$n_z = \frac{(n_p \, 4\pi \, \varepsilon_o \, akT/e^2) \, [B - \alpha_e \, \exp(\alpha_e)]}{[\exp \alpha_e - (m_e/m_z)^{\frac{1}{2}} \, \exp(-\alpha_e)]} > 0 \qquad (4.53)$$

Thus, for given B and K, α_e is unique. Another asymptotic condition for α_e is that

$$\sinh \, [\alpha_e + (1/4) \, \ell n \, (m_z/m_e)] > 0 \qquad (4.54)$$

Therefore, the maximum negative charge that can be collected (besides

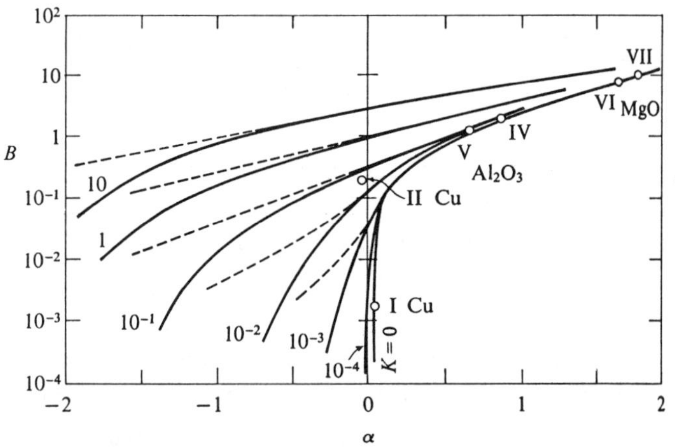

Figure 4.13 Equilibrium of ionization of a gas (argon)-solid suspension (circles show asymptotic states of experiments with various solids; dashed lines for each value of K give the limit of $m_e/m_z = 0$ for any gas)

the limits for electric breakdown [Cobine 1958] by a solid particle is

$$a > -(1/4) \ln (m_z/m_e) \tag{4.55}$$

a condition limited by the rates of collision even in the absence of thermionic emission. The relation between α_e, B, and K for argon is shown in Fig. 4.13 , which gives the trend for all gases. In this case, the values of m_z and m_e for argon give $\alpha_e > -2.8$ as the limiting value of α_e for electron collection (the lowest is $\alpha_e > -0.94$ for hydrogen atoms). Fig. 4.13 shows that for K = 0, the solid particles can only become positively charged [Soo 1967], and this case gives the maximum value of α_e possible for a given B. At higher values of K, depending on its relation to B, the solid particles can become positively or negatively charged (electrons are then effectively collected by solid particles). It is seen that solid particles tend to become negatively charged with the combination of low ionization potential of the gas and high thermionic work function of the solid. The circles in Fig. 4.13 are approximate asymptotic states of the experiments by Soo and Dimick [1965]. The dashed lines for each value of K in Fig. 4.13 give the limit for any gas which produces heavy ions ($m_e/m_z \approx 0$). It is seen that in the range of α_e near and above 0, the value of m_e/m_z does not affect the relation between α_e, B, and K.

It is seen that for Z = 0, $n_e = n_z$, $n_e >> (m_e/m_z)^{\frac{1}{2}} n_z$, and $\bar{r} \sim 0$; A_0 is given by Prob. 1.16. Eq. (4.50), when substituted into Eq. (4.46), with Eq. (1.93), gives, for $\alpha_e = 0$,

$$n_z/n_e = (g_z/g) \exp[-(I - \phi)/kT] \tag{4.56}$$

the equation for surface ionization of an ideal metal given by Langmuir and Kingdon [Reimann 1934] as a reduced form of Eq. (4.51). The same condition, when applied to Eq. (4.46), gives

$$\frac{n_z}{n_e} = 2\left(\frac{g_z}{g}\right) \left(\frac{2\pi m_e kT}{h^2}\right)^{3/4} (2n_b)^{-1/2} \exp\left[- \frac{I-\phi-(\Delta\varepsilon/2)}{kT}\right] \tag{4.57}$$

showing that oxides, for instance, can be more effective in producing surface ionization than metals [Soo 1967]. (Prob. 4.8)

Experimental validation of interactions of a metal (copper) and insulators (alumina and magnesia) with an ionized gas (argon) were conducted in an arc heated subsonic jet. The recombination coefficients of nonequilibrium ionization of various combinations of particles and gas were determined from measurement of ion current with water-cooled probes. The recombination rate constant k_r of an ionized gas alone is defined according to [Loeb 1955]:

$$dn_e/dt = -k_r\, n_e n_z = -k_r\, n_e^2 \qquad (4.58)$$

The effect of ambipolar diffusion is negligible for pressures above 10 mm Hg. Integration of Eq. (4.58) gives

$$k_r = (n_c^{-1} - n_{eo}^{-1})/(t - t_o) \qquad (4.59)$$

In terms of ion current i, for $n_e \simeq n_z$, a corresponding relation is

$$di/dt = -a_2\, i^2 \qquad (4.60)$$

and a_2 is related to k_r according to [Talbot 1960]:

$$a_2 \sim 4\, k_r/e\, A_p\, <U_z> \qquad (4.61)$$

where A_p is the area of the probe face and $<U_z>$ is the average thermal velocity of the ions. When applied to cases with injection of solid particles, the variation of probe current i with time t in the form of a polynomial was postulated according to:

$$di/dt = a_0 - a_1 i - a_2 i^2 \qquad (4.62)$$

where a_0 corresponds to the generation of current or charge carriers between electrodes via thermionic emission or ionization of the vapor of the solid particles; a_1 corresponds to a recombination with the solid particles by three-body collisions; and a_2 has a similar meaning as before. The surface recombination coefficient is thus

$$k_1 = a_1/4\pi \ a^2 \ n_p \tag{4.63}$$

for total particle surface area $4\pi a^2 n_p$.

Experimental results show that for a gas temperature of 1300 K (total temperature 3000 K) and at 45 mm Hg, the recombination rate was 7.3×10^{-7} cm^3/s, and at 800 K (total temperature 2000K), the rate was 1.4×10^{-7} cm^3/s for argon alone. Copper of 10 μm size at a mass flow ratio of 0.28 increased these rates to 2.6×10^{-6} and 7.6×10^{-7} respectively with 2% of the copper evaporated. Alumina of 10 μm at a mass ratio of 0.08 decreases k_r by an order of magnitude, while giving k_1 of 2.5×10^{-3} s^{-1}. Magnesia particles follow similar trend of alumina [Soo & Dimick 1965]. Corresponding states are noted in Fig. 4.13. These results explains the fact that a metallized propellant increases significantly the electron concentration in the rocket plume. Soot particle formed from combustion are invariably charged. It was suggested that soot formation in combustion flames can be reduced by removing particles or nuclei rapidly from the reaction zone with a large magnetic field [Bowser & Weinberg 1974].

4.4 Mobility and Electrical Conductivity

The response of droplets and particles to electric fields is measured by their mobility. From the point of view of electrical control of various combustion processes, Gugan et al. [1964] studied the mobility of charged molecular clusters, droplets, and particles. Their survey indicated that sufficient data have been assembled for ions such as given in the treatise of Loeb [1955]. For small particles such that small ion mobility applies, White [1951] gave the mobility in (cm/s)/(V/cm) as

$$K_p = 0.235 \ [(M + M_i)/M_i]^{1/2} \ [(\varepsilon_{ro}-1)M]^{-1/2} \ (\bar{\rho}_0/\bar{\rho}) \tag{4.64}$$

where M and M_i are the molecular weights of the gas and the ion respectively, ε_{ro} is the dielectric constant, and $\bar{\rho}_o$ is the density of the gas at 0 C and 1 bar, and $\bar{\rho}$ is the density of the gas. For large ions, the Langevin equation gives [Loeb 1955]:

$$K_p = 0.815[Q_i\lambda_i/M<U^2>^{\frac{1}{2}}] \; [(M + M_i)/M_i]^{1/2} \tag{4.65}$$

where Q_i is the ionic charge per kmol, λ_i is the ionic mean free path, and $<U^2>^{\frac{1}{2}}$ is the root-mean-squared velocity of the gas molecule.

For still larger particles, the drag force (Eq. 1.5) on the particle is to be equated to the force due to the electric field, or

$$F = Ee\ Z \tag{4.66}$$

This consideration gives

$$K_p = e\ Z/6\pi\mu a = (q/m)/F \tag{4.67}$$

for the Stokes regime, and

$$K_p \simeq a^{-1}\ (eZ/0.22\pi\bar{\rho}E)^{1/2} \tag{4.68}$$

for the Newton regime. It is seen that Z itself is, in general, a function of the electric field present (Sec. 1.8), thus the mobility is dependent on the electric field.

<u>Electrical conductivity</u>. When applied to high temperature systems, the electrical conductivity of a mixture of charged solid particles, electrons, ions, and gas atoms of suspending gas is often a useful quantity. It was seen that the cross-section for collision between electrons and charged solid particles with coulomb interaction far exceeded that between, say, helium atoms and electrons interacting with an inverse fifth power relation. Due to large Debye shielding distance in this case, combination of effects of diffuse scattering and space charge led to lower electrical conductivity than in an ionized gas of similar electron concentration.

The method of calculating electrical conductivity of a gas consisting of ions, electrons and atoms as presented in Chapman and Cowling [1958] and reduced by Cann [1961] can be simplified for the present approximation as

$$\sigma_e = (3/16)(2\pi/m_e kT)^{1/2} (e^2/Q_{ep}) \qquad (4.69)$$

where σ_e is the electrical conductivity of the mixture and Q_{ep} is the reference cross-section given by

$$Q_{ep} = \pi(e^2 Z/8\pi \ kT \ \varepsilon_o)^2 \ \ell n \ (\alpha_d)_{ep} \qquad (4.70)$$

and the cutoff parameter in coulomb interaction is given by

$$(\alpha_d)_{ep} = 1 + 4(12\pi \ \varepsilon_o \ kT/Ze^2)^2(\varepsilon_o kT/2e^2 n_e) \qquad (4.71)$$

For the electron concentration of a gas-solid mixture as given in Fig. 4.13 for $K = 0$ at various temperatures, one sees that the electrical conductivity decreases with increase in temperature. This is because, from the point of view of coulomb scattering of electrons by solid particles alone, a large reference cross-section suggests higher electrical conductivity as the charge on solid particles decreases. In general,

$$\sigma_e \propto T^{3/2}/n_e \qquad (4.72)$$

Since $n_e \propto T^{\alpha'}$, and $\alpha' > 3/2$, Eq. (4.72) indicates a decrease in electrical conductivity with increase in temperature. However, for a given pressure of the suspending gas, as the influence of coulomb scattering decreases, the elastic scattering of gas atoms becomes significant. The electrical conductivity may then be approximated by

$$\sigma_e = (n_e \ e^2/kT) \ D_{ea} \qquad (4.73)$$

with the diffusivity $D_{ea} \sim \langle U_e \rangle \lambda_{ea}$; $\langle U_e \rangle$ is the mean thermal speed of electrons. Thus, below the temperature range where the effect represented by Eq. (4.72) is prominent, the electrical conductivity increases with increase in temperature, usually below the boiling point of the solid particle.

Non-equilibrium effects. Non-equilibrium in an ionized suspension occurs because of mobility limited diffusion of electrons and ions, and the disparity of particle temperatures. We note that even oxides have substantial conductivity at high temperatures and polarization can be neglected. The equations for mobility limited diffusion currents J of ions and electrons are [Allis and Rose 1954]:

$$-D^+ (dn_i/dr) + n_i K^+ E(r) = J_i(r) \qquad (4.74)$$

$$-D^- (dn_e/dr) - n_e K^- E(r) = J_e(r) \qquad (4.75)$$

where D^+ and D^- are the diffusivities, and K^+ and K^- are the mobilities of positive ions and electrons, with r measured from the center of the particle of radius a. At steady state,

$$J_i(r) = J_e(r) = (a^2/r^2) J_o \qquad (4.76)$$

At the surface of the particle

$$n_{ea} = n_{ep} \exp(-\phi_o/kT_p) \qquad (4.77)$$

where n_{ep} is the free electron density in the solid. The field around the particle is self-consistent. However, for small ion and electron densities compared to that of the neutrals, the radial distribution of the former is determined by the diffusivities and the field of the particle:

$$E = Ze/4\pi \, \varepsilon_o \, r^2 \qquad (4.78)$$

neglecting the space charge due to the field of ions. Further

$$K/D = e/kT \qquad (4.79)$$

Substitution of Eqs. (4.76) to (4.79) into Eqs. (4.74) and (4.75) permits integration of the latter equations to give the non-equilibrium distribution of electrons and ions (Prob. 4.9).

Experimental results showed that particulate matter such as tungsten and alumina tends to acquire negative charges in an ionized gas [Soo et al. 1968].

4.5 Charge Distribution

The above discussions deal with the average conditions of a particle cloud. In reality, perturbation of local charge density by charge collection occurs. Thus, even a monodispersed cloud of particles will not be uniformly charged. For n uncharged droplets per unit volume, at equal positive and negative mobility, distribution of charges of different polarity on the drops with charges of the magnitude of Z electronic charges was given by Fuchs [Coroniti 1965] as

$$n_Z = n_0 \exp[-Z^2 e^2/8\pi \, \epsilon_0 \, a \, kT] \tag{4.80}$$

where n_Z is the number density of particles with Z charges.

It was shown empirically that the charge of a droplet is given by

$$q = 4\pi \, \epsilon_0 \, \phi \, a \tag{4.81}$$

where ϕ is a potential and q ~ 200a electronic charges for a in micrometers, when the surrounding air is neutral, $\phi = -\phi_0$, and the electrokinetic potential ~ 0.3 v. q is then the preponderant negative charge that could be acquired by a drop.

In clouds and fogs, acquisition of negative charges by some particles would leave a substantial concentration of positive ions; thus some particles will be positively charged.

For a cloud with particle size distribution following

$$f(a) = \frac{dn(a)}{da} = \left(\frac{5}{2}\right)^5 \frac{1}{\pi} \left(\frac{\rho_p}{\bar{\rho}_p}\right) \left(\frac{a^2}{a_m^6}\right) \exp\left[-\frac{5a}{a_m}\right] \tag{4.82}$$

where a_m is the mean particle size, the specific charge per unit volume of a droplet is given by

$$q_1 = 5 \, \varepsilon_0 \, \phi/16 \, \pi \, a_m^2 \tag{4.83}$$

Charge distribution among glass particles in a suspension due to surface contact was measured by Turner and Balasubramanian [1976] with particles ranging from 45 to 90 μm. The effect of humidity was also noted. Their results were correlated according to [Green & Lane 1957]:

$$d(N/N_0)/d\xi = \exp(-\xi^2) \tag{4.84}$$

where $d(N/N_0)$ is the fraction of particles with surface charge density ϕ, $\xi = c_1\phi = c_1(q/m)a \, \bar{\rho}_p/3$, c_1 is a constant representing the influence of humidity. Integration of Eq. (4.84) gives the fraction charges with ϕ or more as

$$(N/N_0)\Big|_{\infty}^{\xi} = \mathrm{erfc}(\xi) \tag{4.85}$$

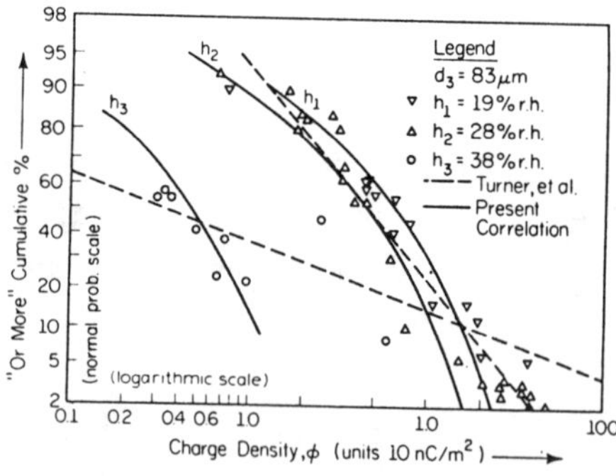

Figure 4.14 Cumulative percentage of particles with charge densities equal to, or greater than, a certain value [Soo 1980]

N/N_0 = 0.5 is given at $\mathrm{erfc}(\xi)$ = 0.5 or ξ = 0.48, giving ϕ_{av} and $(q/m)_{av}$.   The latter amounts to 2.3×10^{-6} C/kg at 19% relative humidity and 0.14×10^{-6} C/kg at 38% relative humidity (Fig. 4.14) [Soo 1981].   Methods of charge elimination at room temperatures include humidity control and use of nuclear ion sources.

4.6 Boundary Conditions of Suspensions.

Rigorous solution of basic equations of multiphase flow is, in most cases, limited to the cases of suspensions of particles that can be approximated by spheres.   Behaviors identifiable with suspensions include deposition, blowing away, splashing of deposited layer by other particles, and slip motion at the wall, besides thermal accommodation (Sec. 3.3).

Many of the boundary conditions arising from phase interactions are directly accountable or can be directly extended from general concepts of fluid dynamics.   It is opportune to treat a boundary condition which is particular with multiphase flow at this point.

It is often necessary to compute the density distribution and wall flux of particulates to determine the effect of erosion or deposition.   This is to be accomplished by solving the diffusion equation (Eq.   2.123) under the influence of field forces and a definitive boundary condition is needed.

When the suspension is dilute, the mixture velocity is identical to that of the fluid phase, and the diffusivity of particle phase p through the mixture is nearly equal to that through the fluid.   Take a two-dimensional semi-infinite system with the wall along the x-direction, z normal to it, and the fluid velocity components conjugate to the x and z directions as U and W.   We further take the fluid velocity as fully developed, that is, $U = U(z)$, $W = 0$.   Then Eqs. (2.123) and (2.124) give:

$$U \frac{\partial \rho_p}{\partial x} = \frac{\partial}{\partial z} D_p \frac{\partial \rho_p}{\partial z} + \frac{\partial}{\partial z} (f_p - f_L) \frac{\rho_p}{F} \qquad (4.86)$$

where f_p is the field force per unit mass normal and toward the wall; f_L is the lift force per unit mass directed outward from the wall due to fluid shear (Sec. 1.5).   With the boundary conditions of U, W, ρ_p, $\partial\rho_p/\partial z = 0$ at $z = \infty$, Eq. (4.86) can be integrated to give

$$\frac{\partial}{\partial x} \int_0^\infty U(z)\ \rho_p\ dz = -D_p\left(\frac{\partial\rho_p}{\partial z}\right)_w - \left(\frac{g}{F}\right)\rho_{pw} + \left(\frac{f_L}{F}\right)\rho_{pw} \qquad (4.87)$$

which is also the change in the total mass flux in the x-direction, caused by diffusion, deposition, and lift from the surface layer; subscript w denote the condition at $z = 0$.

For fine particles, $U_p \sim U$, and the change in total mass flux in the mass balance of the net effect of deposition and entrainment at the wall:

$$\frac{\partial}{\partial x} \int^\infty U_p\ \rho_p\ dz = \sigma_{Lb}\frac{f_L}{F}\ \rho_{pb} + \sigma_L\left(\frac{f_L}{F}\right)\rho_{pw} - \sigma\left(\frac{f_p}{F}\right)\rho_{pw} - \sigma_w\left(\frac{f_w}{F}\right)\rho_{pw}$$
$$(4.88)$$

where σ_{Lb} is the probability of lift from a bed or layer of particles of density ρ_{pb} which may exist at the wall, σ_L is the probability of lift by the fluid field in shear motion, σ is the probability of sticking of particles deposited by the field force, and σ_w is the probability of sticking by adhesive force (Eq. 3.35) per unit mass f_w at the wall and directed toward it.   Equating Eqs. (4.87) and (4.88) gives:

$$-D_p\left(\frac{\partial\rho_p}{\partial z}\right)_w = \sigma_{Lb}\left(\frac{f_L}{F}\right)\rho_{pb} - (1 - \sigma_L)\left(\frac{f_L}{F}\right)\rho_{pw}$$
$$- (1 - \sigma)\left(\frac{f_p}{F}\right)\rho_{pw} - \sigma_w\left(\frac{f_w}{F}\right)\rho_{pw} \qquad (4.89)$$

as a general boundary condition including the effect of diffusion, erosion from a dust layer, lift by shear motion of the fluid on the particles near the wall, deposition by field forces, and deposition by surface adhesion.   [Soo and Chen 1982] This result is in agreement with the boundary condition derived [1972] for pipe flow (Eq. 6.13).

Eq. (4.89) reduces to the boundary conditions of the following cases (neglecting the lift force): (1) In the case of a gaseous pollutant, we get $\partial\rho_p/\partial z = 0$, the "adiabatic" boundary condition given by Sutton [1953]. (2) In the absence of f_w, we get

$$-D_p \left(\frac{\partial\rho_p}{\partial z}\right)_w = (1 - \sigma) \left(\frac{f_p}{F}\right) \rho_{pw} \tag{4.90}$$

which gives for complete sticking, $\sigma = 1$, again $\partial\rho_p/\partial z = 0$, the boundary condition of Williams and Jackson [1962]. When the wall is nonsticking, $\sigma = 0$, we have

$$-D_p \left(\frac{\partial\rho_p}{\partial z}\right)_w = \frac{f_p}{F} \rho_{pw} \tag{4.91}$$

where the depositing force is balanced by a diffusion flux away from the wall and reduces to the "adiabatic" wall condition when field force is absent. (3) When field force is absent, adhesive force gives

$$-D_p \left(\partial\rho_p/\partial z\right)_w = -\sigma_w(f_w/F) \rho_{pw} \tag{4.92}$$

$\sigma_w f_w/F$ corresponds to the deposition velocity of Friedlander [1959]. For very large f_w, $\rho_{pw} \to 0$ at the wall. (4) When both f_p and f_w are present and σ and σ_w are finite, we get

$$-D_p \left(\frac{\partial\rho_p}{\partial z}\right)_w = (1-\sigma)\left(\frac{f_p}{F}\right)\rho_{pw} - \sigma_w \left(\frac{f_w}{F}\right)\rho_{pw} \equiv \omega_m\rho_{pw} - \beta_m\rho_{pw} \tag{4.93}$$

The last equality corresponds to the boundary condition given by Monin [1956] with the meaning of the quantities of ω_m and β_m now identified.

(5) When applied to an erodible bed with $f_w = 0$, we get:

$$-D_p\left(\frac{\partial\rho_p}{\partial z}\right)_w = -(1-\sigma_L) \left(\frac{f_L}{F}\right) \rho_{pw} + (1-\sigma)\left(\frac{f_p}{F}\right)\rho_{pw} \tag{4.94}$$

as the general condition [Graf and Acaroglu 1968], various density distributions may arise depending on the relative magnitudes of σ, σ_L, f_p, and f_L. (Prob. 4.10)

Exercise Problems

4.1 For two walls at 61 mm apart with gap filled by 2 μm iron particles at cloud density of 0.16 kg/m^3. Wall 1 is at 1111 K, reflectivity ρ_1 = 0.1, wall 2 at 278 K, reflectivity ρ_2 = 0.9. Determine the heat flux and the mean temperature of the particles due to scattering of radiation. Compare the result with the case when the particles are replaced by a vacuum, where

$$q = \sigma_r \ (T_1^4 - T_2^4)/[\,(1-\rho_1)^{-1} + (1-\rho_2)^{-1} + 1] \qquad (4.95)$$

Ans.   39.7 kW/m^2, 944K, 8.54 kW/m^2.

4.2 For the same data given in the previous problem except that both walls are at the condition of wall 1 in the way of a two-dimensional channel, with a particle temperature of 540 C at the inlet. Compute the temperature variation of the particles along the channel for a mean particle velocity of 3.05 m/s. Show that the particle temperature reaches 1038 C at 0.23 m from the inlet.

4.3 Assume that the radiation is diffuse, with an index of refraction of 1.25, and surface reflectivity of 0.5, and a mean temperature of 695 K, compute the heat transfer rate for the case in Prob. 4.1 and discuss.

Ans. 284 kW/m^2.

4.4 From Eq. (4.34) determine the asymptotic value of the attenuation parameter for small and large $(2a)^2\omega/\nu$. Calculate for m* = 0.3 and $\overline{\rho}/\overline{\rho}_p$ = 10^{-3}.   Ans. 66(N_{Re}), 0.00135(N_{Re})$^{-\frac{1}{2}}$.

4.5 Derive Eq. (4.38) from basic relations of thermodynamics.

4.6 Compute the ratio of parameter β in Eq. (4.51) of alumina (A_0 = 1.4 amp/cm^{2}K^2, ϕ = 3.77 ev) to that of copper (A_0 = 65, ϕ = 4.5) at 2000 K for similar particle size and number density. Compare this result for the case of $K = 0$ in Fig. 4.13.

Ans. 1.46

4.7 For 1 μm MgO particles, ρ_p = 100 kg/m^3 in an ionized gas of helium (g = 2, g^+ =1, g_e = 2) at 3000 K and 1 kPa. Compute the charge on the particles at equilibrium. $\bar{\rho}_p$ = 3580 kg/m^3, $\Delta\epsilon$ = 10 ev, n_b = 10^4, I = 24.6 ev., ϕ = 3.5 ev.   Ans. 0.

4.8 For 1 μm zirconia particles in argon of n_o =10^{20}/m^3, the particle concentration gives a volume ratio of gas to particle of 1000 (or $n = 2 \times 10^{15}$/m^3), all at 3000 K, ϕ = 3.4 ev, n_{cs} = 10^{26} m^{-3}, I = 15.756 ev , g_z = 6, g = 1. Show that, with the solid particles charged to 2000 holes (that is, with 2000 electrons removed), a reduction of electron concentration to 10^{14}/m^3. From the derivations show that when the particles are neutral initially, the free-electron concentration tends to be raised, while initially negatively charged particles tend to raise the free electron concentration further.   Ans. 1.44 $\times$ 10^{11}m^{-3}.

4.9 For given diffusion current at the radius of a spherical particle J_o, derive relations for non-equilibrium concentration of electrons and ions. By eliminating J_o and taking the case of $T_e \sim T_i$, and far away from the sphere, determine Z, and J_o, for $n_{ea} = n_e - (D^+/D^-)n_i$.

4.10 Discuss the trend of density variation in case (5) of Sec. 4.6 for various combinations of values of f_p, f_L, ρ_{pw}, σ, and σ_L.

Chapter 5

ONE–DIMENSIONAL MOTIONS

5.1 Adiabatic Flow.

Some understanding of the basic nature of the multiphase flow system may be achieved by considering adiabatic flow. For the case of nonshear flow of a nonreactive, uncharged gas-solid suspension, the overall momentum equation takes the form (from Eq. 3.19)

$$\rho(dU_i/dt) + \sum_s K_{ms}\rho_{ps}(dU_{psi}/dt_{ps}) = -\partial P/\partial x_i \tag{5.1}$$

where $d/dt_{ps} = \partial/\partial t + U_{psi}\,\partial/\partial x_i$ along the path of particle ps. The overall energy equation is given by summing Eqs. (2.115) and (2.116) to take the form

$$\rho\frac{d}{dt}[\frac{U_i^2}{2} + cT] + \sum_s \rho_{ps}\frac{d}{dt_{ps}}[\frac{U_{psi}^2}{2} + c_{ps}\,T_{ps}] = \frac{\partial P}{\partial t} \tag{5.2}$$

where c, c_p are the specific heat at constant pressure of the gas and the solid respectively, and the change in enthalpy $dh = c\,dT$. Neglecting the volume occupied by the solid particles and the virtual mass forces, the momentum and energy of a particle of size s given by Eqs. (3.52) and (3.61) are extended to include interactions with other solid phases r according to:

$$(dU_{psi}/dt_{ps}) = F_{sf}(U_i - U_{psi}) + \sum_{(r)} F_{psr}\,(U_{pri} - U_{psi}) \tag{5.3}$$

$$(dT_{ps}/dt_{ps}) = G_{sf}(T - T_{ps}) + \sum_{(r)} G_{psr}(T_{pr} - T_{ps}) \qquad (5.4)$$

dissipations other than those due to phase interactions are neglected in this illustration.

Multiplying Eq. (5.1) by fluid velocity U and subtracting from Eq. (5.2) gives

$$\frac{dP}{dt} = \rho c \frac{dT}{dt} + \sum_s \rho_{ps}[(U_{psi} - K_{ms}U_i) \frac{dU_{psi}}{dt_{ps}} + c_{ps} \frac{dT_{ps}}{dt_{ps}}] \qquad (5.5)$$

Since the entropy of a given volume of suspension is given by:

$$dS_m = \rho c \frac{dT}{T} - \rho \overline{R} \frac{dP}{P} + \sum_s \rho_{ps} c_{ps} \frac{dT_{ps}}{T_{ps}} \qquad (5.6)$$

the rate of entropy generation is given by substituting Eq. (5.5):

$$\begin{aligned}
\dot{S} = \frac{dS_m}{dt} &+ \sum_s [\rho_{ps} c_{ps} G_s (T_{ps} - T)(\frac{1}{T} - \frac{1}{T_{ps}})] \\
&+ \frac{1}{T} \sum_s [\rho_{ps} F_{sf}(U_{psi} - K_{ms}U_i)(U_{psi} - U_i)] \\
&= \sum_s \{\rho_{ps} c_{ps}(\frac{1}{T} - \frac{1}{T_{ps}})[\sum_r G_{psr}(T_{ps} - T_{pr})]\} \\
&+ \sum_s \{\rho_{ps}(U_{psi} - K_{ms}U_i)\frac{1}{T}[\sum_r F_{psr}(U_{psi} - U_{pri})]\} \qquad (5.7)
\end{aligned}$$

It is expected that for a dissipative system such as a gas-solid suspension, $\dot{S} > 0$. The first two summations deal with gas-solid interaction only and are greater than zero ($K_m = 1$ for $U_{pi} < U_i$, $K_m < 1$ for $U_{pi} > U_i$). The last two summations are greater than zero because small particles have smaller inertia and smaller heat capacity than large particles. The detailed proof is left for an exercise (Prob. 5.1).

<u>Speed of sound</u>. Based on Eq. (5.5), the speed of sound and other apparent thermodynamic properties of a suspension can be computed; the former as an alternate to Sec. 4.2 for comparison. Here we illustrate with the case of particles of one size and $K_m = 1$ for simplicity. When the fluid medium is a perfect gas, Eq. (5.5) becomes, for particles of uniform size and mass and $K_m = 1$:

$$-F\rho_p (U_i - U_{pi})^2 + \rho_p c_p G(T-T_p) = (dP/dt) - \rho c(dT/dt) \tag{5.8}$$

The speed of sound a through a suspension is given by

$$\frac{\gamma \bar{R}T}{a_m^2} = \gamma \bar{R}T \, \frac{d(\rho + \rho_p)}{dP} = \gamma [1 - \frac{d\ell n \, T}{d\ell n \, P}] \, \frac{d\rho}{d\bar{\rho}} + \gamma \bar{R}T \, \frac{d\rho_p}{dP}$$

$$= 1 - \frac{\rho_p}{\bar{\rho}_p} \gamma + (\gamma - 1) \, \rho_p [Gc_p(T-T_p) - F(U-U_p)^2] \, (\frac{dP}{dt})^{-1}$$

$$+ \gamma (\frac{1}{\rho} - \frac{1}{\bar{\rho}_p}) \, \frac{d\rho_p}{d\ell n \, P} \tag{5.9}$$

showing that the speed of sound is affected by the dissipative process but not the flow velocities. One readily shows that due to the inertia of the solid particles, the sonic speed of a suspension a_m is always less than $(\gamma \bar{R}T)^{\frac{1}{2}}$, the speed of sound in a clean gas (Prob. 5.2).

5.2 One-dimensional Steady Motion

Studies of one-dimensional steady flow of a gas-solid suspension through a duct of variable area were motivated by the applications to metallized propellant rocketry. Assumptions for the basic formulations are:

1. The motion is one-dimensional and steady; effect of turbulence enters only in characteristic parameters.
2. The solid particles are uniformly distributed over each cross-section, although it is understood that they are suspended by the fluid and interactions exist between components.
3. In each size range, the solid particles are uniform in diameter and physical properties. Variations again can be accounted for with characteristic parameters.

4. The drag on the particles is mainly due to difference between the mean velocities of particles and stream. It is expected that the minimum size of solid particles consists of millions of molecules each (even in the submicron range). Hence the velocity of each solid particle due to its thermal state is extremely low. Slip flow, if occurs, again can be accounted for by an appropriate characteristic parameter.

5. The heat transfer between the gas and the solid is basically due to their mean temperature difference. Effect of fluctuation in temperature will be accounted for by proper characteristic parameters.

6. The solid particles, due to their small size and high thermal conductivity (as compared to that of the gas) are assumed to be at uniform temperatures.

7. The virtual mass force is negligible because of large ratio of densities of materials constituting the particles and the fluid; that due to unsteady flow field is also small.

Through the introduction of densities and pertaining to the particles clouds and the gas excluding the volume occupied by the particles, the continuity relation gives the flow rate of gas as

$$\dot{m} = \rho U A = \alpha_g \bar{\rho} U A = \bar{\rho} U A_g \tag{5.10}$$

where U is the axial flow velocity of the gas in the x-direction and A is the cross-sectional area of the passage. Note that for volume fraction α_g of the gas, the cross-sectional area apportioned to the gas is $A_g = \alpha_g A$. $\rho = \bar{\rho}$ when the volume occupied by solids is neglected. The flow rate of solid particles of species s is

$$\dot{m}_{ps} = \rho_{ps} U_{ps} A = \alpha_{ps} \bar{\rho}_p U_{ps} A = \bar{\rho}_p U_{ps} A_{ps} \tag{5.11}$$

following similar reasoning as that for the gas, A_{ps} is thus the flow area apportioned to particles of species s. The total flow rate of the particles is

$$\dot{m}_p = \sum_s \rho_{ps} U_{ps} A = A \sum_s \rho_{ps} U_{ps} \tag{5.12}$$

Steady-flow condition gives the mass flow ratio $\dot{m}_p^*$ as

$$\dot{m}_p^* = \dot{m}_p/\dot{m} = \sum_s \rho_{ps} U_{ps}/\rho U = \text{constant} \tag{5.13}$$

as well as

$$\dot{m}_{ps}^* = \rho_{ps} U_{ps}/\rho U \tag{5.14}$$

when there is no reaction or phase change, and the mass ratio is given by $\dot{m}_p^* U_{ps}/U$. It is seen that, when $U = U_{ps}$,

$$\dot{m}_p^* = \sum_s \rho_{ps}/\rho = m_p^* \tag{5.15}$$

where m_p^* is the mass ratio. The momentum equations of the particulate phase from Eq. (3.52), neglecting the volume occupied by solid particles and the virtual mass force, are given by

$$U_{ps} dU_{ps}/dx = F_s(U-U_{ps}) + \sum_r F_{sr}(U_{pr}-U_{ps}) \tag{5.16}$$

In deriving the overall momentum equation, we account for the shear stress terms with boundary friction [Shapiro 1953], shear stress at the wall τ_w and Eqs. (2.113) and (2.114), when summed over the phases, becomes

$$\rho U \frac{dU}{dx} + \sum_s \rho_{ps} U_{ps} \frac{dU_{ps}}{dx} = -\frac{dP}{dx} - \frac{\tau_w P}{A} \tag{5.17}$$

where P is the perimeter of the duct at area A, and τ_w is given by the friction coefficient:

$$c_f = \tau_w/(\rho U^2/2) \tag{5.18}$$

When the stress due to impact of particles with the wall is neglected, and the perimeter is related to A via the hydraulic diameter

$$D_H = 4A/P \qquad (5.19)$$

Eq. (5.17), after substitution of Eqs. (5.18) and (5.19) and rearranging, becomes (take $\rho = \bar{\rho}$, and $\bar{\rho} = P/\bar{R}T$):

$$\frac{U}{\bar{R}T}\frac{dU}{dx} + \frac{U}{\bar{R}T}\sum_s \dot{m}^\star_{ps}\frac{dU_{ps}}{dx} + \frac{dP}{Pdx} + \frac{2c_f U^2}{\bar{R}TD_H} = 0 \qquad (5.20)$$

The energy equation of particles including heat input into a particle of size s by convection from the gas and radiation from the wall, but neglecting transfer of heat between particles and wall due to impact, is given by Eq. (3.61) as

$$U_{ps}\frac{dT_{ps}}{dx} = G_{sf}(T-T_{ps}) + \sum_r G_{sr}(T_{pr}-T_{ps}) \qquad (5.21)$$
$$+ \frac{3\,\varepsilon_p \sigma r}{c_p \bar{\rho}_p a_s}\,(T_W^4 - T_{ps}^4)$$

The overall energy equation should account for the heat transfer, if any, at the wall of the amount

$$hP\{T_W - [T + (U^2/2c)]\}$$

over length dx, and by radiation from the wall to the particles, since $dx = U_{ps}dt$ for the s particles, or from the sum of phases of Eqs. (2.115) and (2.116),

$$d(cT + \frac{U^2}{2} + \sum_s \dot{m}^\star_{ps}\,c_{ps}T_{ps} + \sum_s \dot{m}^\star_{ps}\frac{U_{ps}^2}{2})$$
$$= \frac{4(N_{Nu})_{DH}\,\bar{\kappa}A}{D_H^2\,\dot{m}}\,[T_W - (T + \frac{U^2}{2c})]dx + \frac{3\sigma r}{\bar{\rho}_p}\sum_s \frac{\varepsilon_{ps}}{a_s}\,(T_W^4 - T_{ps}^4)\frac{dx}{U_{ps}} \qquad (5.22)$$

The above equations are basic to all one-dimensional gas dynamic systems in steady flow. The latter include nozzle flow, Fanno flow, and shock waves. Detailed treatment is illustrated in this chapter with nozzle flows. Fanno flow or flow of a suspension in a pipe of uniform flow area with wall friction was extensively studied [Trezek & Soo 1966] (Prob. 5.3).

5.3 Adiabatic Flow through a Nozzle

The use of metallized propellant increases the energy per unit mass of a rocket but loses propulsion efficiency by the nature of gas-solid nozzle flow. We first consider the case of flow with particles of one size.

<u>Speed of sound and throat condition.</u> When applied to one-dimensional nozzle flow, Eq. (5.9) gives, for $\dot{m}_p^* = \rho_p U_p/\rho U$, the mass flow ratio of phases

$$\frac{\gamma \bar{R}T}{a_m^2} [1 + \dot{m}_p^* \frac{dU_p}{dU}] = \{1 + \dot{m}_p^* [\gamma \frac{dU_p}{dU} - (\gamma-1) \frac{U_p dU_p}{U dU}$$

$$- (\gamma-1) \frac{c_p dT_p}{U dU}]\} [1 + \dot{m}_p^* \frac{U}{U_p}] + \frac{\gamma \bar{R}T}{U_p^2} [\frac{U_p}{U} - \frac{dU_p}{dU}] \qquad (5.23)$$

taking $\rho \sim \bar{\rho}$. In the case of frictionless adiabatic nozzle handling a pure gas, it is well known that the critical flow is at sonic velocity. Due to internal friction between the phases in the gas-solid system, a different result is expected. The flow through an area A is given by

$$\dot{m}_{mix}/A = \rho U(1 + \dot{m}_p^*) \qquad (5.24)$$

which has a maximum at

$$\rho dU/dx = -U d\rho/dx \qquad (5.25)$$

Eqs. (5.20), (5.22), and (5.23) give, for adiabatic flow without wall friction, a critical flow velocity U_c:

$$\bar{R}T_c/U_c^2 = (\bar{R}T_c/a_m^2)[1 + \dot{m}_p^* \frac{dU_p}{dU}] \qquad (5.26)$$

The Mach number of the gas at the throat based on its sonic speed is therefore

$$(N_M)_g^2 = [1 + \dot{m}_p^* \frac{dU_p}{dU}]^{-1} \qquad (5.27)$$

Since $U_p < U$ in an accelerating flow, both phases are subsonic at the throat because of the dissipation in the mixture even in the absence of wall friction. (Prob. 5.4)

When wall friction is significant, the momentum equation in one-dimension takes the form

$$\frac{dP}{Pdx} = - \frac{1}{RT} \left[U \frac{dU}{dx} + \dot{m}_p^* \, U \frac{dU_p}{dx} + 2 \frac{C_f}{D_H} U^2 \right] \tag{5.28}$$

where C_f is the friction factor and D_H is the hydraulic diameter. The Mach number of the gas phase at the throat is now

$$(N_M)_g^2 = \left[1 + \dot{m}_p^* \frac{dU_p}{dU} + 2 \frac{C_f}{D_H} U \left(\frac{dU}{dx}\right)^{-1} \right]^{-1} \tag{5.29}$$

indicating a still lower Mach number than in Eq. (5.27).

The above considerations further show that for the nozzles of similar inlet, throat, and exit areas operated at similar inlet temperature, pressure and exit pressure, the throat and sonic speeds for a given gas-solid mixture are different for different lengths an shapes of nozzles.

<u>Nondimensionalizing of basic equations</u>. The functional dependence as indicated in Eqs. (5.27) and (5.29) shows that non-dimensionalizing Eqs. (5.13), (5.16), (5.17), (5.20), (5.21), and (5.22) in terms of Mach number is inappropriate. A logical choice is to use a reference velocity defined in terms of the stagnation temperature T : $U_o = (2cT_o)^{\frac{1}{2}}$ [Soo 1967]. For length of the nozzle L, the nondimensional variables (superscript *) are:

$$A^* = A(U_o \rho_o / \dot{m}) = A(U_o P_o / \dot{m} \bar{R} T_o), \quad x^* = x/L$$

$$T^* = T/T_o, \quad T_p^* = T_p/T_o, \quad T_w^* = T_w/T_o,$$

$$P^* = P/P_o, \quad U^* = U/U_o, \quad U_p^* = U_p/U_o \tag{5.30}$$

Eq. (5.10) becomes

$$A^*U^*P^* = T^* \tag{5.31}$$

Eq. (5.16) rewritten in dimensionless form becomes

$$U_p^*(dU_p^*/dx^*) = E^*(U^* - U_p^*) \tag{5.32}$$

where a drag parameter E^* is introduced

$$E^* = FL/U_o \tag{5.33}$$

Eq. (5.21) becomes

$$U_p^* dT_p^*/dx^* = B^*(T^* - T_p^*) + D^*(T_w^{*4} - T_p^{*4}) \tag{5.34}$$

where B^* is a convection heat transfer parameter,

$$B^* = 3(N_{Nu})_p \bar{\kappa} L/2 c_p \bar{\rho}_p a^2 U_o \tag{5.35}$$

and D^* is a radiative heat transfer parameter,

$$D^* = 3\epsilon_p \sigma_r T_o^3 L/c_p \bar{\rho}_p a U_o \tag{5.36}$$

Eq. (5.20) takes the form

$$\frac{dP^*}{P^*} + \frac{2c}{\bar{R}} \frac{U^*}{T^*} dU^* + \frac{2\dot{m}_p^* c_p}{\bar{R}T^*} U^* dU_p^* + \frac{4cC_f U^{*2}}{\bar{R}D_H T^*} dx^* = 0 \tag{5.37}$$

Eq. (5.22) becomes

$$d(T^* + U^{*2} + \frac{\dot{m}_p^* c_p}{c} T_p^* + \dot{m}_p^* U_p^{*2})$$

$$= B_w^* A^* [T_w^* - (T^* - U^{*2})] \, dx^*$$

$$+ \frac{c_p}{c} D^* \frac{\dot{m}_p^*}{U_p^*} (T_w^{*4} - T_p^{*4}) \, dx^* \tag{5.38}$$

where B^*_w is a convective heat transfer parameter for the wall

$$B^*_w = 4(N_{Nu})_{D_H} \,\overline{\kappa}\overline{L}\overline{R}T_o/D_H^2 \, c \, U_o P_o \tag{5.39}$$

These five equations, (5.31), (5.32), (5.34), (5.37), and (5.38), are the governing equations for a gas-solid suspension derived under the assumptions outlined previously. The basic nonlinearity is noted in Eq. (5.32).

<u>Adiabatic nozzle flow without wall friction.</u> When radiation, wall friction and heat transfer at the wall are negligible, and when the Stokesian drag and hence $(N_{Nu})_p = 2$ can be assumed, Eqs. (5.34), (5.37), and (5.38) take simplified forms:

$$U^*_p dT^*_p/dx^* = B^*(T^* - T^*_p) \tag{5.40}$$

$$\frac{1}{P^*}\frac{dP^*}{dx^*} + \frac{2cU^*}{\overline{R}T^*}\frac{dU^*}{dx^*} + \frac{2\dot{m}^*_p c \, U^*}{\overline{R}T^*}\frac{dU^*_p}{dx^*} = 0 \tag{5.41}$$

and Eq. (5.38) integrates to

$$T^* + U^{*2} + \frac{\dot{m}^*_p \, c_p}{c} T^*_p + \dot{m}^*_p \, U^{*2}_p = 1 + (\dot{m}^* c_p/c) \tag{5.42}$$

with the initial conditions at $x^* = 0$ as

$$T = T_o, \; T^* = 1; \; T_p = T_o, \; T^*_p = 1$$

$$U = 0, \; U^* = 0; \; U_p = 0, \; U^*_p = 0 \tag{5.43}$$

These five equations, (5.31), (5.32), and (5.40) to (5.42) are now the governing equation for adiabatic nozzle flow with the simplifications as stated. They constitute a set of ordinary, nonlinear, first order, differential equations. In a numerical solution for a given nozzle area distribution, the critical flow condition at the throat must be accounted for.

<u>Uniform acceleration of the gas.</u>   An analytical solution for the above five simultaneous equations is available for the case of uniform acceleration of the gas:

$$U^* = \alpha x^* \tag{5.44}$$

where α is a positive arbitrary constant less than 1. Substitution into Eq. (5.32) gives

$$U_p^* = \frac{1}{2} \left[(E^{*2} + 4E^*\alpha)^{\frac{1}{2}} - E^* \right] x^* = \beta x^*. \tag{5.45}$$

showing that the particles always lag behind the gas. Eqs. (5.40) to (5.42) are linear and readily solved.   Note that the area distribution along x^* cannot be prescribed but is given by Eq. (5.31) from U^*, P^*, and T^*.   However, an understanding of multiphase flow can be gained from this example Eqs. (5.42) and (5.40) may be combined to give:

$$\frac{dT_p^*}{dx^*} + \frac{B^*}{U_p^*} \left(1 + \frac{\dot{m}_p^* c_p}{c}\right) T_p^* = - \frac{B^*}{U_p^*} \left[U^{*2} + \dot{m}_p^* U_p^{*2} - \left(1 + \frac{\dot{m}_p^* c_p}{c}\right) \right] \tag{5.46}$$

which belongs to a standard form for solution by introducing an integrating factor

$$\exp \left\{ [1 + (\dot{m}_p^* c_p / c)] \int (B^*/U_p^*) dx^* \right\}$$

which, with Eqs. (5.44) and (5.45) and the initial condition of $x^* = 0$, $T_p^* = 1$, to give

$$T_p^* = 1 - (B^*/\beta)(\alpha^2 + \dot{m}_p^* \beta^2) \left[(B^*/\beta)\left(1 + \frac{\dot{m}_p^* c_p}{c}\right) + 2 \right]^{-1} x^{*2} \tag{5.47}$$

Equations (5.42), (5.44), and (5.47) give T^* for the gas phase

$$T^* = 1 - \alpha^2 x^{*2} (c/c_e) \tag{5.48}$$

where c_e is the apparent specific heat at constant pressure of the suspension:

$$\frac{c_e}{c} = [\frac{B\star}{\alpha}(1 + \frac{\dot{m}_p^\star c_p}{c}) + 2(\frac{\beta}{\alpha})][1 + \dot{m}_p^\star \frac{\beta^2}{\alpha^2}]^{-1}[\frac{B\star}{\alpha} + 2(\frac{\beta}{\alpha})]^{-1} \qquad (5.49)$$

It is interesting to note that specific heat, normally a thermo-dynamic property, is dependent on the transport process in the case of a gas-solid suspension. Non-equilibrium exists as seen in the disparity of velocity and temperature of phases. The limiting cases are:

(a) When the phases are at equilibrium, $E\star \to \infty$, $\beta/\alpha \to 1$, $B\star \to \infty$, $c_e = (c + \dot{m}_p^\star c_p)/(1 + \dot{m}_p^\star)$, like a heavy gas-light gas mixture.

(b) When there is no heat transfer between the phases, we have: $E\star \to \infty$, $\beta/\alpha \to 1$, $B\star \to 0$, $c_e = c/1 + \dot{m}_p^\star$, with the solid particles contributing to inertia only.

(c) The case of $E\star \to 0$, $\beta/\alpha \to 0$, $B\star \to \infty$, gives, $c_e = c + \dot{m}_p^\star c_p$, with solid particles left behind at a low velocity and heating the gas phase. The expansion of the gas tends to constant temperature; however, the volume occupied by the particles can no longer be neglected.

(d) The case of a clean gas is given by: $E\star \to 0$, $\beta/\alpha \to 0$, $B\star \to 0$, $c_e = c$.

The pressure distribution $P\star(x\star)$ is given by Eq. (5.41) as:

$$\ell n P\star = (c_e/\overline{R})[1 + \dot{m}_p^\star(\beta/\alpha)]\ell n[1 + \alpha^2 x\star^2(c/c_e)] \qquad (5.50)$$

showing that presence of solid particles retards the drop in pressure. Eq. (5.47) may be written to include $T\star$:

$$T_p^\star = T\star + 2(\beta/\alpha)(c/c_e)\alpha^2[(B\star/\alpha) + 2(\beta/\alpha)]^{-1}x\star^2 \qquad (5.51)$$

showing that the particle temperature always stays higher than the gas temperature. In many cases, where $E\star$ and $B\star$ are both greater than 1000, the approximation of $U = U_p$, and $T = T_p$, should be quite accurate.

The speed of sound of the mixture, a_m from Eq. (5.23), for the case of uniform acceleration, gives

$$a_m^2 = \gamma \bar{R} T (1 + \dot{m}_p^* \frac{\beta}{\alpha}) (1 + \dot{m}_p^* \frac{\alpha}{\beta})^{-1} [\gamma - (\gamma-1) \frac{c}{c_e} + \gamma \dot{m}_p^* \frac{\beta}{\alpha}]^{-1} \tag{5.52}$$

The throat Mach number is given by Eq. (5.27) for the present case as

$$(N_M)_g = (1 + \dot{m}_p^* \frac{\beta}{\alpha})^{-\frac{1}{2}} \tag{5.53}$$

With the above values of P^*, T^*, and U^*, $A^*(x^*)$ is given by Eq. (5.31).   In spite of the inflexibility of this analytical solution, it is exact and provides a check solution to any computer program, with calculated $A^*(x^*)$ from the above as input to determine U^*, etc. [Hultberg and Soo 1965], with critical flow at the throat as an additional condition.   For the case with a distribution in particle sizes, the general summation procedure over each sixe range as given in Sec. 5.2 can be applied.   When a continuous distribution is applicable as illustrated in Sec. 1.9 an integration procedure can be used [Soo 1967]; noting that the size distribution function of a particle cloud in motion is different from the results of static measurement.

It is difficult to draw conclusions from this solution because the area distribution cannot be held constant for a parametric study.   For a typical example of: $c = 1$ kJ/kg K (air), $c_p = 0.491$ kJ/kg K (magnesia), $\bar{R} = 0.287$ kJ/kg K, $\dot{m}_p^* = 0.30$ kg solid/kg gas, $E^* = 100$, $B^* = 10$, and $\alpha = 0.9$, the solutions for $A^*(x^*)$, $P^*(x^*)$, $T^*(x)$, and $N_M(x^*)$ are illustrated in Figs. 5.1 to 5.4.   In the latter 3 figures, the trends of various values of $\dot{m}_p^*$ are also shown ($\dot{m}_p^* = 0$ for gas alone); especially notable are the Mach numbers at the throat in Fig. 5.4.

For the same numerical example, the general character of c/c_e is shown in Fig. 5.5. The speed of sound of the mixture a_m as influenced by the transport processes is now expressed in terms of its ratio to the speed of sound a_g in the gas alone: (Prob. 5.5)

$$a_m^2/a_g^2 = [(\frac{B^*}{E^*})(\frac{\beta}{\alpha})(\frac{c + \dot{m}_p^* c_p}{c}) + 2(1 - \frac{\beta}{\alpha})] \ \{(\frac{B^*}{E^*})(\frac{\beta}{\alpha})[(1 + \dot{m}_p^* \frac{\beta}{\alpha})(\frac{c + \dot{m}_p^* c_p}{c})$$

$$- (\frac{\overline{R}}{c - \overline{R}})(1 + \dot{m}_p^* \frac{\beta^2}{\alpha^2})] + 2(1 - \frac{\beta}{\alpha}) \ [(\frac{c}{c - \overline{R}}) \ (1 + \dot{m}_p^* \frac{\beta}{\alpha})$$

$$- (\frac{\overline{R}}{c - \overline{R}}) \ (1 + \dot{m}_p^* \frac{\beta^2}{\alpha^2})]\}^{-1} \tag{5.54}$$

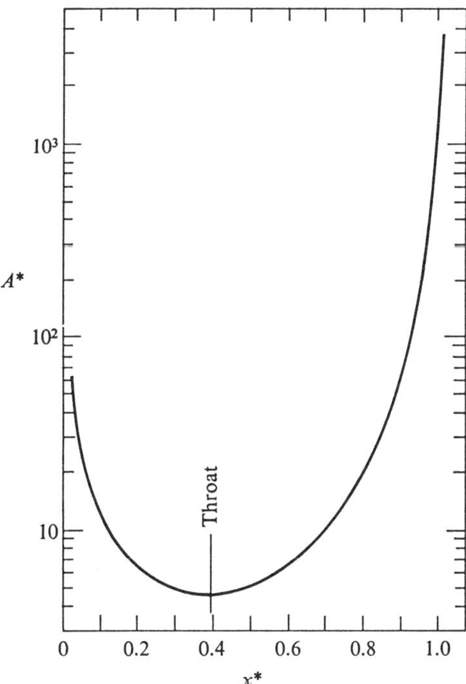

Figure 5.1 Area distribution of nozzle designed for linear acceleration with $\dot{m}_p^* = 0.3$ [Hultberg & Soo 1965]

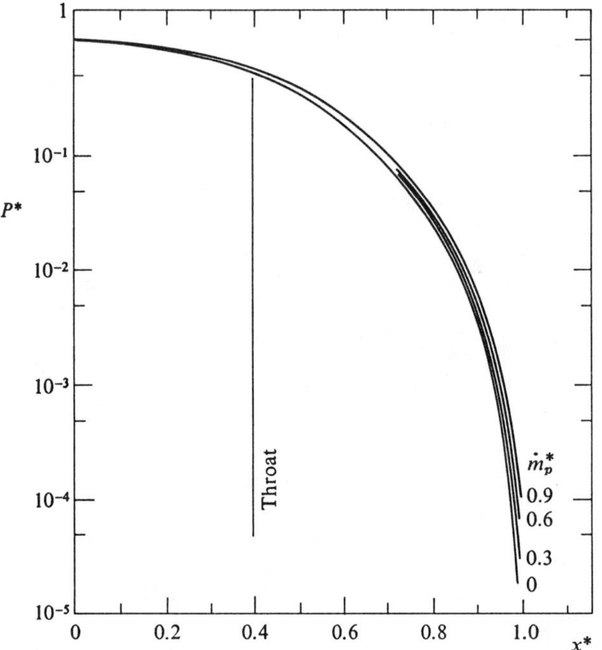

Figure 5.2   Pressure distribution in nozzle designed for linear
acceleration at $\dot{m}_p^* = 0.3$ [Hultberg & Soo 1965]

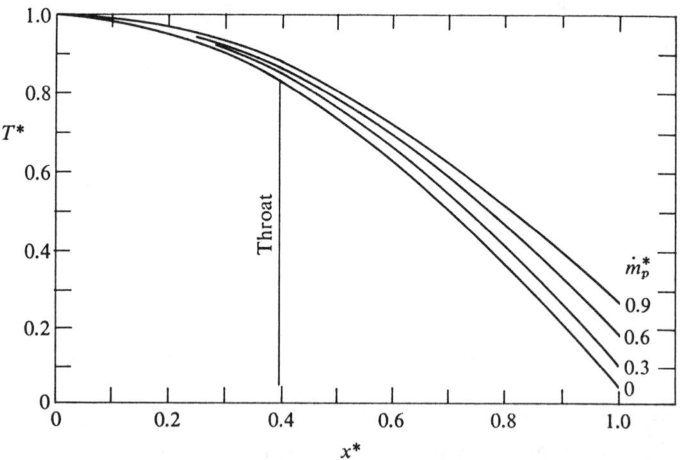

Figure 5.3   Gas-temperature distribution [Hultberg & Soo 1965]

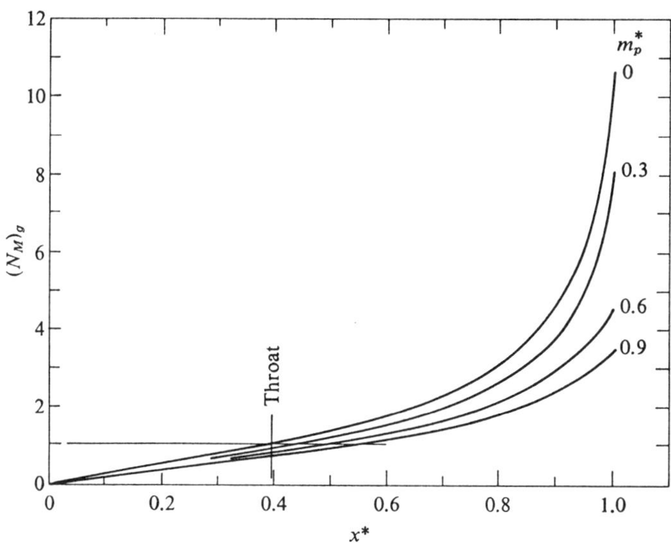

Figure 5.4   Mach number of the gas phase [Hultberg & Soo 1965]

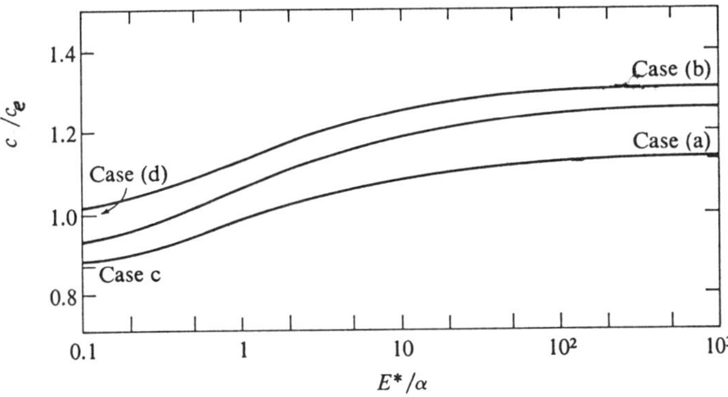

Figure 5.5   Apparent specific heat of a suspension as affected
by transport processes ($\dot{m}_p^* = 0.3$ MgO in air)

This relation is shown in Fig. 5.6 for the same numerical example. For large drag and small heat transfer, the solid contributes only as added molecular weight. If the heat transfer rate is also large, the situation of a heavy gas mixture arises. If the drag is very small, the solid particles do not contribute to the speed of sound. (Prob. 5.6, 5.7, 5.8)

One notes that it is not just a matter of small particles and dense gas that will produce small momentum and thermal lag between the phases. It calls for a large heat-transfer parameter: B* (Eq. 5.35), and a large drag parameter : E* (Eq. 5.33). In the Stokesian

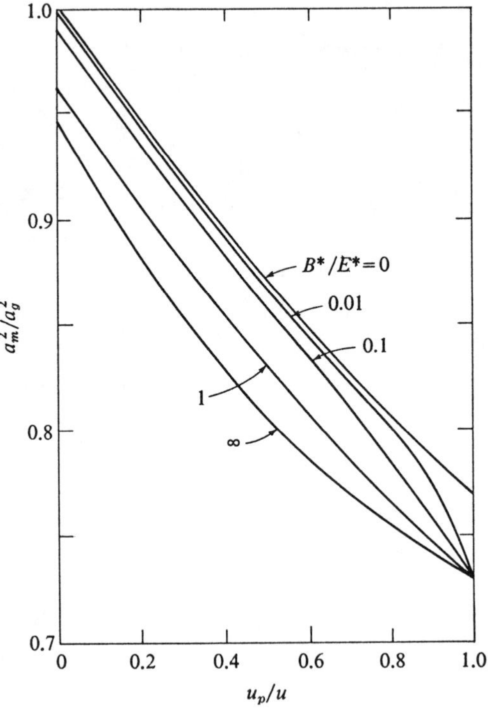

Figure 5.6   Velocity of sound through a gas-solid suspension
$(\dot{m}_p^* = 0.3)$

range, the ratio

$$B^*/E^* = (2/3)\ (c_\rho/c_p \bar{\rho}_p) N_{Pr}$$

being usually small, thermal lag is usually more significant than momentum lag. For particles sizes smaller than the mean free path λ of the gas, modifications for slip motion (Eq. 1.8) and free molecular flow ($C_D = 2$) have to be applied.

These results indicate that, even without wall friction, the shape and length of the nozzle affects its thrust. For a nozzle operating with very low exit pressure, the expansion should be as complete as possible, because the particles do not contribute to pressure thrust. When applied to a rocket, the specific impulse is given by

$$I = (\alpha\ \sqrt{2cT_o}/g_c)\ [1 + \dot{m}_p^*(\beta/\alpha)][1 + \dot{m}_p^*]^{-1} \qquad (5.55)$$

where g_c is the gravitational constant to give I in seconds. Hence, for the same exit velocity from the nozzle, the specific impulse of a gas-solid system is always nearly the same as that of a pure gas of similar composition as the gas phase, regardless of the exit velocity and concentration of the solid phase. For the same impulse, the inlet pressure to the nozzle usually has to be higher when solid particles are present. While thermal lag is greater than velocity lag in most of the actual cases, the effect of thermal lag on the performance of a rocket is far less than that of the velocity lag.

Experimental validation. Together with the computer program for given area distribution of a nozzle, experimental study with a converging nozzle and a Mach 2.3 nozzle was conducted in a blow-down system with glass particles in air [Hultberg & Soo 1965]. Instrumentation included pressure transducers for the gas and shadowgraph determination of concentration of solids. High concentration of solid particles along the wall due to electrostatic effect and modification of velocity distribution due to boundary layer effect was noted.

Experiments on metallized (aluminum) propellant rockets were made by Brown and McArty [1962]. Average particle size of 2 to 3 micrometer were measured and long nozzles with gradual contours produced higher particle exit velocities than short nozzles.

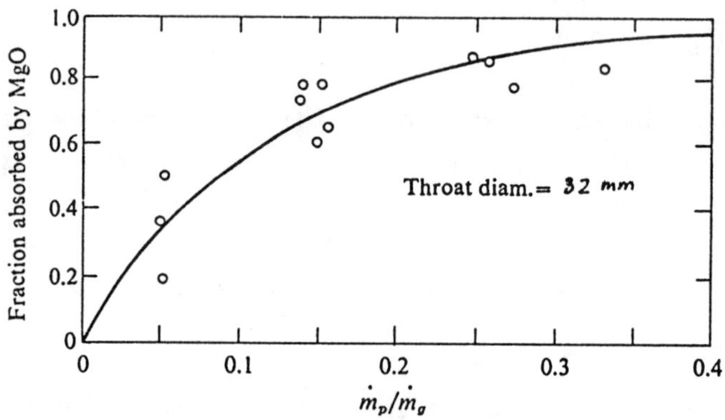

Figure 5.7   Absorptivity of rocket exhaust due to presence of MgO cloud; measured at 5911 A [Carlson 1962]

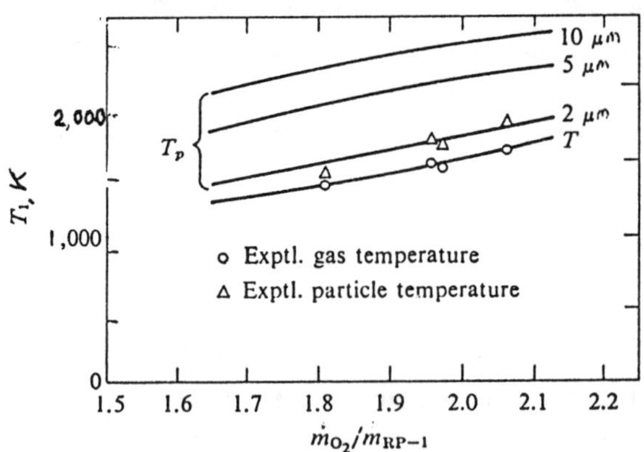

Figure 5.8   Experimental and theoretical gas and particle temperature at mass-flow ratio of particle to gas of 0.275 [Carlson 1962]

Measurements on a rocket of 4500 N thrust and 27 bar chamber pressure, burning a slurry of magnesium in RP-1 fuel, with oxygen as oxidizer by Carlson [1962] gave absorptivity of the MgO cloud at the exhaust and particle and gas temperatures as shown in Figs. 5.7 and 5.8. Particle size distribution around 0.5 micrometer was measured from velocity lags.

5.4 Gas-liquid Systems

Gas-liquid systems include liquid droplets in a vapor and vapor bubbles in liquid. Both have significant applications; they include droplet formation during expansion of a gas that may lead to condensation shock and the bubble formation and growth during critical flow of a liquid.

<u>Condensation.</u> The use of a supersonic nozzle makes possible experimental investigation of homogeneous nucleation and condensation because it allows a maximum relaxation rate in the simplest possible manner compared to other methods of sudden expansion. Static pressure measurements along the nozzle serve as a sensitive indicator of heat release by condensation and density of the gas can be measured by interferometry [Duff and Hill 1964]. The nucleation rate equation for condensation from supersaturated vapor was given by Frenkel [1946]:

$$\Gamma = (\frac{P}{kT})^2 \frac{m}{\bar{\rho}} (\frac{2\sigma}{\pi m})^{\frac{1}{2}} \exp [-4\pi\sigma r^{*2}/3kT] \qquad (5.56)$$

where Γ is the nucleation rate in nuclei/m^3s, σ is the surface tension of the drop, $\bar{\rho}$ the density of the vapor, r^* the Kelvin-Helmholtz initial radius at which a liquid is at equilibrium with its vapor, and m the molecular mass. The drop temperature is given by, for N_g complexes consisting of g simple molecules of similar structure, and $\xi = (N_g)^{1/g}$,

$$\frac{\xi}{2\overline{R}T} \, [u_{fg} + \frac{1}{2} \overline{R}T_p] \, [1 - \frac{P_\infty}{P} \, \exp \, (\frac{2\sigma}{\rho \overline{R} \, T_p}) \, (\frac{T}{T_p})] = \frac{T_p}{T} - 1 \qquad (5.57)$$

where u_{fg} is the potential energy in the absence of rotation and vibration, T_p is the droplet temperature, P_∞ is the flat surface vapor pressure; this equation gives the droplet temperature T_p. The rate of droplet growth is given by consideration of incident and evaporating molecules in terms of

$$\frac{dr}{dt} = \frac{2P}{\overline{\rho} \, \sqrt{2\pi \overline{R}T}} \, [\frac{\overline{R} \, (T_p - T)}{u_{fg} + (1/2)\overline{R}T_p}] \qquad (5.58)$$

The gas dynamic relations are now, for $U \sim U_p$, $\overline{\rho}_p >> \overline{\rho}$,

$$\frac{dP}{dx} = \frac{P\gamma(N_M)^2}{(N_M)^2 - 1} \, [(\frac{h_{fg}}{cT} - \frac{1}{1 + m_p^\star}) \, \frac{dm_p^\star}{dx} - \frac{1}{A} \frac{dA}{dx}] \qquad (5.59)$$

where (N_M) is the Mach number of the vapor phase, h_{fg} is the latent heat, c is the specific heat at constant pressure of the vapor, $m_p^\star$ is the mass ratio of liquid; the velocity U is given by:

$$\frac{1}{U} \frac{dU}{dx} = - \frac{1}{\gamma(N_M)^2} \, \frac{1}{P} \frac{dP}{dx} \qquad (5.60)$$

with the contribution of the droplets neglected, and

$$\frac{dT}{dx} = T(\frac{\gamma - 1}{\gamma}) \, \frac{1}{P} \frac{dP}{dx} + \frac{h_{fg}}{c} \frac{dm_p^\star}{dx} \qquad (5.61)$$

the change in $m_p^\star$ is given by Eqs. (5.56) and (5.58).

The computed results of Hill [1966] in comparison to the experimental results of Binnie and Green [1942] on steam is shown in Fig. 5.9, in which the effect of supersaturation is clearly indicated. Liquid-vapor surface accommodation coefficient is near unity. The condensation coefficient ξ is given by mass flux $\dot{m}_c$, for $\dot{m}_c = \xi P/(2\pi \overline{R}T)^{\frac{1}{2}}$ over a wide range up to 1. The cross-sectional area of the stream A is expressed in relation to a reference area A*. The predicted nucleation rate of carbon dioxide is given in Fig. 5.10, showing that greater supersaturation is required at lower

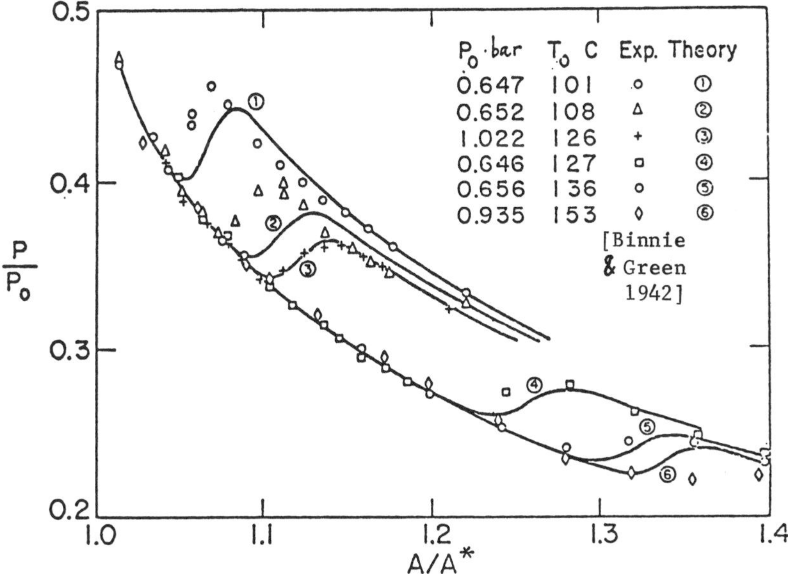

Figure 5.9   Comparison of experimental results to theoretical pressure profile [Hill 1966]

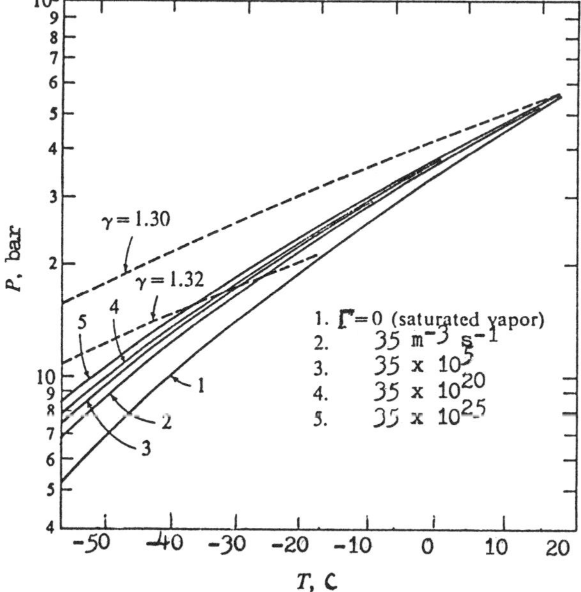

Figure 5.10 Predicted nucleation rates for CO_2 [Duff 1964]

temperatures to obtain a given nucleation rate. The supersaturation ratios steadily increase prior to the occurrence of rapid condensation.

Critical vapor-liquid flow. Extensive studies were made on critical flow of liquids near saturation because of applications to nuclear power systems. A study based on fundamental relations of multiphase flow was the one by Ardron [1978]. Taking into account the usual presence of impurities, the rate of formation of bubble nuclei per unit volume of liquid is given by [Hirth & Pound 1963]:

$$dn_b/dt = v\ n_s \exp\ [-\ \psi(\Delta G)/kT_\ell) \qquad (5.62)$$

where n_s is the effective density of heterogeneous nucleation sites in the liquid and $\bar{v}$ is a frequency factor $(2\sigma\bar{R}/\pi k)^{\frac{1}{2}}$ of impingement flux of liquid molecules; the factor $\psi(\leq 1)$ depends on the surface geometry and contact angle with the solid. ΔG is the Gibbs energy of formation of the nucleus in a pure liquid,

$$\Delta G = 16\ \pi\ \sigma^3/3\ [P - P_g(T_\ell)]^2 \qquad (5.63)$$

where $P_g\ (T_\ell)$ is the saturation pressure based on the liquid temperature. The liquid superheat must exceed a critical value before sufficient nucleation can take place. Since for low vapor density, the Clapeyron equation takes the form:

$$\ell n\ P = -h_{fg}\ (\bar{\rho}_g/P) + constant \qquad (5.64)$$

where h_{fg} is the latent at T_g, the saturation temperature at the given P. In a linearized form, Cole [1974] gave the incipient boiling superheat as, with dn_b/dt of unity as a reference,

$$\theta_c = T_\ell - T_g(P) = \frac{T_g}{\bar{\rho}_g\ h_{fg}}\ [\frac{16\ \pi\ \sigma^3\ \psi}{3\ k\ T_\ell\ \ell n\ (n_s\ \bar{v})}]^{\frac{1}{2}} \qquad (5.65)$$

Eliminating ψ between Eqs. (5.62) and (5.65), and $\theta_c \ll T_\ell$, n_s

$<< \bar{v}$, the average nucleation rate is given by [Cole 1974]:

$$\frac{dn_b}{dt} = n_s \, \bar{v} \, \exp\{ - \frac{\bar{\rho}_g^2 \, h_{fg}^2 \, (\ell n \, \bar{v}) \, \theta_c^2}{T_\ell^2 \, [P - P_g(T_\ell)]^2} \} \tag{5.66}$$

The bubble growth rate is given by

$$\frac{d}{dt} \, [\frac{4\pi}{3} \, a^3 \, \bar{\rho}_g] = \frac{q_\ell}{h_{fg}} \, 4 \, \pi \, a^2 \tag{5.67}$$

where q_ℓ is the heat flux at the bubble wall. The temperature and latent heat at the bubble wall or interface are those based on the saturation at pressure P. Taking into account the diffusive heat flux at time t in the wall of a bubble created at t', a linearized form is:

$$q_\ell(t,t') = \frac{\bar{\kappa}_\ell}{\sqrt{\pi \, D_{t\ell}(t-t')}} \, [2 \, \theta_c - \theta_c(t')] \tag{5.68}$$

$\bar{\kappa}_\ell$ is the thermal conductivity and $D_{t\ell}$ is the thermal diffusivity of the liquid. The superheats are $\theta_c(t)$ and $\theta_c(t')$, and $a(t, t')$ is the bubble radius. The evaporation rate for all the bubbles per unit volume Γ_g is:

$$\Gamma_g(t) = h_{fg}^{-1} \int_o^t \, (\frac{dn_b}{dt}) \, [1-\alpha_g(t')] \, q_\ell(t,t') \, 4\pi a^2(t,t')dt' \tag{5.69}$$

for the continuity equation. The time coordinate is taken as based on the bubble velocity, or $dx = U \, dt$. Eqs. (2.111) and (2.112) for one-dimensional motion with variable area are now illustrated by:

$$U_g \, \bar{\rho}_g \, \frac{d\alpha_g}{dx} + \alpha_g \, U_g \, \frac{d\bar{\rho}_g}{dx} + \alpha_g \, \bar{\rho}_g \, \frac{dU_g}{dx} = \Gamma_g - \alpha_g \, \rho_g \, U_g \, \frac{d\ell nA}{dx} \tag{5.70}$$

Ardron [1978] further assumed similar velocity of interface, that of the vapor at the interface, and that of the liquid phase; as well as the corresponding pressures. The wall friction was taken as due to the liquid phase alone, and application of the momentum equations (2.113) and (2.114) now includes the effect of virtual mass force. (Prob. 5.9)

Solution of the basic equations were made via a computer program. Computations showed best fit to experimental results was given by n_s of $10^6/m^3$, and θ_c = 3 C. Illustrative calculations for cases of a tapering inlet and a radiused inlet, each followed by a length of pipe of constant diameter of 20 mm. The calculated critical flow rate of initially saturated water at 70 bar showed that the choice of n_s is insensitive to variation within a factor of 2. Comparison of calculated results to the experimental results of Zaloudek [1964] for critical flow through a conical tapered pipe (254 mm long, 12.8 mm diameter, with 20 deg. taper over 36 mm long) for stagnation pressures between 20 and 100 bar is shown in Fig. 5.11, for the various enthalpy of liquid h_0 in the reservoir. The case of radiused entry (12.7 mm diameter, up to 1.8 m long) is shown in Fig. 5.12 for a comparison to the experimental results of Sozzi and Sutherland [1975]. Fig. 5.12 shows also the inaccuracy of isentropic equilibrium model. Since most of the critical flow analyses were conducted on a one dimensional system, Ardron's study showed the significance of the geometry of the entry, that is, the three-dimensional effect.

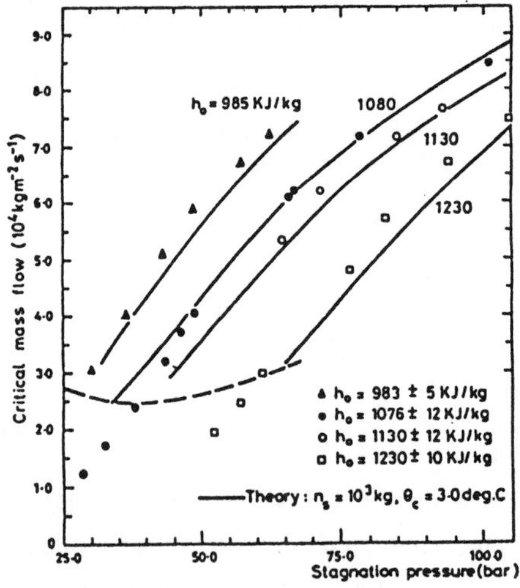

Figure 5.11 Comparison of theory with Zaloudek's [1964] data. Pipe with a tapering inlet.

<u>General liquid-gas flow</u>. The above shows that relatively accurate
analyses are feasible for either a dilute suspension of droplets or
bubbles, but various semi-empirical approximation procedures have to
be used in formulating the general flow regimes.

Critical velocity of low-quality steam was measured by
Dryndrozhik [1975] using convergent nozzles of 9, 10, 11 mm
diameters. Agreement was seen with Henry et al. [1970] on slip
ratios. For steam velocity of 150 to 250 m/s, liquid velocity was 20
to 110 m/s. Readers are referred to the treatises of Wallis [1969]
and Govier and Aziz [1972] for extensive details and empirical
correlations on general one-dimensional gas-liquid flow. One notes
that the descriptions of the flow regimes other than bubbles and
droplets have been qualitative and restricted to pipe flows. In
between, there are slug flow (nonspherical bubble with size close to
the pipe diameter), churn flow ("intermediate" range in Fig. 2.2),
annular (radial stratified flow with liquid along the pipe wall) and
annular mist flow (break-off of droplets from the liquid layer along
the wall). Ways of correlating these flow regimes have been
suggested.

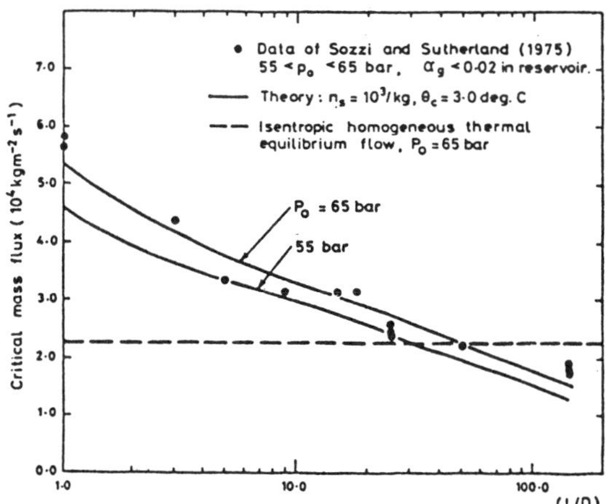

Figure 5.12 Comparison of theory with the measured critical
            mass flux of saturated water in radiused entry
            12.7 dia. pipes.

Efforts toward evaluating the transfer integrals (Chapter 2) for various flow regimes are still in the developing stage. The primary quantity is seen to be the area of interface per unit volume A_k/v. The latter quantity for spherical dispersed phase or radius a_k and volume fraction α_k is given by:

$$A_k/v = 3\,\alpha_k/a_k \tag{5.71}$$

and for annular flow in a pipe of radius R, α_g is given by, for vapor phase g,

$$A_g/v = 4\alpha_g^{1/2}/2R \tag{5.72}$$

while other flow configurations fall in between. In an effort to see some generality of relations of A_k/v to flow configurations and parameters, two correlation parameters were tested, using the A_k/v data collected by Ishii and Mishima [1980].

(1) Volumetric rate of flow of the gas past the interface per unit volume of mixture, $(A_g/v)U_g$ in s^{-1}. This rate is large for small interfacial areas per unit volume. For vertical upward flow (Fig. 1.11):

Flow Regimes	Ranges	No. of data points
Bubble flow	$(A_g/v)U_g < 100\ s^{-1}$	
Bubble to slug flow	$100 < (A_g/v)U_g < 2.2 \times 10^3$	(7)
Slug to churn flow	$2.2 \times 10^3 < (A_g/v)U_g < 1.7 \times 10^4$	(19)
Churn to annular	$1.7 \times 10^4 < (A_g/v)U_g < 3 \times 10^4$	(17)
Annular to mist flow	(dependent on α_ℓ of liquid)	

(2) A dimensionless parameter given by A*, for $\alpha_f = 1 - \alpha_g$,

$$A^* = 2R\left(\frac{A_g}{v}\right)\left(\frac{U_g}{\alpha_\ell U_\ell}\right)\left(\frac{\bar{\rho}_g}{\rho_\ell}\right) = 2R\left(\frac{A_g}{\alpha_g v}\right)\left(\frac{\alpha_g \bar{\rho}_g U_g}{\alpha_\ell \bar{\rho}_\ell U_\ell}\right) \tag{5.73}$$

as a measure of transfer of momentum across the interface.

Correlation of the same data gives:

Flow Regimes	Ranges	No. of data points
Bubble flow	$A* < 0.1$	
Bubble to slug flow	$0.1 < A* < 0.2$	(7)
Slug to churn flow	$0.2 < A* < 1$	(19)
Churn to annular	$1 < A* < 10$	(17)

These ranges are distinct, in spite of the spread of numerical values of $A*$. For horizontal pipe flow the transitions were shown to be not so distinct:

Bubble to slug flow	$0.05 < A* < 0.1$
Churn to annular flow	$A* \sim 50$

It is further noted that information on droplets was given only in connection with annular mist flow. Droplet flow may lead to small $A*$ for diffusive motion, but large $A*$ when there is significant streaming of droplets. For the purpose of computation, various correlations for friction and heat transfer are outlined in various handbooks (see, for instance, Hetsroni [1982]). Other elaborations include effects of chemical reaction and radiation, modifications in the generation and heat source terms are readily made.

5.5 Unsteady flow

Studies in connection to nuclear reactor safety led to investigations on unsteady multiphase flow or two-phase flow. One readily sees from Sec. 2.2 and 2.7 that, in the case of pure stratified flow, inertia force is not transferred between the phases (interface momentum transfer integral equals to zero) and hence wave interaction is absent. This is not the case for wavy stratified flow (Prob. 2.10). This inertial coupling becomes significant when one phase is dispersed in the other, whence there is a common speed of sound (Sec. 4.2). These physical conditions exhibit mathematical significance when using the method of characteristics (see, for

instance, Shapiro [1953]) to solve problems of one-dimensional
unsteady motion of a two-phase system.

For the case of incompressible fluids of phases 1 and
2, $\rho_1 = \alpha\bar{\rho}_1$, $\rho_2 = (1 - \alpha)\bar{\rho}_2$, and $\rho_2 \gg \rho_1$, the continuity and
momentum equations of phases (Sec. 2.7) can be expressed in general
forms:

$$\bar{\rho}_1 \frac{\partial \alpha}{\partial t} + \bar{\rho}_1 U_1 \frac{\partial \alpha}{\partial x} + \bar{\rho}_1 \alpha \frac{\partial U_1}{\partial x} = \Gamma \tag{5.74}$$

$$-\bar{\rho}_2 \frac{\partial \alpha}{\partial t} - \bar{\rho}_2 U_2 \frac{\partial \alpha}{\partial x} + \bar{\rho}_2(1-\alpha) \frac{\partial U_2}{\partial x} = -\Gamma \tag{5.75}$$

$$\bar{\rho}_1 \alpha[1 + \beta \frac{\bar{\rho}_2}{\bar{\rho}_1}] \frac{\partial U_1}{\partial t} + \bar{\rho}_1 \alpha[1 + \beta \frac{\bar{\rho}_2}{\bar{\rho}_1}] U_1 \frac{\partial U_1}{\partial x}$$

$$- \beta \bar{\rho}_1 \alpha \frac{\bar{\rho}_2}{\bar{\rho}_1} U_1 \frac{\partial U_2}{\partial x} + \omega_1 \alpha \frac{\partial P}{\partial x} + \omega_2 P \frac{\partial \alpha}{\partial x}$$

$$= - (U_1 - U_2) \Gamma + \rho_1 \alpha F(U_2 - U_1) \tag{5.76}$$

$$\bar{\rho}_2 (1-\alpha) \frac{\partial U_2}{\partial t} + (1-\omega_1\alpha) \frac{\partial P}{\partial x} - \omega_2 P \frac{\partial \alpha}{\partial x}$$

$$+ \bar{\rho}_2(1-\alpha) U_2 \frac{\partial U_2}{\partial x} = - \bar{\rho}_1 \alpha F(U_1 - U_2) \tag{5.77}$$

where P_k is retained for the sake of generality and β is a coefficient introduced for the virtual mass force ($1/2 < \beta < 1$, Eq.
2.107) which physically accounts for the inertial interactions between the phases; ω_1 and ω_2 are coefficients related to the following
phase configurations: molecular mixture, $\beta = 0$, $\omega_1 = \omega_2 = 1$; dilute
dispersed system, $1/2 < \beta < 1$, $\omega_1 = \omega_2 = 0$; and pure stratified flow
system, $\beta = 0$, $\omega_1 = 1$, $\omega_2 = 0$. Equations (5.74) to (5.77) can be
expressed in terms of the operator, $A\, \mathbf{U}_t + B\, \mathbf{U}_x = f$ where

$$\mathbf{U} \;=\; \begin{vmatrix} \alpha \\ P \\ U_1 \\ U_2 \end{vmatrix} \tag{5.78}$$

and f includes all the inhomogeneous terms. The normals (n_1, n_2) to
the characteristics are given by the determinant

$$(An_1 + Bn_2) = 0 \tag{5.79}$$

which gives

$$-\,\bar{\rho}_1^2\,\bar{\rho}_2\,\alpha n_2^2 [\,(n_1 + U_1 n_2)^2 (1-\alpha)(1-\omega_2\alpha)\left(1+\beta\,\frac{\bar{\rho}_2}{\bar{\rho}_1}\right) + (n_1 + U_2 n_2)^2\,\frac{\bar{\rho}_2}{\bar{\rho}_1}\,\alpha\omega_1(1-\alpha)$$

$$+\,\beta(1-\omega_1\alpha)\,U_1 n_2\,(n_1 + U_2 n_2)\,\alpha\,\frac{\bar{\rho}_2}{\bar{\rho}_1} - (1-\alpha)\omega_2\,\frac{P}{\bar{\rho}_1}\,n_2^2\,] = 0 \tag{5.80}$$

Eq. (5.80) gives four characteristic curves. Two have normals to n_2
$= 0$, or $\mathbf{n} = (1, 0)$, and $\mathbf{n} = (1, 0)$.   The normals (n_1, n_2) to the
characteristics are determined by arranging Eq. (5.80) in the form:

$$a\,n_1^2 + b\,n_1 n_2 + c\,n_2^2 = 0 \tag{5.81}$$

n_1/n_2 has real roots and the characteristics are real
for $b^2 - 4ac \geq 0$.

Case 1: In the case of a molecular mixture, the term in Eq. (5.80)
gives, besides $n_2^2 = 0$ for $\beta = 0$, $\omega_1 = \omega_2 = 1$

$$n_1 = n_2\,[-\,U_2 \pm (P/\alpha\bar{\rho}_2)^{\frac{1}{2}}\,] \tag{5.82}$$

which is always real, $(P/\alpha\bar{\rho}_2)^{\frac{1}{2}}$ is the wave velocity.

Case 2: For a dilute dispersed system, $\beta > 1/2$, $\omega_1 = \omega_2 = 0$, Eq. (5.80) reduces to:

$$n_1^2(1-\alpha) - n_1 n_2 U_1(2-\alpha-\alpha\kappa_1) -$$

$$n_2^2[(1-\alpha)U_1 + \alpha U_2 - \alpha\kappa_1 U_2]U_1 = 0$$

where $\kappa_1 = [1 + \beta(\bar{\rho}_2/\bar{\rho}_1)]^{-1}$, giving real characteristics. (Prob. 5.10.) A limiting case is for $\beta = 0$, where the motion of phase 1 is due to the drag force exerted by phase 2 only (Sec. 5.6).

Case 3: When there is no inertial coupling while the pressure gradients exist in both phases, Eq. (5.81) becomes

$$(n_1 + U_1 n_2)^2 + (n_1 + U_2 n_2)^2 \frac{\alpha}{1-\alpha} \left(\frac{\bar{\rho}_2}{\bar{\rho}_1}\right) = 0 \qquad (5.83)$$

The characteristics are complex, which physically means that the common set of characteristics does not exist in the case of pure stratified flow where each phase has its own set of characteristics and speed of sound. The corresponding interface momentum transfer integral to the above virtual mass force term remains to be determined for various flow regimes, an empirical correlation shows different degrees of inertial interactions between phases ranging from highly dispersed flow to pure stratified flow [Soo 1980].

5.6 Shock Waves in Dusty Gas

The nature of passage of a dusty gas through a shock wave is interesting when determining losses due to over expansion of a suspension in a nozzle or the strength of a nuclear blast. For the case of a standing shock wave, similar sets of basic equations as in one-dimensional nozzle flow of a suspension apply to normal shock except that the continuity equation is replaced by [Kriebel 1963]:

$$(\rho_1 + \rho_{p1})\, U_1 = (\rho_\infty + \rho_{p\infty})\, U_\infty \qquad (5.84)$$

for the upstream of shock at condition 1 and far downstream when equilibrium is again reached at condition with subscript. The effect of large Reynolds number due to relative motion was accounted for in the time constants for momentum and heat transfer to particles. Typical results of machine computation are shown in Fig. 5.13 for upstream Mach number of 2.0 and mass-flow ratio of particle to gas of 0.4, mean particle radius of 1.5 μm, c = 2.51 kJ/kg K, c_p = 1.42 kJ/kg K, $\bar{\kappa}$ = 0.2735 W/m K, $\bar{\mu}$ = 6.41 x 10^{-5} kg/m s, $\bar{\rho}_p$ = 3850 kg/m^3, for a distribution in particle sizes represented by a/a_m (subscript m for the mean radius) of 1/3, 1 and 5/3 (subscripts 1, 2, 3 in Fig. 5.13). Fig. 5.13 shows the nature of relaxation of particles after traversing through the shock front. Realistic modifications of these results should include the irreversibility in the conversion of kinetic energy of particles to static pressure (Eq. 3.19); a reduced pressure rise in the shock process is expected. (Prob. 5.11). An analysis of reactive shock waves in a carbon particle-oxygen mixture was given by Elperin et al. [1986].

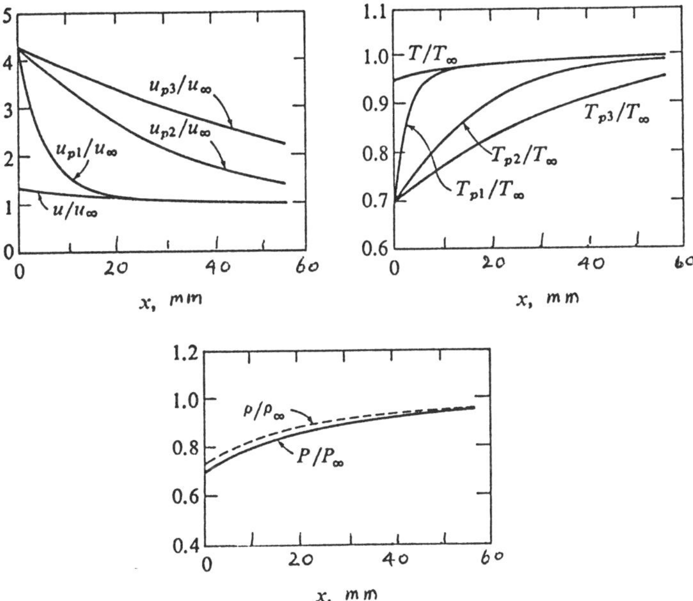

Figure 5.13 Typical shock structure of dusy gas [Kriebel 1963]

<u>Method of characteristics.</u>

Deviation of particle flux line and gas trajectories in the x, t coordinates is expected in one-dimensional unsteady motion of a suspension since particles cannot follow the rapid changes in velocity and temperature that are produced by wave motion. Additional relaxation processes are expected. Extension of the method of characteristics (see, for instance, Shapiro [1953]), to a suspension provides additional insight into the basic interactions as well as to applications such as blast waves in a dusty gas [Rudinger & Chang 1964].

When applied to one-dimensional unsteady motion, the basic equations can be rewritten in the form:

$$a_g \frac{\partial U}{\partial x} + 2 \frac{da_g}{dt} + \frac{a_g}{\rho} \frac{dP}{dt} = - \frac{a_g}{A} U \frac{\partial A}{\partial x} \tag{5.85}$$

$$\frac{d\rho_p}{dt_p} = - \rho_p \frac{\partial U_p}{\partial x} - \frac{\rho_p U_p}{A} \frac{\partial A}{\partial x} \tag{5.86}$$

$$\rho \frac{dU}{dt} + \rho_p \frac{dU_p}{dt_p} + \frac{\partial P}{\partial x} = \rho X + \rho_p X_p \tag{5.87}$$

$$\frac{dU_p}{dt_p} = F(U - U_p) + X_p \tag{5.88}$$

$$\rho(U \frac{dU}{dt} + \frac{\alpha}{\gamma-1} a_g \frac{da_g}{dt}) - \frac{\partial P}{\partial t} + \rho_p (U_p \frac{dU_p}{dt_p} + c_p \frac{dT_p}{dt_p})$$

$$+ \rho(q + UX) + \rho_p(q_p + U_p X_p) \tag{5.89}$$

$$\frac{dT_p}{dt_p} = G(T - T_p) + \frac{q_p}{c_p} \tag{5.90}$$

where A is the flow area of the duct, x is the coordinate along the duct, a_g is the speed of sound based on the gas phase only as an approximation; X, X_p are force per unit mass acting on the gas the particles respectively; q, q_p are heat added per unit mass of gas and the particles respectively, γ is the ratio of specific heats, $\overline{R}$ is the gas constant per unit mass of gas; c, c_p are the specific heats at constant pressure of the gas and the particles. Eqs. (5.86), (5.88) and (5.90) are compatibility conditions for $dx/dt = U_p$ along the particle flux line together with T_p, ρ_p. The entropy s is defined along the gas trajectories $dx/dt = U$,

$$\frac{ds}{dt} = \frac{1}{T} \{[F(U - U_p)^2 - G(T - T_p)] \frac{\rho_p}{\rho} + q\} \qquad (5.91)$$

For the method of characteristics, the wave equations are:

$$\frac{dx}{dt}\Big|_Q^P = U \pm a_g \qquad (5.92)$$

for the P (right travelling) waves (upper sign) and the Q (left travelling) waves (lower sign) (Fig. 5.14) and the state equations take the form: (Prob. 5.12).

$$dU\Big|_Q^P = \mp \frac{2}{\gamma-1} da_g + [\mp \frac{a_g U}{A} \frac{\partial A}{\partial x} \pm X \mp \frac{\rho_p}{\rho} \frac{dU_p}{dt}$$
$$+ \frac{a_g^2}{\gamma R} \frac{\partial s}{\partial x} + \frac{a_g}{R} \frac{ds}{dt}] \, dt \qquad (5.93)$$

Solution of a given problem can be obtained from solving Eqs. (5.86), (5.88), and (5.90) to (5.93), using numerical method of characteristics. Two cases were computed:

Case A. Centered expansion wave. A constant area duct is initially filled with a suspension of glass particles of $2a = 10$ μm or 4 μm in air ($\gamma = 1.40$, $c_p/c = 1.125$) at atmospheric conditions, initially $\rho_{po}/\rho_o = 0.3$. One end of the duct is suddenly opened to a vacuum; a centered expansion wave is created. Initially, the characteristics of the expansion wave form a fan of straight lines as a pure gas in frozen flow because the particles have yet to respond to the change. Subsequently, interaction between the phases makes the characteristics curved. A few P waves are shown in Fig. 5.14 for 10 μm particles. Two of the crossing Q waves are shown in dotted lines. As denoted by the shaded triangle, the condition at N is computed from known conditions at A and B (obtained previously) via intersecting waves by Eq. (5.92) and states of intersecting waves according to Eq. (5.93) with terms in the latter given by Eqs. (5.86), (5.88), (5.90), and (5.91).   The particle and gas trajectories starting from a common point is shown in Fig. 5.14, The dashed line denoting wave and gas trajectory at equilibrium flow was given for:

$$a_{ge}^2 = a_{go}^2 (1 + m_p^\star \frac{c_p}{c}) \; (1 + m_p^\star)^{-1} \; (1 + \gamma \, m_p^\star \frac{c_p}{c})^{-1} \tag{5.94}$$

where $m_p^\star = \rho_{po}/\rho_o$ is the initial mass ratio of particles to gas, a $= \gamma \, \overline{R}T$, and $T_p = T$, $U_p = U$ at equilibrium.   For the example in Fig. 5.14, T_p - T up to 25 C was obtained for an initial temperature of 300 K, a_{ge}/a_{go} = 0.836.   Note that a_g in Eq. (5.85) is a matter of

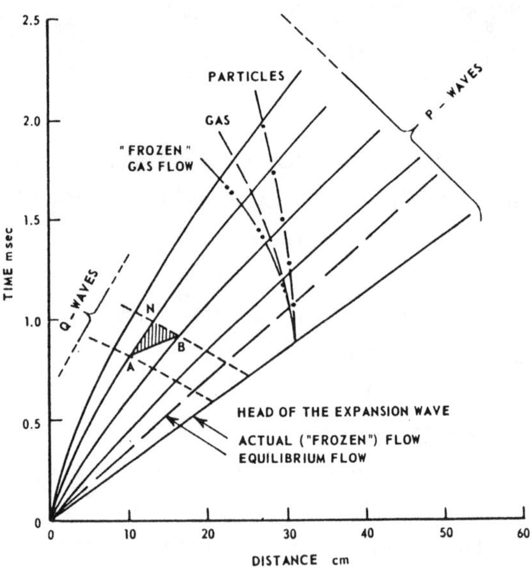

Figure 5.14 Some characteristics of the centered expansion wave
            that is produced if a suspension of spherical glass
            particles of 10 μm diameter in air is suddenly
            discharged into a vacuum. The initial mass-loading
            ratio is 0.3 [Rudinger and Chang 1964]

definition to eliminate ρ in the continuity equation of the gas phase by substituting $a_g^2 = \gamma \overline{R} T$ and $\rho = P/\overline{R}T$, but a_g in Eq. (5.91) and (5.92) should have been the actual speed of sound as given by Eq. (5.9) to respond to the local ρ_p, T, T_p, U, U_p, and P.

Case B.   Shock wave generated by an impulsively driven piston.   A semi-infinite, constant area duct is filled initially with a suspension as in Case A.   A piston at one end is impulsively accelerated to a constant velocity at 0.442 a_{go} giving an initial velocity of shock wave of 1.3 a_{go}.   This shock wave is instantaneously formed, producing a gas velocity equal to the piston velocity.   Thereafter, the interaction between the phases modifies the gas flow and the fact that gas velocity at the piston must be equal to the piston velocity leads to a change in shock strength which decreases to 1.2 a_{go} after travelling 100 cm.   The waves are illustrated in Fig. 5.15.   Particles overtaken by the piston are assumed lost rather than trying to deal with the collision process. The particle and gas trajectories in Fig. 5.14 illustrates the lag in

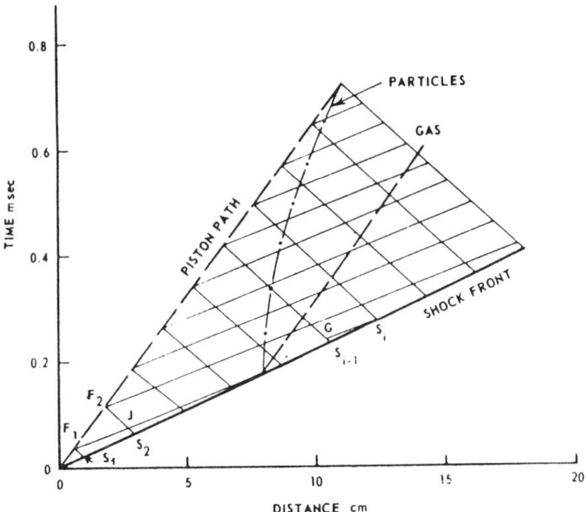

Figure 5.15 Shock wave produced by an impulsively accelerated piston moving in a suspension of spherical glass particles of 10 μm diameter in air at an initial mass-loading ratio of 0.3. Initial shock velocity is 1.3 a_o [Rudinger and Chang 1964]

particle motion. Similar comments on the magnitude of speed of sound applies here.

<u>Shock wave in a gas-liquid system</u>. Using an air bubble-water system, Eddington [1970] produced a mixture with speed of sound as low as 20 m/s. Normal and oblique shocks were obtained at velocities below 100 m/s. The range of experiments covered highly dispersed droplet, bubble, and froth flow regimes. Measured isentropic stagnation pressure recovery follows mixture conservation relation. This again confirms the existence of characteristics in an inertially coupled system.

5.7 Settling in an External Field.

Batch settling of an initially homogeneous and quiescent suspension in the gravitational field can be formulated according to Sec. 2.7, neglecting diffusion and thermal effects:

$$\partial \rho / \partial t + \partial \, \rho W / \partial z = 0 \tag{5.95}$$

$$\partial \rho_p / \partial t = \partial \, \rho_p \, W_p / \partial z = 0 \tag{5.96}$$

$$\rho_p \frac{\partial W_p}{\partial t} + \rho_p \, W_p \frac{\partial W_p}{\partial z} = - \rho_p g + \rho_p F(W - W_p) - \frac{\rho_p}{\rho_p} \frac{\partial P}{\partial z} \tag{5.97}$$

$$\rho \frac{\partial W}{\partial t} + \rho W \frac{\partial W}{\partial z} + \rho_p \frac{\partial W_p}{\partial t} + \rho_p \frac{\partial W_p}{\partial z} = - \frac{\partial P}{\partial z} - (\rho_p + \rho) \, g \tag{5.98}$$

Eq. (5.98) being the momentum equation of the mixture; with velocities W, W_p and coordinate z in the opposite direction of gravitational acceleration g. These equations apply to the case of solid particles of uniform size and we have neglected viscous forces due to dilatation and the virtual mass force. We further take the case where the fluid is incompressible such that

$$\rho = \bar{\rho}\,\epsilon, \quad \rho_p = \bar{\rho}_p(1 - \epsilon) = \bar{\rho}_p\,\alpha \tag{5.99}$$

ϵ being the fraction void.

It is seen that Eqs. (5.95) and (5.96) account for the fact that, as solids of finite volume settle, the fluid is displaced upward. Eq. (5.97) accounts for the forces acting on the solid particles due to buoyancy, fluid drag, and pressure gradient in the gravitational field. Eqs. (5.95) and (5.96) with Eq. (5.99) give:

$$W_p = -W\epsilon/(1-\epsilon) \tag{5.100}$$

meaning that, where the volume occupied by the particles counts, as the particles settle downward, the fluid moves upward. For this case, where the volume fraction of particles becomes large as the particles settle, the inverse relaxation time F for momentum transfer from fluid to particle takes the form given by Eq. (3.3), whose last term can be neglected because of slow motion.

For a complete solution Eq. (5.96) can be expressed in terms of ϵ. A second equation can be obtained by reducing Eqs. (5.97) and (5.98) with elimination of P and substitution of Eq. (5.100) to give an equation in terms of W_p and ϵ. (Prob. 5.13) It can be reduced to a simple form for the case $\bar{\rho}_p \gg \bar{\rho}$ and when $F/\epsilon^2 = F'$ constant, and inelastic collision of particles with the bottom of the vessel and settled particles:

$$\frac{\partial W_p}{\partial t} + W_p \frac{\partial W_p}{\partial z} = -g - F'W_p = \frac{dW_p}{dt} \tag{5.101}$$

For an initially (t = 0) homogeneous suspension of height z and volume fraction solid $\alpha_0 = 1 - \epsilon_0$, and complete settling to α_f at some time later, the height of solids then extend to $z_f = z_0\alpha_0/\alpha_f$. The settling velocity W_p and fluid velocity W are given by:

$$-\frac{W_p F'}{g} = (1-e^{-F't}) = \frac{WF'(1-\alpha_0)}{\alpha_0 g} \tag{5.102}$$

and the position of a particle starting at z_{01} at $t = 0$ is given by:

$$\frac{z}{z_0} = \frac{z_{01}}{z_0} - (\frac{g}{F'z_0})t + \frac{g}{F'^2 z_0}(1 - e^{-F't}) \tag{5.103}$$

These relations are shown in Fig. 5.16 for $\alpha_0 = 0.2$, $\alpha_f = 0.5$, $g/F'z_0 = 1$, and $F' = 10\ s^{-1}$. In this simple case the area AOB in Fig. 5.16 retains a constant α of 0.20; in this area the particles settle downward at velocity W_p while the fluid is displaced upward at velocity W. Above ABC we have clean fluid and $W = 0$, below ABC the particles settle to $\alpha = 0.50$ and $W = W_p = 0$. The time for complete settling is 0.694 s. This diagram shows that at $t = 0.5$, for instance, the top 40 % of the overall height of the column is filled with clean fluid, below which we have 16.5% of the column of a suspension of $\alpha = 0.2$, and the bottom 13.5% is filled with a settled bed of $\alpha = 0.5$. This illustrates the simplest nontrivial situation. In reality, F as given by Eq. (3.3), scattering motion of particles, and diffusion effects must be accounted for in a numerical solution. Empirical correlations were given by Lapidus and Elgin [1957] for solid-liquid systems, by Wallis [1961] for gas-liquid systems, Pratt et al. [1953] for liquid-liquid systems, and Miles et al. [1945] for foams.

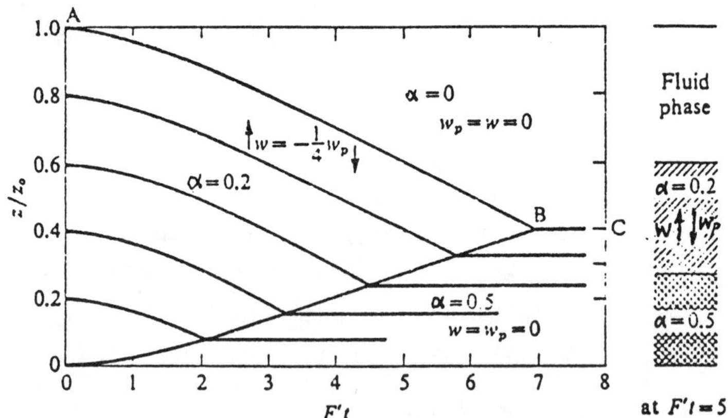

Figure 5.16 Simple sedimentation, $F' = F/\epsilon^2 = 10$, $\alpha_0 = 0.2$, $\alpha_f = 0.5$, $g/F'z_0 = 1$.

Note that Eqs. (5.95) to (5.98) give imaginary characteristics according to Case 3 of Sec. 5.5. The elliptic nature of the differential equations is physically represented by the exponential decay of the motion without wave behavior.

5.8 Settling of Charged Dusts.

A case of one-dimensional unsteady motion is the interaction of gravitational and electric fields in the settling of charged aerosols. The simplified system is defined in Fig. 5.17, which shows a uniform non-reactive gaseous suspension of charged aerosols of uniform size (radius a, mass m), charge q, and uniform cloud density ρ_{po} is suddenly brought between two grounded conductor plates of infinite extent at distance z_o apart normal to the direction of gravity of acceleration g. In this case, settling will take place at the surface at both the upper and the lower plates, but at different rates [Soo 1970]. Changes in the system with time are computed with the assumption of a dilute suspension such that the motion of the gas phase is neglected. The continuity and momentum equations in Sec. 2.7 give:

$$\partial \rho_p / \partial t + \partial \rho_p W_p / \partial z = 0 \qquad\qquad (5.104)$$

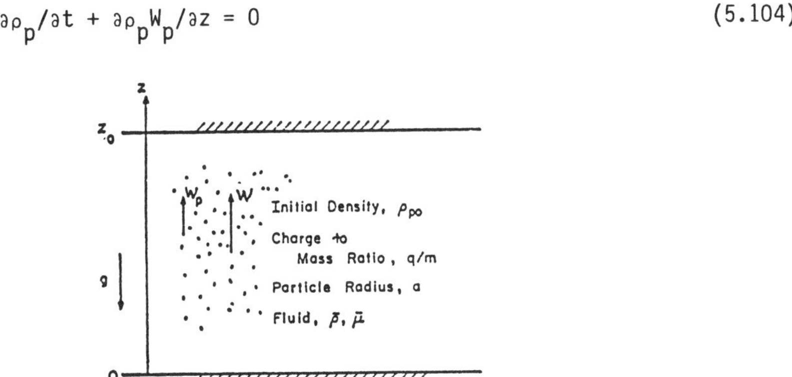

Figure 5.17 Settling of a uniformly charged dust suspension
between two grounded plates at distance z_o apart.

$$(\partial W_p/\partial t) + W_p(\partial W_p/\partial z) = -g + (q/m)\, E_z + F(-W_p) \tag{5.105}$$

where W_p is the particle velocity. The electric field E_z is given by writing the Poisson equation in the form:

$$\partial E_z/\partial z = \epsilon_o^{-1}\, (q/m)\, \rho_p \tag{5.106}$$

The initial conditions are:

$$t = 0, \quad \rho_p = \rho_{po}, \quad W_p = 0 \tag{5.107}$$

$$\partial W_p/\partial t = -g$$

The solution of these equations is facilitated by applying the Mises transformation with the introduction of a "stream function" ψ such that

$$\rho_p = -\partial \psi_p/\partial z, \quad \rho_p W_p = \partial \psi_p/\partial t \tag{5.108}$$

whereby Eq. (5.104) is satisfied. Changing the independent variables from z and t to ψ_p and t, Eqs. (5.105) and (5.106) become

$$(\partial/\partial t)\,[W_p\, \exp(Ft)] = -g\, \exp(Ft) + (q/m)\, E_z\, \exp(Ft) \tag{5.109}$$

and

$$\partial E_z/\partial \psi_p = -\epsilon_o^{-1}(q/m) \tag{5.110}$$

Elimination of E_z between Eqs. (5.109) and (5.110) gives

$$(\partial^2/\partial \psi_p\, \partial t)\,[W_p\, \exp(Ft)] = -\epsilon_o^{-1}(q/m)^2\, \exp(Ft) \tag{5.111}$$

with the initial condition changed to, with $\psi_p = 0$ at $z = z_0/2$,

$$t = 0, \quad \psi_p = \rho_{po}\left(\frac{z_0}{2} - z\right), \quad W_p = 0, \quad \left(\frac{\partial W_p}{\partial t}\right)_{\psi_p} = -g \tag{5.112}$$

Solution of Eq. (5.111) subjecting to Eq. (5.112) gives

$$-W_p = [(q/m)^2/\epsilon_o F] \ \psi_p \ [1 - \exp(-Ft)] + gt \ \exp(-Ft) \qquad (5.113)$$

or, in dimensionless form, with $Ft = t^*$, $(\psi_p/\rho_{po} \ z_o) = \psi_p^*$,

$$-(W_p/Fz_o) = - W_p^* = (N_{ev})[1-\exp(-t^*)]\psi_p^* + (N_m)_g t^* \exp(-t^*) \qquad (5.114)$$

with dimensionless quantities of electroviscous number

$$N_{ev} = \rho_{po}(q/m)^2/\epsilon_o F^2 \qquad (5.115)$$

and the gas-particle momentum transfer number in the gravitational field

$$(N_m)_g = g/z_o \ F^2 \qquad (5.116)$$

The predominance of the influence of either the electrostatic effect or the gravity effect is seen via the relative magnitude of (N_{ev}) and $(N_m)_g$, or whether the ratio

$$(N_{ev})/(N_m)_g = \rho_{po}(q/m)^2 \ z_o/\epsilon_o \ g \qquad (5.117)$$

is much greater or much smaller than 1. Both effects could be significant in the case of atmospheric dust (Prob. 5.14).

To determine and ρ_p and ψ_p, we substitute Eq. (5.114) with Eq. (5.108) into Eq. (5.104) to get, for $\rho_p^* = \rho_p/\rho_{po}$,

$$(\partial\rho_p^*/\partial t^*) + (N_{ev}) \ [1 - \exp(-t^*)](\partial/\partial z^*)(\rho_p^* \int \rho_p^* \ dz^*)$$

$$- (N_m)_g \ t^* \ \exp(-t^*) \ (\partial\rho_p^*/\partial z^*) = 0 \qquad (5.118)$$

with $z^* = z/z_o$. In Eq. (5.118), the second term gives the change in density due to space charge effect, and the third term gives the change in density due to gravity effect. Eq. (5.118) is not readily solved analytically, but some understanding of the problem can be gained by considering a few limiting conditions.

(a) When the particles are not charged, $(q/m) = 0$, Eq. (5.118) takes the form

$$\partial \rho_p^*/\partial t^* = (N_m)_g \, t^* \exp(-t^*) \, (\partial \rho_p^*/\partial z^*) \tag{5.119}$$

Its pertinent solution for the given conditions is

$$\rho_p^* = 1 \quad \text{at } z^* \leq 1 - (N_m)_g \, [1 - (1+t^*) \exp(-t^*)] \tag{5.120}$$

$$\rho_p^* = 0 \quad \text{at } z^* > 1 - (N_m)_g \, [1 - (1+t^*) \exp(-t^*)] \tag{5.121}$$

Noting that at small t^*, Eq. (5.120) simplifies to

$$z^* \leq 1 - (N_m)_g \, t^{*2} \tag{5.122}$$

which accounts for the start of the descent. This compares to

$$z^* \leq 1 - (N_m)_g \, t^* \tag{5.123}$$

which corresponds to the case in Sec. 5.7 of long time batch settling.

(b) For $N_{ev} \gg (N_m)_g$ at sufficiently large t^*, Eq. (5.118) reduces to

$$(\partial \rho_p/\partial t_1^*) + (\partial/\partial z^*) \, (\rho_p \int \rho_p^* \, dy) = 0 \tag{5.124}$$

with t_1^* given by:

$$t_1^* = [t^* + \exp(-t^*)] \, (N_{ev}) \tag{5.125}$$

Solution by series expansion gives:

$$\rho_p^* = (1 + t_1^*)^{-1} + (z^* - \tfrac{1}{2})^2 (1 + t_1^*)^{-4} + \ldots \tag{5.126}$$

$$\psi_p^* = -(z^* - \tfrac{1}{2}) (1+t_1^*)^{-1} - (1/3)(z^* - \tfrac{1}{2})^3 (1+t_1^*)^{-4} + \ldots \tag{5.127}$$

and

$$W_p^* = -(N_{ev}) [1 - \exp(-t_1^*)] \psi_p^* \tag{5.128}$$

as illustrated in Fig. 5.18. The E field is given by Eq. (5.110) whose integration gives

$$E^* = \epsilon_o E/(q/m) \rho_{po} z_o \psi_p^* \tag{5.129}$$

(c) When $N_{ev} \sim (N_m)_g$, both the electrostatic effect and gravity effect are prominent. A complete solution of Eq. (5.118) must embody the above features. Experimental work on settling of dust at both top and bottom of horizontal plates was reported by Owen [1960] showing more collection at the bottom plate than at the top

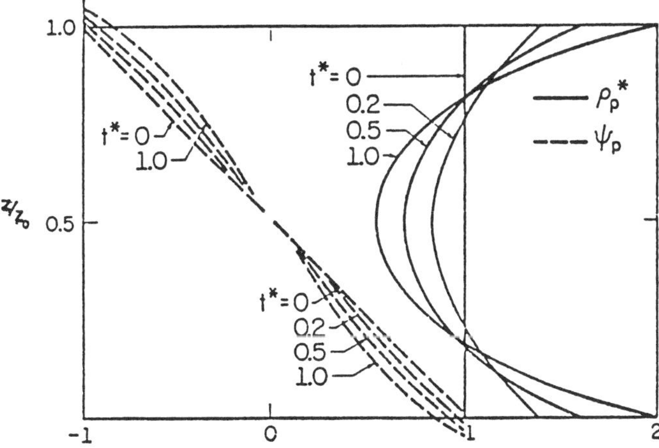

Figure 5.18 Behavior of settling between two walls, $N_{ev} \gg (N_m)_g$.

plates.

An application of the above case is that of electric field in a
rain cloud [Shishkin 1965]. (Prob. 5.15)

Another case of one-dimensional settling is that of atmospheric
fallout. Bannister and Davis [1962] treated the descent of small
particles or heavy molecules through an exponential (density)
atmosphere; both gravitational fall and molecular diffusion were
accounted for. This effect of diffusion will be dealt with in the
case of settling plume in Chapter 7.

EXERCISE PROBLEMS.

5.1   Show, from considerations of relative magnitudes of inverse
      relaxation times F's and G's for a multiphase system, that the
      last two terms of Eq. (5.6) are always greater than zero.

5.2   Show that, for particles suspended in a perfect gas, and $T =
      T_p$, $U = U_p$, and a constant mass ratio m_p^* of solid to gas, the
      sonic speed in a suspension is always less than that in the
      pure gas.

5.3   Determine the relation between total pressure and other
      properties of a suspension for $K_m \sim 0$, and for low
      velocities.

5.4   Derive Eqs. (5.23) and (5.26).

5.5   From Eq. (5.46) derive Eq. (5.48).

5.6   From Eq. (5.23) derive Eq. (5.54).

5.7   Follow the numerical example above Eq. (5.54), compute the
      speed of sound for $T = 300$ K.

5.8   Write down the necessary equations for one-dimensional pipeflow of a gas-solid suspension with wall friction. Compare the number of independent equations and dependent variables [Trezek & Soo 1966].

5.9   Complete the formulation for a two-fluid model of critical vapor-liquid flow in a pipe. Identify the basic equations needed for solution of the problem with initially saturated water. [Ardron 1978]

5.10 Perform an analysis based on method of characteristics as in Sec. 5.5, show that the characteristics are real for a dilute suspension with virtual mass force.

5.11 Apply the one-dimensional equations to the stagnation region of a body in hypersonic flight at velocity U_0 in a dust cloud, noting that immediately behind the shock wave, the relative velocity of the particles is still at U_0. Compute the velocity of impact of the particles. The temperature rise of the gas behind the shock wave is assumed to be $T_s - T_0$, T_0 is the temperature in the free stream; compute the temperature variation of the particles. State the condition in which the particles would have evaporated before reaching the surface. [Soo 1973a]

5.12 Derive Eq. (5.93).

5.13 Derive Eq. (5.101) by eliminating P from Eqs. (5.97) and (5.98) with substitution of Eq. (5.100) and simplify according to the assumptions.

5.14 Compute the ratio of electroviscous and the momentum transfer (gravity) numbers for the case: particle cloud density 0.01 kg/m^3, charge-to-mass ratio 10^{-4} C/kg, cloud thickness 1 m.

Compute each number for the case of particle cloud density 1 kg/m^3, $q/m = 10^{-4}$ C/kg, $a = 5$ μm at room condition or $F = 3000$ s^{-1}, for $z_0 = 1$ m.   Ans.   1, 10^{-4}, 10^{-6}.

5.15   Compute the electric field of a rain cloud of $\rho_p = 0.001$ kg/m^3, $a = 10$ μm, $\bar{\mu} = 1.74 \times 10^{-5}$ kg/m s, the particle charge is given by Eq. (4.81), with an electrokinetic potential (ϕ) of 0.33×10^{-11} C/m, atmospheric conductivity 4×10^{-14} mho/m.   This occurs because of charge separation in the rain cloud.

Ans. 10^3 V/m.

Chapter 6

PIPE FLOW OF A SUSPENSION

Fully developed pipe flow has been a favorite configuration in the studies of fluid dynamics because of the absence of the nonlinear inertial terms in the momentum equation by its definition (see, for instance, Schlichting [1979]). Such a simplification facilitates solutions for various boundary conditions which further the understanding of the basic interactions. Measurements and correlation of momentum transfer in fully developed pipe flow are useful because of applications in estimating the power requirements for transferring fluids or suspensions over a distance. Applications are also found in tubular heat exchangers.

In this chapter, we shall deal with fundamentals concerning interactions in dilute suspensions, leaving the treatment of dense systems to Chapter 8. Also we shall leave out the design aspects of pipe flow transport of particulates, where the task consists of determining empirically the minimum transport velocity to minimize the power consumption, followed by determining the pipe diameter, the empirical friction factor, and the needed power input. These procedures are given in various handbooks (see, for instance, Hetsroni [1982], and Cheremisinoff and Gupta [1983]).

Extensive experimental studies predated the formulation of a dilute suspension (defined in Sec. 2.7 and 3.4). Gravity was shown to be important in the flow of a suspension of sand in water by Newitt et al. [1962]. In the case of gas-particle suspension in pipe flow, the particle charge effect is significant while the gravity effect may be small for high velocity flow in small pipes [Soo et al. 1964, Soo and Trezek 1966].

6.1 Experimental Studies on Dilute Suspensions.

While the above section has indicated that the case of pipe flow of a dilute suspension is readily formulated based on Chapter 2 and the transport properties were given in subsequent chapters, evolution actually has taken a different path. It was measurements on pipeflow that provided the suggestions and validations for much of the concepts which are given in the following sections. Chronologically, experimental results on the velocity and density distributions of the phases were obtained first (for example, Soo et al. [1964]) while their prediction came later (for example, Soo [1969]).

Several reviews on measurement techniques on multiphase flow were given in Hetsroni [1982] and an extensive survey was given in Soo [1982] according to mechanical, optical, nuclear, acoustical, and electrical methods. The averaged density, velocity, and mass flow of, for instance, of a particulate suspension have a triangular relation, namely, determination of any two of these three quantities will give the third.

<u>Velocity of particle phase</u>. Photographic methods for determining the mean velocity of particles in pipe flow are illustrated by Chandok and Pei [1971]. They used the photographic technique with three flashes ("dot-streak-dot") to obtain velocities of particles of glass (150 μm to 500 μm diameters) in a 10 cm. diameter pipe with air flow at up to 20 m/s mean velocity. A special design using fiber optics to collimate the light from three flash units gave a single illumination slit. Both two-dimensional photography and stereophotogrammetry were used to evaluate the axial, lateral, and transverse velocity components of particle motion. The latter method also gave particle concentrations. Their measured particle velocity profile in vertical pipe flow is shown in Fig. 6.1, obtained from a least square fit to a power law.

<u>Mass flow</u>. All sampling techniques extracting particles at isokinetic condition (similar inlet fluid velocity to the probe as the local fluid velocity in the system) give quantities which are in

direct proportion to mass flow. The electrostatic probe (Sec. 3.3) [Cheng and Soo 1970], and the isokinetic sampling system [Mitchell and Engdahl 1963], for instance, give mass flow of particulate suspensions.

The isokinetic sampling system serves as a convenient primary standard. The local flow velocity of the fluid in a particulate suspension is readily determined by a Shapiro probe (Fig. 6.2) originally calibrated by a pitot static probe in a clean gas stream [Cheng et al. 1970, Vaitkus et al. 1971, Perez-Blanco et al. 1980]. Matching of inlet velocity to a sampling probe (Fig. 6.3) is readily carried out with a sampling system as shown in Fig. 6.4, by adjusting needle valves 1 and 2 when the switching valve connects the probe through filter 2, the matching filter. Once matched, we take two samples m_{p1} and m_{p2} via thimbles in filter 1 over two different time

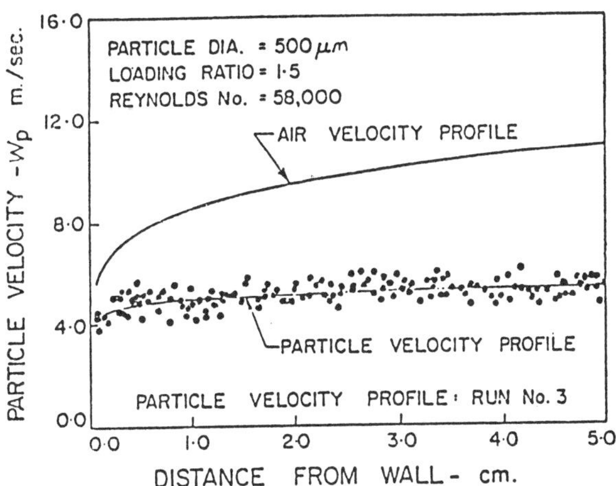

Fig. 6.1   Particle velocity profile from photo-optical method (Glass particles in air, 10 cm. inside diameter of pipe). [Chandok & Pei 1971]

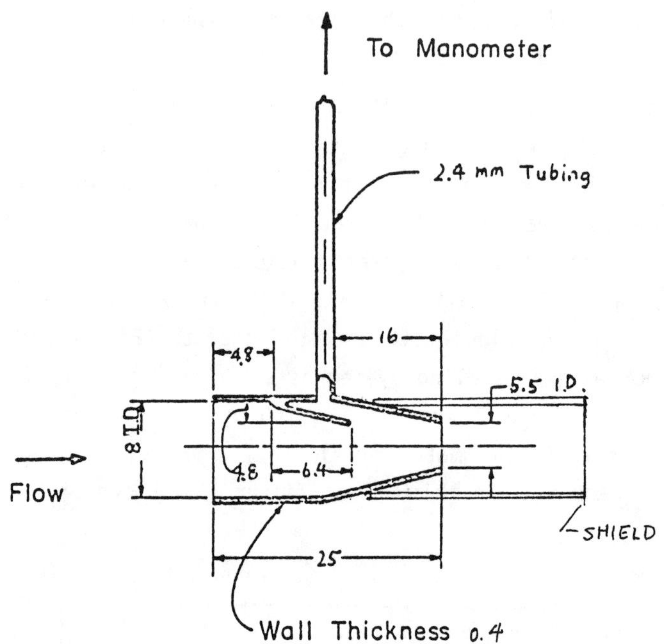

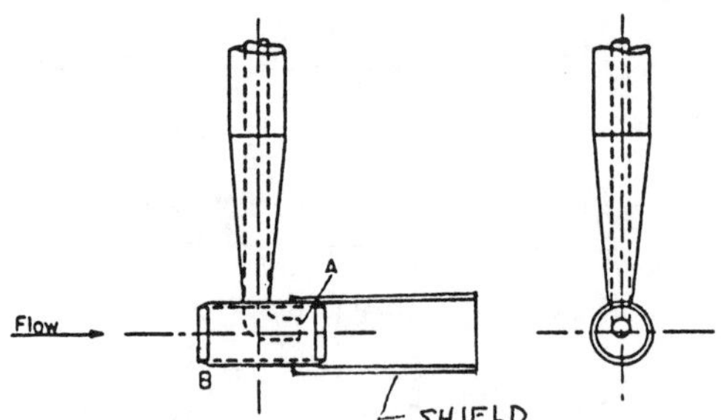

(b)   Inverted Kiel Probe [Perez-Blanco et al. 1980]

Fig. 6.2 Shapiro probe [Cheng et al. 1970, Vaitkus et al. 1971]

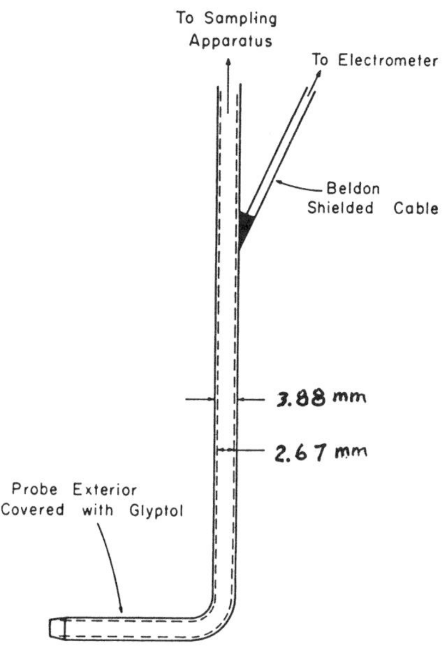

Fig. 6.3 Sampling probe [Soo et al. 1969]

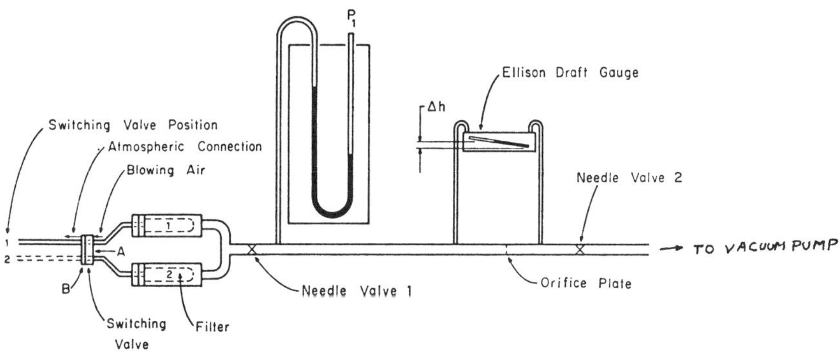

Fig. 6.4 Schematic diagram of sampling apparatus [Soo et al. 1969]

durations Δt_1 and Δt_2 gives:

$$\rho_p U_p = (m_{p1} - m_{p2})/A(\Delta t_1 - \Delta t_2) \qquad (6.1)$$

where A is the effective flow area at the probe [Soo et al. 1969]. The effect of the hold-up volume of the sampling system is thus eliminated.

The electrostatic probe or ball probe (Fig. 6.5) was used earlier [Soo et al. 1964], with the probe theory developed subsequently [Cheng and Soo 1970]. When the ball probe is inserted in a dusty gas, charge transfer occurs by impact. For probe dimensions smaller than the mean free path of the particles, single scattering can be assumed. For given probe potential, the probe current is given by Eq. (3.96).

Particle density. Methods of measurement of local density of particles in a suspension are limited to electrical capacitance [Daniel & Brackett 1951] and optical methods [Soo et al. 1969]. The latter applied the principle of fiber optics. Designs of the optical probes are as shown in Fig. 6.6, utilizing the principle of light attenuation between a light source and a photocell via fiber optics. Calibration of such a probe was made with the isokinetic sampling system at the center of a pipe flow system where the fluid

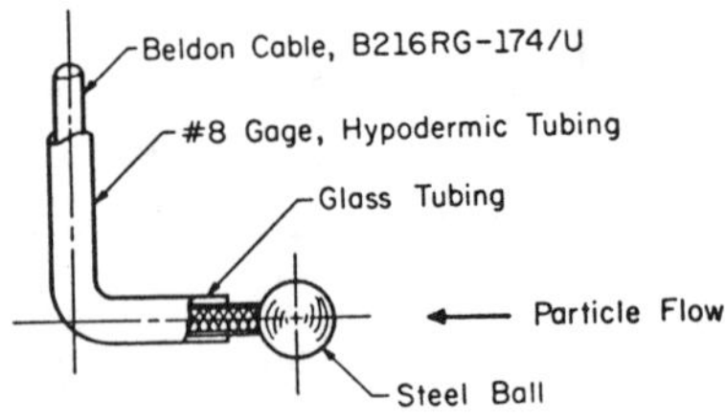

Fig. 6.5 Electrostatic ball probe [Cheng et al. 1970]

velocity and the particle velocity were nearly equal. Techniques based on both scattering and attenuation of light have been used. Measurement of density distribution was made from the total output of the phototube in each case using a bucking circuit. Over a range the probe current is proportional to the local particle density.

Typical results of measurements of particle mass flow by iso-kinetic sampling, particle density by a modified optical probe, and the calculated particle velocity distribution in the vertical plane of horizontal pipe flow are shown in Fig. 6.7 [Ni 1986]. The data in Fig. 6.7 were obtained from measurements made on a suspension of 50 μm alumina in air flowing through a 127 mm diameter pipe. The centerline air velocity was 20.7 m/s (W_o), with a solid-to-gas mass flow ratio of 0.2 kg/kg. Fig. 6.7(a) gives the particle mass flow $\rho_p W_p$, (b) its density ρ_p, (c) its velocity W_p. It is seen that the gravity effect is prominent as well as slip motion of particles at the wall (Sec. 3.5), while a turbulent velocity profiles of air

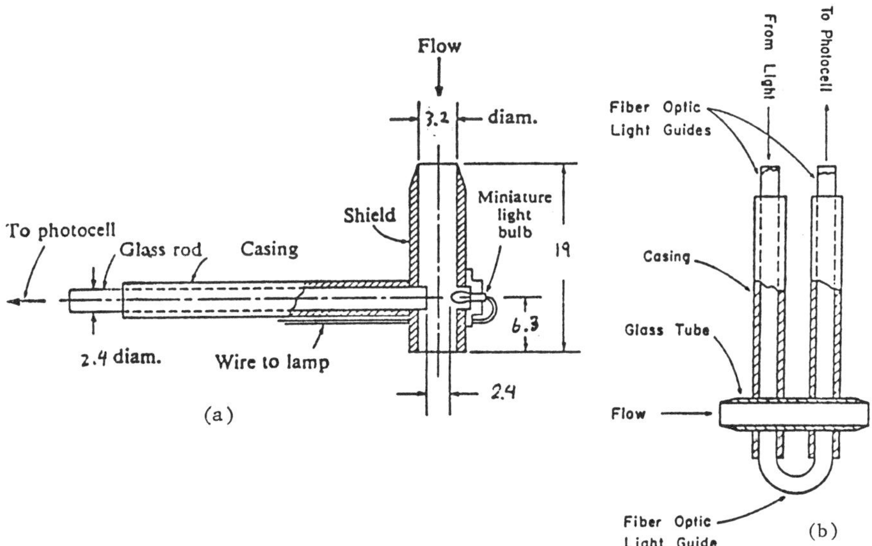

Fig. 6.6 Fiber optic density probe
  (a) Probe with built-in light source (dimensions in mm)
  (b) Two-path riber optic density probe [Soo et al. 1969]

was measured. Earlier measurements showed that the particle velocity
profiles were nearly parabolic [Soo et al. 1964] (dashed line in Fig.
6.7 (c)).

By defining the mass flow ratio of particles to gas as the ratio
of total mass of solid particles conveyed by a gas through the pipe
per unit time to that of the total flow rate of gas in mass per unit
time, $\dot{m}^*$; and the mass ratio based on integrated density of the
phases, m^*; Fig. 6.8 was obtained by Cheng et al. [1970]. The effect
of different velocities of the phases is seen; these two ratios are
identical only for gaseous mixtures. Also shown is that greater lag
of particle phase occurs at a lower air velocity than at a high air
velocity.

<u>Electrostatic charge on particles</u>. Together with mass flow measure-
ments, we can determine the charge to mass ratios of particles. This
can be done with a shielded Belden cable soldered to the sampling
probe with the probe shielded from ground (Fig. 6.3). A current was
measured when the sampling probe discharged to ground through an

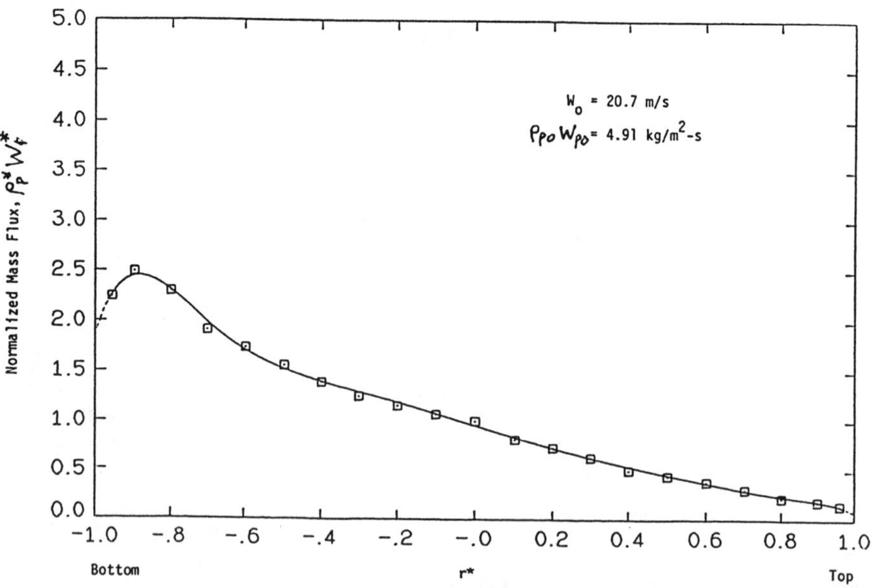

(a) Mass Flux Distribution along Vertical Diameter.

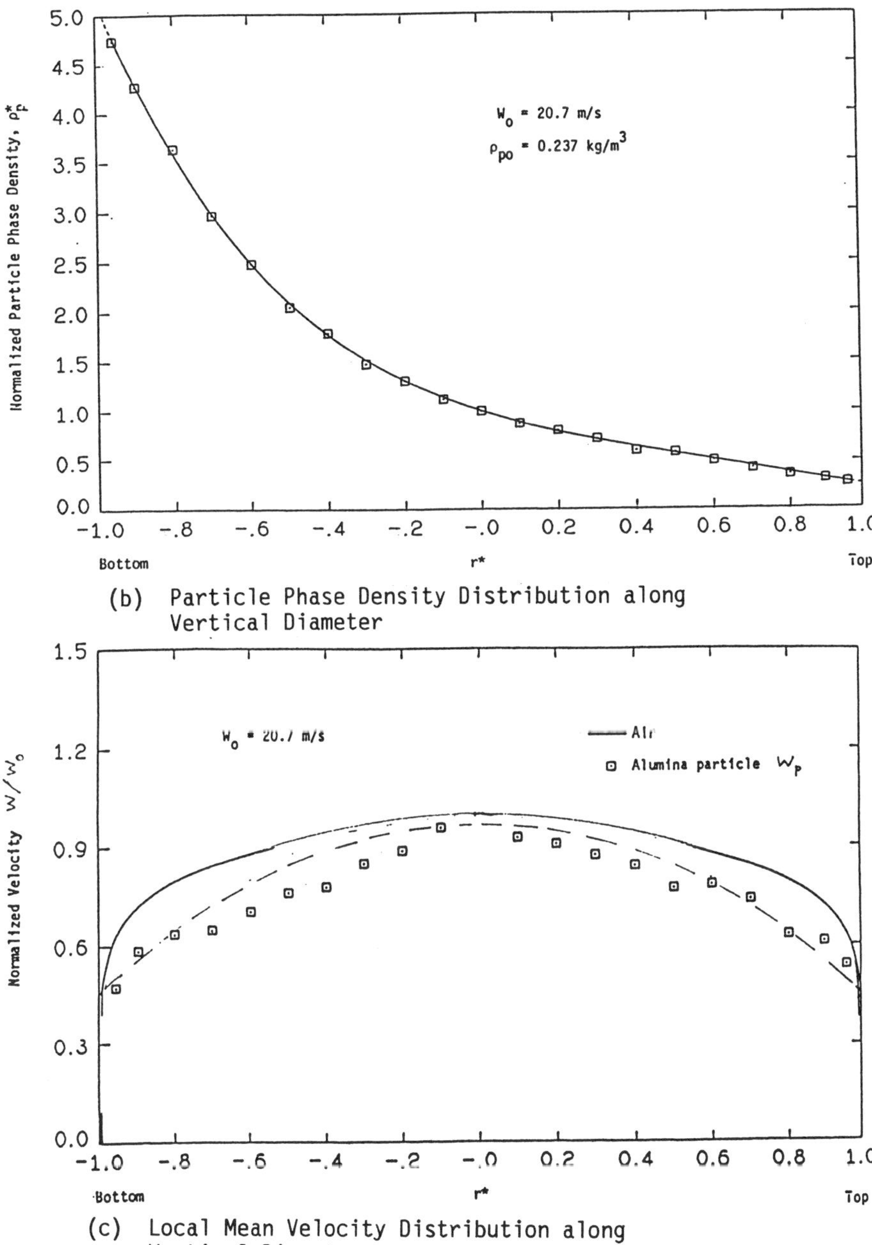

(b) Particle Phase Density Distribution along
     Vertical Diameter

(c) Local Mean Velocity Distribution along
     Vertical Diameter

Fig. 6.7 Measured mass flow and density distributions of particles
          in the vertical plane of fully developed horizontal pipe
          of a suspension. Starred quantities are ratios refer to
          quantities along pipe axis [Ni 1986]

electrometer. Since the mass flow is known, the average charge to
mass ratio (q/m) can be estimated by the probe current in amp. by
the mass collected per unit time in kg/s to give C/kg. Measurements
were made where the probe potential was floated between ±20 V. The
current under this condition did not change appreciably. Measurement
of particle charge can be made accurately using a Faraday cage [Min
et al. 1963]. Magnitudes of particle charges due to conveyance by
air through a pipe at room condition and below 35 m/s varied from
10^{-5} C/kg to 10^{-3} C/kg. depending on pipe and particle materials
(Sec. 1.8 and Eq. 3.91).

6.2 Basic Relations of Pipe Flow

The above experimental efforts are typical among those which
constitute the basis of our understanding of multiphase flow, and
rigorous computation of velocity and density distributions of the
particle phase became possible after recognizing [Soo 1969]:

(1) In a dilute suspension, transfer of momentum of the particle
phase is by diffusion, that is, its shear stress in pipe flow is
given by Eq. (3.97).

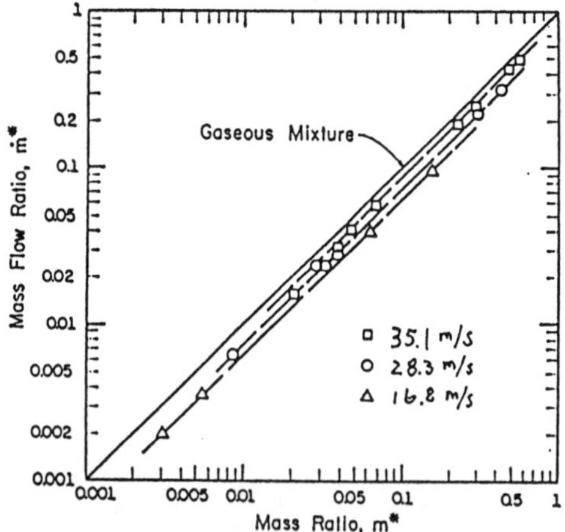

Fig. 6.8 Mass ratio versus mass flow ratio 16 μm coal dust in air.
          [Cheng et al. 1970]

(2) The slip motion is governed by the interaction length (Eq. 1.72), and the slip velocity at the wall is given by Eq. (3.122) for particle radius a much smaller than the pipe radius.

(3) The relation of diffusion under the influence of field forces (fluid flow field, gravity, and electromagnetic fields) (Eqs. 2.123 and 2.124) is applicable.

(4) In a dilute suspension, the motion of the fluid phase is unaffected by the presence of the particle phase. This condition is still valid when deposition of particles occurs when the thickness of the layer of deposit is much smaller than the pipe radius. This further means that the motion of the particle phase corresponds to viscous slip motion and motion of particles are not correlated to each other even when the fluid is turbulent (Sec. 3.5).

(5) The maximum density of the particle phase is that of its packed bed.

Further details will be elaborated later. It suffices to say that these are the basic relations in addition to those in fluid mechanics of a single phase which we need in making rigorous calculations. The basis of the above items are experimental facts.

The present knowledge of the dynamics of suspensions is such that a meaningful set of analytical formulations is limited to the cases of a dilute suspension on the one hand and a sliding bed of solid particle on the other. The reason is that in a dilute suspension, the scale of motion is large but the particle-particle interactions are weak; in a sliding bed, these interactions are strong but the displacements are small.

General relations [Soo & Tung 1972]. Unless otherwise stated, the present formulation deals with a fluid phase in fully developed motion and a particle phase in a general quasi-stationary motion by neglecting time dependence and inertia effect. The flow system consists of a circular pipe with its axis making an angle with the direction of gravity of acceleration g as shown in Fig. 6.9. In Fig. 6.9, r, ϕ, and z are the radial, azimuthal, and axial coordinates; U, V, and W are the conjugate components of velocity of the fluid; and U_p, V_p, and W_p are those of the particle phase. We shall treat only the case of a single species of monodispersed suspension of spherical particles of radius a, although extension to include a distribution in particle size is readily accomplished.

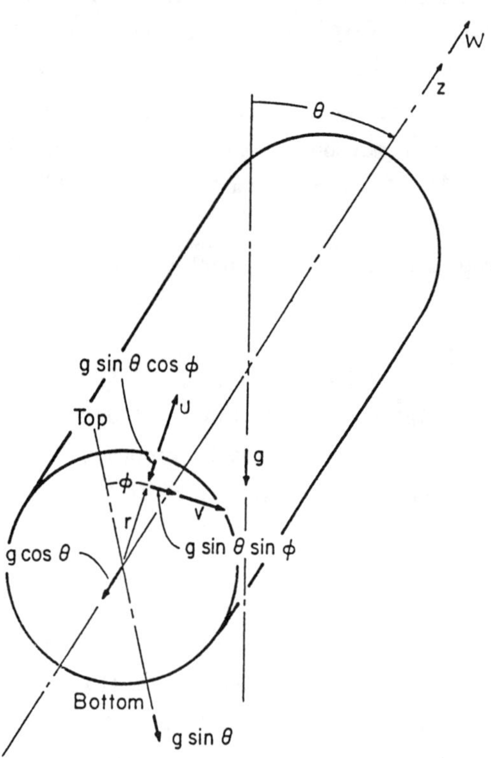

Fig. 6.9   Coordinate system and components of gravitational acceleration

Application of Eqs. (2.113), (2.114), and (2.123) calls for identifying (use subscript p for the particle phase and no subscript for the fluid phase), for the fluid phase, $\tau = \tau_{zr}$, and Eq. (2.114) is reduced to, for the fluid phase under negligible gravity effect:

$$-(dP/dz) + \rho\, g\, \cos\theta + r^{-1}(d/dr)(r\tau) = 0 \tag{6.2}$$

where τ is the shear stress of the fluid with $\rho \sim \bar{\rho}$. For turbulent flow, let $y = R - r$,

$$W = W_o (y/R)^{1/7} \tag{6.3}$$

for $y_s < y < R$ with $y_s = 60(W\, R\, \bar{\rho}/\bar{\mu})^{-7/8} R$, the thickness of the laminar sublayer. Under this condition, gravity effect contributes only to the change in static pressure along the axis of the pipe. For laminar flow, Poiseuille motion exists in the fluid phase.

For the particle phase, when the particle-fluid interaction length L_p is not too small when compared to R, we have, from Eq. (3.97),

$$\tau_{pzr} = \rho_p\, D_p (\partial W_p/\partial r)$$

$$\tau_{pz\phi} = \rho_p\, D_p\, r^{-1}(\partial W_p/\partial \phi) \tag{6.4}$$

The flux of particles due to field and fluid forces is now given, for negligible inertia effect, by Eq. (2.113) as:

$$\rho_p(U_p - U) = (f_p\, \rho_p/F) + F^{-1}\nabla \cdot \tau_p = J_p \tag{6.5}$$

Where deposition may occur, the conservation of mass has to be accounted for by the diffusion equation for a dilute suspension given by Eq. (2.123), with substitution of Eq. (6.5), to give, for a quasi-stationary state,

$$U \cdot \nabla \rho_p = \nabla \cdot (D_p \nabla \rho_p) - F^{-1} \nabla \cdot \rho_p\, f_p \tag{6.6}$$

when the divergence of the divergence of shear stress can be neglected (Prob. 6.1). For the pipe flow system, the components of field forces and the fluid force f_L due to shear lift (Eq. 1.36 or 1.37) are (neglecting the ϕ, z components of shear lift force):

$$f_{pr} = -[1 - (\overline{\rho}/\overline{\rho}_p)] \, g \, \sin\theta \, \cos\phi + (q/m) \, E_r + f_L \qquad (6.7)$$

$$f_{p\phi} = [1 - (\overline{\rho}/\overline{\rho}_p)] \, g \, \sin\theta \, \sin\phi + (q/m) \, E_\phi \qquad (6.8)$$

$$f_{pz} = -[1 - (\overline{\rho}/\overline{\rho}_p)] \, g \, \cos\theta + (q/m) \, E_z \qquad (6.9)$$

where q is the charge, m is the mass of each particle, the E's are the components of electric field $E = -\nabla V_e$, V_e being the electric potential, and E is given by the Poisson equation

$$\nabla^2 V_e = -\rho_p (q/m)/\varepsilon_0 \qquad (6.10)$$

where ε_0 is the permittivity of free space and E_z is small for gradual deposition.

The boundary condition for particle density is given by the conservation of total flow:

$$\frac{\partial}{\partial z} \int_0^{2\pi} \int_0^R \rho_p W_p \, r \, dr d\phi = -\sigma R \int_0^{2\pi} U_{pR} \, \rho_{pR} \, d\phi - \sigma_w \, 2\pi R \, \rho_{pR}(f_w/F)$$

$$+ \sigma_w' \, 2\pi R \, \rho_{pb}(f_L/F) \qquad (6.11)$$

where the terms were explained in Sec. 4.6. Eq. (6.11) is actually the basis via which Eq. (4.89) was originally derived.

Since we assume the motion of the fluid phase as fully developed ($\partial U/\partial z = 0$), we have, for $E_z = 0$,

$$\frac{\partial}{\partial z} (\rho_p W_p) = \frac{\partial}{\partial z} [\rho_p(W_p - W)] + W \frac{\partial \rho_p}{\partial z}$$

$$= [W - F^{-1} (1 - \frac{\overline{\rho}}{\overline{\rho}_p}) \, g \, \cos\theta] \frac{\partial \rho_p}{\partial z} \qquad (6.12)$$

Substitution of Eqs. (6.12) and (6.6) into Eq. (6.11) gives, with $U_{pR} \sim f_{pR}/F$:

$$D_p(\partial\rho_p/\partial r)|_R = -(1-\sigma)\ (1 - \frac{\bar{\rho}}{\rho_p})\ (\frac{g}{F})\ \sin\theta\ \cos\phi\ \rho_{pR}$$

$$+ [(1-\sigma)\ \rho_{pR} + \sigma_w'\rho_{pb}]\ (f_L/F)$$

$$+ (1-\sigma)\ (q/m)\ F^{-1}\ E_r\ \rho_{pR} - \sigma_w\ \rho_{pR}(f_w/F) \qquad (6.13)$$

which, for $\sigma = 0$, $\sigma_w = 0$, $\sigma_w' = 0$, reverts to the non-depositing boundary condition. Existence of ρ_{pb} means the presence of a fixed or moving bed, or the condition of saltation. (Prob. 6.2, 6.3) Additional boundary conditions are:

$$V_e(R,\phi) = V_R(\phi) \qquad (6.14)$$

for a conducting pipe, the surface at radius R is a Gaussian surface such that V_e is uniform while $V_e = 0$ at an interior point with $V_e < 0$ everywhere else; and the slip boundary condition:

$$W_{pw} = -L_p\ d\ W_p/dr\Big|_{r=R-a} \qquad (6.15)$$

In the following, we treat specific cases using these dimensionless quantities:

$$r^* = r/R,\ z^* = z/R,\ \rho_p^* = \rho_p/\rho_{p1} = \rho_p^*(r^*,z^*,\phi)$$

$$W^* = W/W_o = W^*(r^*),\ W_p^* = W_p/W_o = W_p^*(r^*,z^*,\phi)$$

$$V^* = (q/m)\ V_c/D_p F,\ E^* = -\nabla^* V^* \qquad (6.16)$$

where W_o is the fluid velocity at the center of the pipe, ρ_{p1} is the density of particles at the initial condition ($z^* = 0$) and at $r = 0$ of fully developed motion. We also define the following parameters:

$$\alpha = (\rho_{p1}/4\varepsilon_o)(q/m)^2\ (R^2/D_p F) \qquad (6.17)$$

$$\beta = R^2 F/D_p \tag{6.18}$$

$$\gamma = 2[1 - (\bar{\rho}/\bar{\rho}_p)] R^2 g \cos\theta/D_p W_o \tag{6.19}$$

$$\eta = 2[1 - (\bar{\rho}/\bar{\rho}_p)] R g \sin\theta/D_p F \tag{6.20}$$

and the momentum transfer number N_m and the particle Knudsen number N_{Kp}:

$$N_m = W_o/R F \tag{6.21}$$

$$N_{Kp} = L_p/R \tag{6.22}$$

and an adhesive force parameter

$$\lambda = R f_w/F D_p \tag{6.23}$$

f_w acts only on particles at the immediate vicinity of the wall (Eq. 3.35). From Eqs. (1.36) and (1.37), we correlate the shear-lift parameter as the ratio of the characteristic terminal velocity due to lift force per unit mass to the diffusion velocity of a particle. For shear-lift in the sublayer of turbulent flow, Eq. (1.37) gives:

$$\zeta_2 = f_{Lo} R/F D_p = 0.04(a/R)^{1/2} (R W_o/D_p) \tag{6.24}$$

with the characteristic lift force per unit mass f_{Lo} given by the lift force per unit mass as:

$$f_L = f_{Lo} \left|\frac{\partial W^*}{\partial r^*}\right|^{1/2} \frac{(W_p^* - W^* + \Delta W^*)}{|W_p^* - W^* + \Delta W|^{1/2}} \tag{6.25}$$

where $\Delta W^* = |\partial W^*/\partial r^*| (a/R)$, and $|\partial W^*/\partial r^*| = 60^{-6/7} (W_o R/\bar{\nu})^{3/4}$ in the laminar sublayer of thickness y given by $y_s/R = 60(W_o R/\bar{\nu})^{-7/8}$. For laminar shear motion based on Eq. (1.36),

$$\zeta_1 = 0.343 (R W_o/\bar{\nu})^{1/2} (a/R) (W_o R/D_p) \tag{6.26}$$

and $|\partial W^*/\partial r^*| = 2W_0/R$.

The above parameters α, β, etc. can be expressed in terms of dimensionless parameters of interactions of two effects, they are:

$$N_{ED} = (\rho_{p1}/4\epsilon_0)^{1/2} \; (q/m) \; (R^2/D_p) \tag{6.27}$$

which is the ratio of displacement by electrostatic repulsion to that by diffusion, the diffusion response number

$$N_{DF} = (D_p/F)/R^2 = \beta^{-1} \tag{6.28}$$

which is the ratio of relaxation time $(1/F)$ to diffusion time, the well known Froude number correlating inertia force to gravity force:

$$N_{Fr} = W_0/(2Rg)^{1/2} \tag{6.29}$$

and the momentum transfer number correlating relaxation time to transport time:

$$N_m = W_0/RF \tag{6.30}$$

In terms of these dimensionless numbers:

$$\alpha = N_{ED}^2 \; N_{DF} \tag{6.31}$$

$$\gamma = N_{Fr}^{-2} \; N_{DF}^{-1} \; N_m[1 - (\bar{\rho}/\bar{\rho}_p)] \; \cos\theta \tag{6.32}$$

etc. which account for multiple interactions. The use of these two-interaction parameters has been shown to facilitate scale-up from tested designs [Perez-Blanco et al. 1980].

With these correlations, the momentum equation of the particle phase in pipe flow reduces from Eq. (2.113) to the form:

$$- \frac{1}{2} \gamma \rho_p^* + \frac{1}{r^*} \frac{\partial}{\partial r^*} \left(r^* \rho_p^* \frac{\partial W_p^*}{\partial r^*} \right) + \frac{1}{r^{*2}} \frac{\partial}{\partial \phi} \left(\rho_p^* \frac{\partial W_p^*}{\partial \phi} \right)$$

$$+ \beta \rho_p^* (W^* - W_p^*) = 0 \tag{6.33}$$

where the inertia effect is neglected as stated earlier and because the velocity distribution in the particle phase is independent of the mass flow ratio [Soo & Tung 1971]. We also choose the case of gas-solid suspension, such that $\overline{\rho_p} \gg \overline{\rho}$, unless an exception is stated. The diffusion equation now reduces from Eq. (2.123) to:

$$N_m \left(\beta W^* - \frac{1}{2} \gamma \right) \frac{\partial \rho_p^*}{\partial z^*} = \frac{1}{r^*} \frac{\partial}{\partial r^*} r^* \frac{\partial \rho_p^*}{\partial r^*} + \frac{1}{r^{*2}} \frac{\partial^2 \rho_p^*}{\partial \phi^2} - \frac{1}{r^*} \frac{\partial}{\partial r^*} r^* \rho_p^* E_r^*$$

$$+ \frac{1}{2} \eta [\cos \phi \frac{\partial \rho_p^*}{\partial r^*} - \frac{\sin \phi}{r^*} \frac{\partial \rho_p^*}{\partial \phi}] + \frac{1}{r^*} \frac{\partial}{\partial r^*} r^* \rho_p^* L^* \tag{6.34}$$

where

$$L^* = \zeta \left| \frac{\partial W^*}{\partial r^*} \right|^{1/2} \frac{(W_p^* - W^* + \Delta W^*)}{|W_p^* - W^* + \Delta W^*|^{1/2}} \tag{6.35}$$

with ζ_1 given by Eq. (6.26) for relative motion in the laminar range and ζ_2 given by Eq. (6.24) for that in the turbulent range. The Poisson equation is now given by:

$$\frac{1}{r^*} \frac{\partial}{\partial r^*} r^* \frac{\partial V^*}{\partial r^*} + \frac{1}{r^{*2}} \frac{\partial^2 V^*}{\partial \phi^2} = -4\alpha \rho_p^* \tag{6.36}$$

The boundary conditions are:

$$\phi = \pm \frac{\pi}{2}, \quad \partial \rho_p^* / \partial r^* \big|_o = 0 \tag{6.37}$$

and from Eq. (6.13),

$$\frac{\partial \rho_p^*}{\partial r^*} = -(1-\sigma) \frac{1}{2} \eta \cos \phi \, \rho_p^* - \sigma_w \lambda \rho_p^* - (1-\sigma) \frac{\partial V^*}{\partial r^*} \rho_p^* \tag{6.38}$$

for $\pi/2 < \phi < 3\pi/2$. Since the top of the pipe cannot have particles falling into the suspension, the boundary condition becomes

$$\left.\frac{\partial \rho_p^*}{\partial r^*}\right|_1 = -\frac{1}{2} \eta \cos\phi \; \rho_p^* - \sigma_w \lambda \; \rho_p^* - (1-\sigma) \frac{\partial V^*}{\partial r^*} \; \rho_p^* \qquad (6.39)$$

for $-\pi/2 < \phi < \pi/2$. Note that the reverse has to be specified when $\bar{\rho}_p < \bar{\rho}$, such as in the case of buoyant particles or bubbles.

6.3 Fully Developed Pipe Flow

Fully developed motion excludes deposition of the particulate phase under field force $\mathbf{f}_p$; that is, the fluxes due to diffusion $(-D_p \nabla \rho_p)$ and drift under field force $\mathbf{J}_F$ must be equal, or

$$\mathbf{J}_F + (-D_p \nabla \rho_p) = 0 \qquad (6.40)$$

Further, Eq. (6.5) gives

$$\mathbf{J}_F = \rho_p (\mathbf{U}_p - \mathbf{U}) = \rho_p \mathbf{f}_p/F \qquad (6.41)$$

In dimensionless forms, Eq. (6.34) is replaced by, for $\lambda = 0$ and $\rho_{p1} = \rho_{po}$,

$$\partial \ln \rho_p^*/\partial r^* = -(1/2) \eta \cos\phi - (\partial V^*/\partial r^*) - L^* \qquad (6.42)$$

$$\partial \ln \rho_p^*/r^* \; \partial\phi = (1/2) \eta \sin\phi - (\partial V^*/r^* \partial\phi) \qquad (6.43)$$

It is readily seen that when the effect of L^* is negligible, these simultaneous equations integrate to:

$$\ln \rho_p^* = -(1/2) \eta \; r^* \cos\phi - V^* + c_1 \qquad (6.44)$$

and the constant of integration $c_1(\phi) = c_1(r^*) = c_1$, a constant. For a pipe made of an electrical conductor, $V_R(\phi)$ – constant and the gravity effect gives the relation of particle density at the top $\rho_p(R, 0)$ and at the bottom $\rho_p(R,\pi)$ in physical quantities as, for $R \gg a$,

$$\ln \frac{\rho_p(R,\pi)}{\rho_p(R,0)} = \frac{2 \; g \; \sin\theta \; R}{D_p \; F} \left(1 - \frac{\bar{\rho}}{\bar{\rho}_p}\right) \qquad (6.45)$$

Therefore, the significance of the effect of gravity depends on this dimensionless quantity and the magnitude of $\rho_p(R,\pi) - \rho_p(R,0)$ depends on actual densities and θ. Note that this is a basic means via which particle diffusivity in pipe flow can be determined. The values of D_p so determined from measurements [Cheng et al. 1970] (Fig. 1.8) amount to 0.5 to 1.2 times of the turbulent diffusivity of the fluid phase D in pipes of 51 mm and 127 mm diameters, and ducts of 76 mm to 304 mm square sections, with particles of glass, coal, and magnesia. In addition to other means of determination (Sec. 1.7), detailed measurements are needed.

The dimensionless total flow rate in terms of the characteristic mass flow ratio is given by:

$$\dot{m}^* = \frac{(\rho_{po}/\rho)}{2\pi} \int_o^{2\pi} \int_o^1 \rho_p^* \, W_p^* \, r^* \, dr^* \, d\phi / \int_o^1 W^* \, r^* \, dr^* \qquad (6.46)$$

and the average density of particle cloud is represented by the characteristic mass ratio is given by:

$$m^* = \frac{(\rho_{po}/\rho)}{\pi} \int_o^{2\pi} \int_o^1 \rho_p^* \, r^* \, dr^* \, d\phi \qquad (6.47)$$

We also have $W^* = y^{*1/7}$, with Reynolds number $N_{Re} = RW_o \, \overline{\rho/\mu}$ in the turbulent range.

Interesting cases of non-depositing pipe flow in a turbulent fluid are:

(1) Negligible gravity and shear effects ($\eta = 0$, $\gamma = 0$, $\zeta = 0$). Distribution in particle density is given by: (Prob. 6.4)

$$\rho_p^* = [1 - (\alpha/2) \, r^{*2}]^{-2} \qquad (6.48)$$

and the velocity distribution computed is given in Fig. 6.10 for N_{Kp} = 0.1 to 1, and β = 4 and 40. The influence of N_{Kp} on the slip velocity at the wall, and N_{DF} on the velocity lag at the axis are readily seen, also shown is the influence of electrostatic charges of particle on particle density distribution in terms of α.

(2) $\alpha = 0$, $\eta = 0$, $\gamma = 0$, $\zeta = 0$; the density of particles is uniform and the velocity distribution is given by Eq. (6.33).

(3) $\alpha = 0$, finite η, $\gamma = 0$ (horizontal pipe), $\zeta = 0$, a case illustrated in Fig. 6.11, with comparison to experimental data of Wen [1966].

(4) $\alpha = 0$, $\eta = 0$, finite γ (vertical pipe), $\zeta = 0$, cases of cocurrent and countercurrent flows are illustrated in Fig. 6.12, the profiles are symmetrical.

(5) $\alpha = 0$, finite η and γ, $\zeta = 0$, combined case of (3) and (4).

(6) Finite α, η, γ; $\zeta = 0$, combined case of electrical and gravity effects [Nieh et al. 1985].

(7) $\alpha = 0$, $\gamma = 0$, finite η, ζ, case is illustrated in Fig. 6.13. ζ is based on Eqs. (6.24) and (6.26).

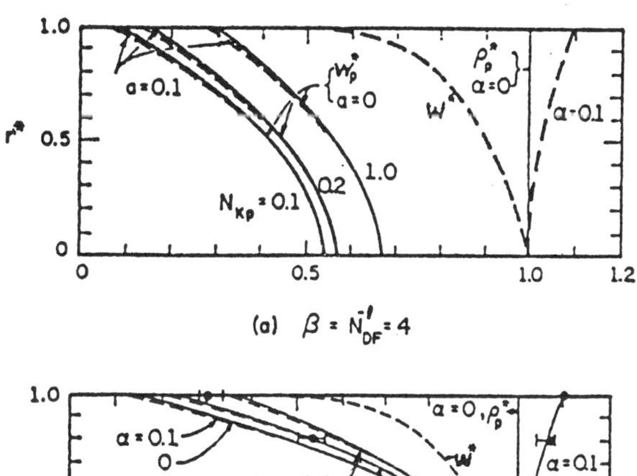

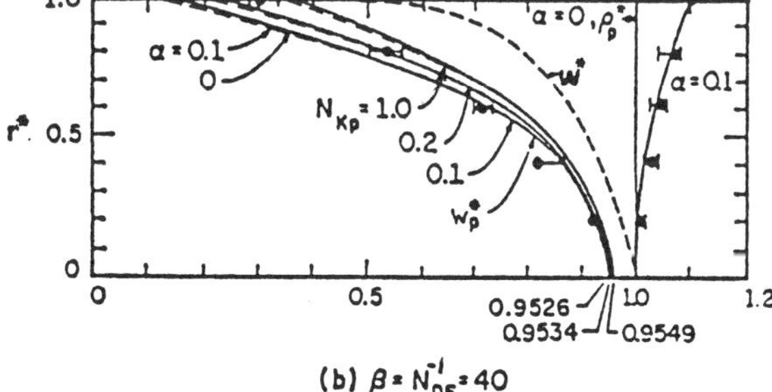

Fig. 6.10 Velocity and density profiles in fully developed turbulent pipe flow of a suspension (O Measured value for solid-gas mass ratio of 0.45, W_o = 42 m/s, 127 mm, pipe; - Range for other mass ratios.)

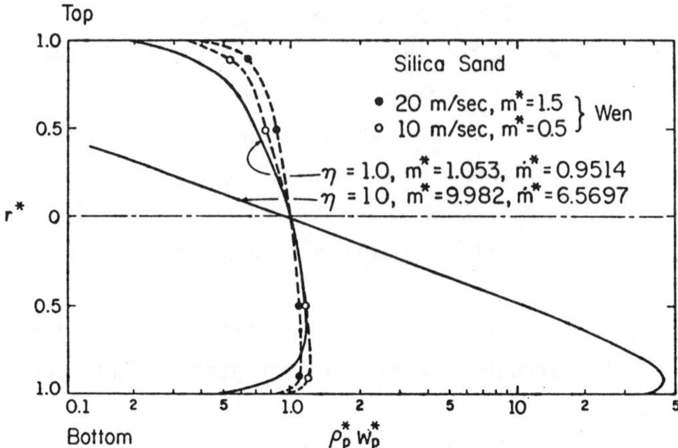

Fig. 6.11 Mass flow distribution at various η in horizontal pipe flow (γ = 0), β = 100, N_{Kp} = 0.1--with comparison to experimental data of Wen [1966].

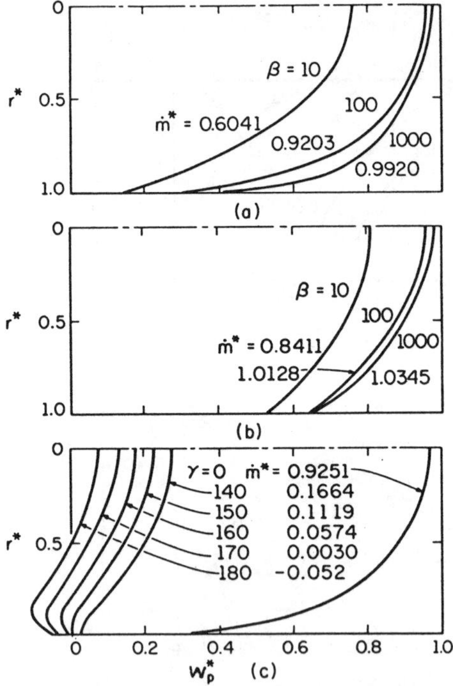

Fig. 6.12 Vertical pipe flow (η = 0) at negligible electric charge effect (α = 0), m* − 1
(a) Effect of β, N_{Kp} = 0.1, γ = 1.0
(b) Effect of β, N_{Kp} = 1.0, γ = 1.0, m* = 1.0
(c) Effect of γ, N_{Kp} = 0.1, β = 100
[Soo & Tung 1971]

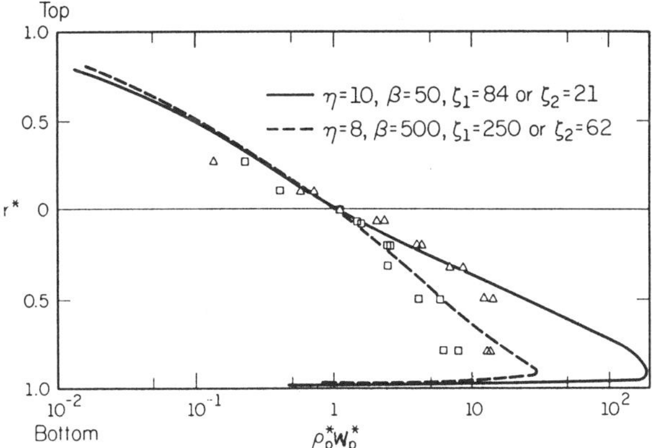

(a) Mass flow distribution in vertical plane of coarse
     sand in water ($\eta = 0$, air = 1/35.8, $N_{Kp} = 0.1$,
     $N_{Re} = 10^4$) as compared to experimental results of
     Newitt et al. (mean volume fraction solid:
     $\triangle$ 0.0050. $\square$ 0.0648

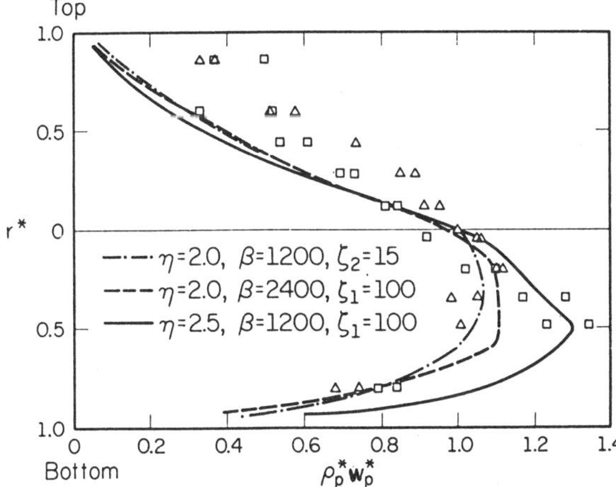

(b) Mass flow distribution in vertical plane of perpex
     in water ($\eta = 0$, R = 1 13.5, $N_{Kp} = 0.1$,
     $N_{Re} = 10^2$) as compared to experimental results of
     Newitt et al. (mean volume fraction solid:
     $\triangle$ 0.0983 $\square$ 0.0295).

Fig. 6.13 Mass flow distribution in vertical plane [Soo & Tung 1972]

<u>Residence time</u>. The method as outlined is readily applicable to the design of transfer-lines in catalytic cracking units [Saxton & Worley 1970].   It is readily seen that for chemical processes, the mean residence time of particles over a length of pipe L is given by $\int_0^L dz/W_p$, where

$$\overline{W}_p = \int_0^{2\pi} \int_0^R \rho_p W_p \, r \, drd\,\phi \Big/ \int_0^{2\pi} \int_0^R \rho_p \, r \, drd\,\phi \qquad (6.49)$$

that is, the mean residence time over length L is given by the total hold-up in L divided by the total flow rate.

<u>Laminar flow</u>. A number of studies have been made on the viscous flow of suspensoid sols and macromolecular solutions; an important application is in understanding the flow of blood [McDonald 1960] and the development of a thrombus as in coronary thrombosis. Segre and Silberberg [1962] performed experiments on laminar flow of solid particles of neutral buoyancy in water and found that the maximum density of particles occur at nearly 0.6 pipe radius, in direct contradiction to the prediction of Einstein [1906] and Jeffery [1922].   This controversy appears to be resolved by taking into account the shear lift effect and the phase velocities.

With a parabolic velocity profile $W^*(r^*)$ of the fluid phase in fully developed flow, Eq. (6.34) now takes the form of:

$$d \, \ell n \, \rho_p^*/dr^* = -L^* \qquad (6.50)$$

and Eq. (6.33) takes the form:

$$\frac{1}{r^*} \frac{d}{dr^*} \left(r^* \, \rho_p^* \, \frac{dW_p^*}{dr^*} \right) + \beta \, \rho_p^* \, (W^* - W_p^*) = 0 \qquad (6.51)$$

with ζ_1 given by Eq. (6.26). and $\Delta W^* = -(dW^*/dr^*)(a/R) = 2(a/R)r^*$.

The simplest case is for small β, that is, the effect of diffusion is negligible, thus $W_p \rightarrow W$ and Eq. (6.51) gives

$$\ell n \, \rho_p^* = - \, \zeta_1(a/R)^{1/2} \, r^{*2} \qquad (6.52)$$

with the maximum concentration at the axis of the pipe as predicted by Einstein and Jeffery. The parameter L* shows that a has to be sufficiently large for the distribution as measured by Segre and Silberberg to arise. A comparison to a numerical solution of Eqs. (6.50) and (6.51) is shown in Fig. 6.14, with fluid velocity follows the Poiseuille profile, the profiles of particle velocity show slip at the wall (depending on N_{Kp}), and the density profile of particles shows a maximum near 0.6 radius [Soo 1969, Soo & Tung 1974]. An alternate treatment based on effective viscosity (Eq. (3.112)) was given by McTigue et al. [1986].

Drag reduction. A decrease in friction factor with the presence of solid particles in gas at mass flow ratios up to 2 was observed by Halstrom [1953]. Studies have been made to explain this phenomenon. From synthesizing the relations of viscous sublayer thickness, the wall shear stress, structure of eddies and perturbations, and the particle inertia, Jokati and Tomita [1973] gave a maximum diameter of particle to produce drag reduction:

$$(2a)_{max} = (\nu^2/g)^{1/3}/2 \tag{6.53}$$

A rigorous explanation remains to be found.

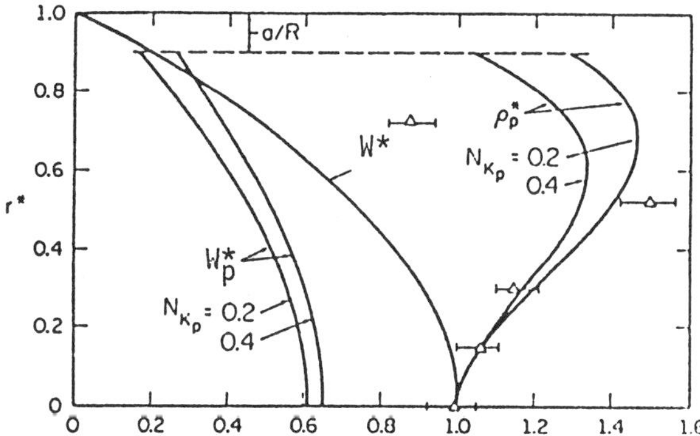

Fig. 6.14 Density and velocity distribution of a suspension of neutrally buoyant spheres in laminar motion (ζ_1 = 4, β = 8, a/R = 0.1, and N_{Kp} = 0.2, 0.4 Experimental points of $\rho_p^*(\triangle)$ were given by Sergé and Silberberg for N_{Re} = 7.3, 2a = 1.71 mm, W_0 = 6.8 m/s)

6.4 Sedimentary Flow

Stationary sedimentary flow of a dilute suspension can be treated within the present frame work by solving Eqs. (6.33) and (6.34) with appropriate boundary conditions. For turbulent flow of charged suspensions without gravity gravity effect, we first treat the effect of particle charge on deposition with γ, η, ζ equal to zero. The following cases can be identified:

(1) Diffusion effect alone in a dilute suspension. $\alpha = 0$ and deposition is by sticking at the wall alone. The is the case studied by Friedlander and Johnstone [1957]. Here we are concerned with a very dilute suspension of, say, room dust. Eq. (6.34) becomes

$$\beta \, N_m \, W^* \, \frac{\partial \rho_p^*}{\partial z^*} = \frac{1}{r^*} \, \frac{\partial}{\partial r^*} \, r^* \, \frac{\partial \rho_p^*}{\partial r^*} \qquad\qquad (6.54)$$

with $W^* = W^*(r^*)$ for turbulent motion and the boundary conditions are: $r^* = 0$, $\partial \rho_p^*/\partial r^* = 0$; $r^* = 1$, $\partial \rho_p^*/\partial r^* = -\sigma_w \lambda \, \rho_p^*$; and λ may include the van der Waals force and the electrostatic force. $\beta N_m = W_o R/D_p$ is a Peclet number. Other than the dependence in the form of $W^*(r^*)$, the solution is straight forward. For turbulent slug flow, we may take $W \sim 1$ and we have

$$\rho_p^* = \sum_k C_k \, \exp[- k^2(z^*/\beta \, N_m)] \, J_0(kr^*)$$

$$\sim \exp(-2\sigma_w \, \lambda \, z^*/\beta N_m) \, [1 - \frac{1}{2} \, \sigma_w \, \lambda \, r^{*2} + \frac{1}{16} \, \sigma_w^2 \, \lambda^2 \, r^{*4} ...] \quad (6.55)$$

where J_0 is the zeroth order Bessel function of the first kind, c_k is the Fourier coefficient for an eigenvalue $k = (2\sigma_w \lambda)^{-1/2}$. The approximation is for small $\sigma_w \lambda$, and $\sigma_w f_w/F$ is the deposition velocity defined by Friedlander and Johnstone, now expressed in terms of material and surface properties. An example is the adhesion and deposition of quartz particles to glass [Corn 1961]. This force amounts to 0.01 dyne/μm of particle size. A deposition velocity of 10 cm/s in air suggests $\sigma_w = 0.5 \times 10^{-4}$. (Prob. 6.5)

(2) Electrostatic force alone. When the electrostatic force due to space charge effect is significant, we can neglect both the effect of radial diffusion and λ. We get, from Eq. (6.34),

$$\frac{\beta \, N_m W^*}{4\alpha} \frac{\partial \rho_p^*}{\partial z^*} = - \frac{\partial}{\partial (r^*)^2} \, [\rho_p^* \int \rho_p^* \, d(r^{*2})] \tag{6.56}$$

For $W^* \sim 1$, the distribution of particle density is given by:

$$\rho_p^* = (1 + \xi)^{-1} + r^{*4} \, (1 + \xi)^{-4} + \ldots \tag{6.57}$$

with $\xi = z^*(4\alpha/\beta N_m)$. Similarity to the solution in Sec. 5.8 is noted.

(3) Simultaneous action of diffusion and electrostatic repulsion. Eqs. (6.33) and (6.34) have to be solved numerically [Soo & Tung 1971]. Although any form of initial condition can be accounted for, the change from a fully developed condition at $z = 0$ is sufficiently interesting. Fig. 6.15 shows the change in particle density at the wall along the length of the pipe. The change in density at the center of the pipe and density at the wall in the axial direction are readily seen. The particle velocity was determined from fluid velocity given by the 1/7th velocity law.

(4) Attainment of packed or moving bed density. A limiting condition is when α is large enough so that the density of a packed or moving bed ρ_{ps} of volume fraction solid α_s is reached at the wall; that is we have

$$\rho_{pR}/\rho_{p1} = (1 - \tfrac{\alpha}{2})^{-2} = \rho_{ps}/\rho_{p1} = \alpha_s/\alpha_1 \tag{6.58}$$

when sedimentation occurs because of slowing down of a wall layer. For $\alpha_s \gg \alpha_1$, such a condition occurs when

$$\alpha_1 > 2/\bar{\alpha} \tag{6.59}$$

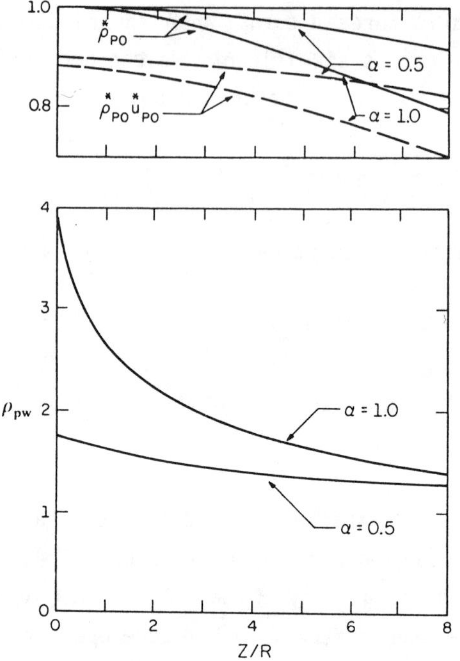

Fig. 6.15 Electrostatic sedimentation β = 20, N_{Kp} = 0.2, N_m = 2.0, σ = 0.2, λ = 0.

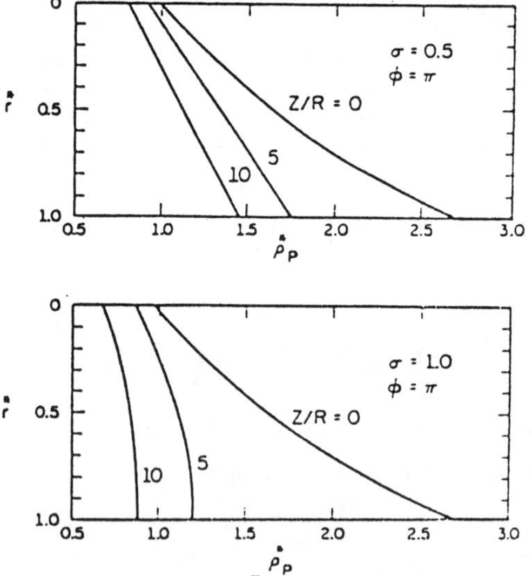

Fig. 6.16 Density distribution when deposition occurs due to gravity and diffusion β = 10, γ = 1, η = 2, N_{Kp} = 0.2, N_m = 2.0, λ = 0.

where $\bar{\alpha} = (\bar{\rho}_p/4\varepsilon_0)\,(q/m)^2\,(R^2/D_pF)$.   In this case, σ is that for particle-particle sticking and f_w is negligible.   When the bed of deposits is stationary, the rate at which its thickness is built up is given by

$$d\;\delta_s/dt = \rho_{pR}\;U_{pR}/\rho_{ps} \qquad\qquad (6.60)$$

At a steady state, we may have a deposit layer at the wall of thickness δ_s moving along the z-direction.

(5)   Gravity flow alone, negligible λ.   In this case, $\eta \to \infty$, σ does not play role.   We introduce a coordinate ξ' such that

$$d\;\xi' = \frac{1}{2}\,\eta\;dz^*/N_m\left(\beta - \frac{1}{2}\,\gamma\right)$$

$$= [\frac{W_oF}{g}\,(1 - \frac{\bar{\rho}}{\rho_p})^{-1} - \cos\theta]^{-1}\,\sin\theta\;dz^* \qquad (6.61)$$

and, taking $W^* \sim 1$ for turbulent flow, Eq. (6.34) becomes (Prob. 6.6)

$$\frac{\partial\rho_p^*}{\partial\xi'} = \cos\phi\,\frac{\partial\rho_p^*}{\partial r^*} - \frac{\sin\phi}{r^*}\,\frac{\partial\rho_p^*}{\partial\phi} = \frac{\partial\rho_p^*}{\partial y} \qquad (6.62)$$

In this case, ρ_p^* remains constant except that, starting from fully developed motion, the top boundary of the particle phase falls according to the case of batch settling; that is, the top surface of the suspension is at

$$y = 2R - R\xi' \qquad\qquad (6.63)$$

with $y = 0$ at $\phi = \pi$, $r = R$, for $\theta \neq 0$ ($\theta = 0$ for a vertical pipe) sedimentation occurs with a packed bed of α_s building up from the bottom of the pipe.   This solution is similar to the one in Sec. 5.7, but the previous time coordinate is replaced by the space coordinate.

(6) Simultaneous action of diffusion and gravity effect.   Solution can be obtained numerically.   With fully-developed flow as initial condition and 1/7the turbulent velocity law of the fluid

phase, computations were carried out for $\lambda = 0$. Figure 6.16 shows the change in density distribution. (Prob. 6.7)

(7) Attainment of packed bed density at the bottom of the pipe. This case is analogous to case (4). Sliding bed motion may occur in both the axial and peripheral directions.

The above cases show how sedimentation may get started. Once started, three more cases call for a detailed treatment of the time dependence:

(a) Transient deposition may continue until the pipe is completely plugged.

(b) Transient deposition together with bed motion.

(c) Formation of dunes and repeated piling up of deposits until a large pressure difference in the fluid is built up to blow it away. This causes a pulsating motion in the pipe. This wave motion is produced by a variation in fluid velocity due to flow restriction by the dunes.

6.5 Flow of Suspension of Fibers

Experiments were performed with dilute suspension of different papermaking fibers in the range of concentration between 0.1 and 1.0 per cent by weight [Daily and Bugliarello 1959]. Laminar flow was observed to leave two regions: a central core with all the fibers gathered in an entangled structure and a very thin peripheral annulus of clear water of thickness d across which the velocity drops from the constant value in the core. Friction factors measured show a trend similar to other suspensions. The thickness of clear-water annulus d was found to be related to the wall shear stress:

$$\tau_w = \frac{\bar{\mu}\,W}{d} = \frac{f\,\rho\,W^2}{8} \qquad (6.64)$$

d increases with flow rate and decreases with increase in concentration. Depending on pipe size, d is to the order of 30 μm and for velocities up to 0.9 m/s in 51 mm and 19 mm diameter pipe, d varies from 30 μm to 3 mm. Daily and Bugliarello obtained

$$d \sim 200 \, \bar{v}/W \tag{6.65}$$

while Forgacs et al. [1958] obtained a coefficient of 310. The peculiar nature of flow of a fiber suspension is seen in the typical friction-loss curves in Fig. 6.17 for sulfite (paper pulp) stock recommended in the Standards of Hydraulic Institute [1930] for design purpose. It is seen that for a 4 percent pulp at bone-dry consistency, the head loss decreases as flow quantity is increased from 3 to 4.55 m^3/min. It is also seen that with 1 per cent pulp the head loss is less than pure water over the whole flow range. This is a case of drag reduction by suspended materials (Sec. 6.3).

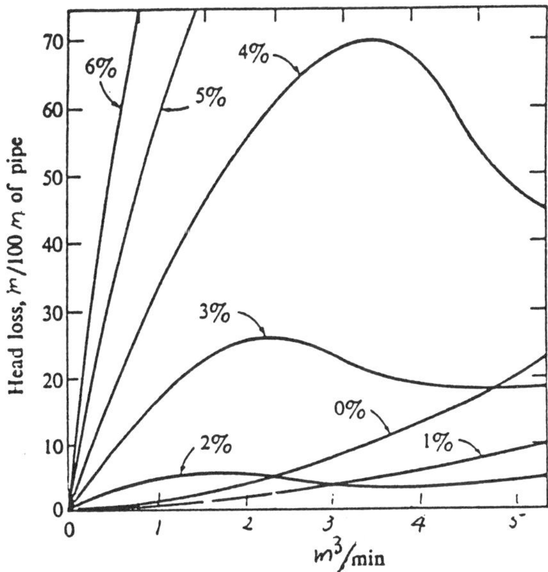

Fig. 6.17 Friction loss of sulfite (paper) stock through 153 mm I.D. cast iron pipe, percentage pulp based on bone-dry consistenty [Standards of Hydraulic Institute 1930]

6.6 Heat Transfer in Pipe Flow

Assuming constant physical properties and uniform particle density, and identical particle and fluid velocities in fully developed flow, the time-volume averaged energy equations for pipe flow of a dilute suspension can be expressed as: (Fig. 6.18) (unprimed velocities and temperatures are averages)

$$W \frac{\partial T_p}{\partial z} + G(T_p - T) - G_r(T_w^4 - T_p^4) + \frac{1}{r}\frac{\partial}{\partial r} r <U'T'> = 0 \qquad (6.66)$$

for the particle phase, since $U = V = 0$, $<W_p' T_p'>$ is independent of z and $<V_p' T_p'> = 0$; and

$$\rho c\, W \frac{\partial T}{\partial z} + \rho_p\, c_p\, W \frac{\partial T_p}{\partial z} = \frac{1}{r}\frac{\partial}{\partial r}\left(r\bar{\kappa}\frac{\partial T}{\partial r} - \rho c r <U'T'>\right) \qquad (6.67)$$

for the mixture, where is the density of the gas phase ($\rho \sim \bar{\rho}$), ρ_p is the density of the cloud of particles, T is the temperature of the fluid and T_p is that of the particles. The inverse relaxation times for heat transfer by convection and by radiation are:

$$G = \frac{h_p\, A_p}{\bar{\rho}_p\, v_p\, C_p} \qquad\qquad G_r = \frac{\sigma_r\, A_p\, \varepsilon_p}{\bar{\rho}_p\, v_p\, C_p} \qquad (6.68)$$

from Eq. (1.20) and considering the particle-wall interaction only in the radiative transfer, where h_p is the heat transfer coefficient

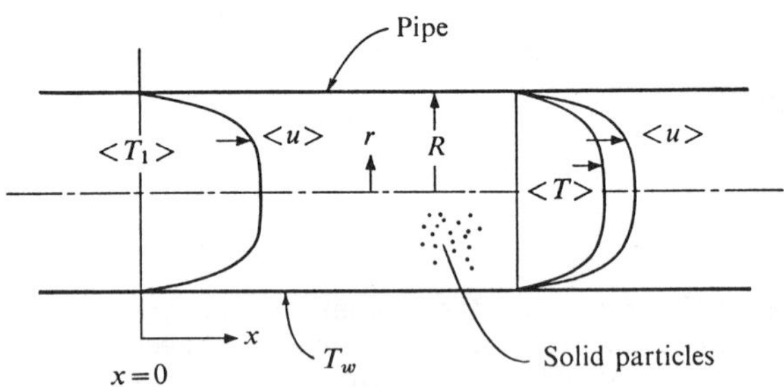

Fig. 6.18 Two-phase flow in a circular pipe [Tien 1961]

from fluid to particle, A_p is the surface area, v_p is the volume of the solid particle, c_p is the specific heat of the solid, σ_r is the Stefan-Boltzmann constant of radiation, ε_p is the emissivity of the surface of the solid. The heat transfer coefficient h_p is not well known when a particle is suspended in a turbulent stream.

The energy equation (6.67) can be further written as:

$$W \frac{\partial T}{\partial z} + \frac{\rho_p}{\rho} W \frac{\partial T_p}{\partial z} = \frac{1}{r} \frac{\partial}{\partial r} \left[r \left(\frac{\bar{v}}{N_{pr}} + \varepsilon_H \right) \frac{\partial T}{\partial r} \right] + \beta_2 (T_p - T) \qquad (6.69)$$

where

$$\beta_2 = \frac{n_p \, h_p \, A_p}{\rho C} = \frac{c_p \, \rho_p \, G}{C\rho}$$

$\bar{v}$ is the kinematic viscosity, and ε_H is the eddy diffusivity for heat. Eq. (6.66) can be simplified since its last term can be neglected because

$$0 <U'T'> \; > \; 0 <U'T_p'> \; \simeq \; <U_p'T_p'>$$

from a statistical point of view, since T' is directly related to the motion of a fluid element while T_p' is not, the correlation between U' and T' is much larger than that between U' or U_p' and T_p'. Hence Eq. (6.66) becomes

$$W \frac{\partial T_p}{\partial z} + G(T_p - T) - G_r(T_w^4 - T_w^4) = 0 \qquad (6.70)$$

Equations (6.69) and (6.70) are now the governing equations for determining the average temperatures T and T_p, with the boundary conditions:

$$r = 0, \; \partial T/\partial r = 0, \; \partial T_p/\partial r = 0; \; z = 0, \; T = T_o, \; T_p = T_o;$$

$$r = R, \; T = T_w, \; T_p = T_w; \; z \to \infty, \; T = T_w, \; T_p = T_w.$$

Neglecting radiation and introducing the following dimensionless quantities:

$$\theta = \frac{T - T_w}{T_0 - T_w} \qquad\qquad \theta_p = \frac{T_p - T_w}{T_0 - T_w}$$

$$\eta = \frac{r}{R}, \quad \xi = \frac{z}{N_{Re} \, N_{pr} \, R}, \quad f(\eta) = \frac{W}{W_{avg}} \; , \quad g(\eta) = 1 + \frac{\varepsilon_H \, N_{pr}}{\overline{\nu}}$$

the governing equations become, respectively,

$$f \, \frac{\partial \theta_p}{\partial \xi} + \beta_3 (\theta_p - \theta) = 0 \tag{6.71}$$

and

$$f \, \frac{\partial \theta}{\partial \xi} = \frac{2}{\eta} \frac{\partial}{\partial \eta} \left[\eta g \, \frac{\partial \theta}{\partial \eta} \right] + \beta_4 (\theta_p - \theta) \tag{6.72}$$

with

$$\beta_3 = \left(\frac{\rho C}{\rho_p C}\right) \left(\frac{4R^2 \, h_p \, A_p}{2 \, \overline{\kappa} \, V_p}\right) \qquad\qquad \beta_4 = \frac{n_p \, h_p \, A_p (2R)^2}{2 \, \overline{\kappa}}$$

The boundary conditions are now

$$\eta = 0, \qquad \frac{\partial \theta}{\partial \eta} = 0, \qquad \frac{\partial \theta_p}{\partial \eta} = 0$$

$$\eta = 1, \qquad \theta = 0, \qquad \theta_p = 0$$

$$\xi = 0, \qquad \theta = 1, \qquad \theta_p = 1 \tag{6.73}$$

$$\xi \to \infty, \qquad \theta = 0, \qquad \theta_p = 0$$

The last boundary condition in Eq. (6.73) suggests a solution of Eqs. (6.71) and (6.72) by separation of variables, or

$$\theta = \sum_{n=0}^{\infty} C_n \, R_n \, \exp(-\lambda_n^2 \, \xi) \tag{6.74}$$

and

$$\theta_p = \sum_{n=0}^{\infty} C_n \, R_{pn} \, \exp(-\lambda_n^2 \, \xi) \tag{6.75}$$

where $R_n(\eta)$ and $R_{pn}(\eta)$ satisfy the relations:

$$R_n - R_{pn}(1 - \frac{\lambda_n^2}{\beta_3} f) = 0 \tag{6.76}$$

and

$$(\eta \, g \, R_n')' + \frac{1}{2} \beta_4 \, \eta (R_{pn} - R_n) + \frac{\lambda_n^2}{2} \eta \, f \, R_n = 0 \tag{6.77}$$

with the boundary conditions in the forms:

$$R_n'(0) = 0, \; R_{pn}'(0) = 0$$

$$R_n(1) = 0, \quad R_{pn}(1) = 0 \tag{6.78}$$

Substitution of Eq. (6.76) into Eq. (6.77) gives

$$(\eta \, g \, R_n')' + [(1 - \frac{\lambda_n^2}{\beta_3})^{-1} - 1 + \frac{\lambda_n^2 \, f}{\beta_4}] \frac{\beta_4}{2} \eta \, R_n = 0 \tag{6.79}$$

where the boundary conditions of $R_n'(0)$ and $R_n(1)$ are given in Eq. (6.78).

For a dilute suspension, $\lambda_n^2 \, f/\beta_3 \ll 1$, we further let

$$\bar{\lambda}_n^2 = [1 + (\beta_4/\beta_3)] \, \lambda_n^2 \tag{6.80}$$

Eq. (6.79) is reduced to a Sturm-Liouville system: [Tien 1961]

$$(\eta \, g \, R_n')' + \frac{1}{2} \bar{\lambda}_n^2 \, f \, \eta \, R_n = 0 \tag{6.81}$$

with linear homogeneous boundary conditions $R_n'(0) = 0$ and $R_n(1) = 0$. Note that Eq. (6.81) is reduced to that of a single phase fluid when $n_p = 0$ or $\beta_4 = 0$. In other words, the present solution accounts for the increase in heat capacity of the fluid only.

The heat flux to the two-phase flow at the wall is given by:

$$J_q = \bar{\kappa} \left.\frac{\partial T}{\partial r}\right|_{r=R} = \frac{-2\kappa \ (T_1 - T_w)}{T} \sum_n A_n \exp(-\lambda_n^2 \xi) \tag{6.82}$$

where

$$A_n = - C_n \ R_n'(1)/2 \tag{6.83}$$

and the coefficients C_n are given by the orthogonality condition:

$$C_n = \int_0^1 f \ \eta \ R_n \ d\eta / \int_0^1 f \ \eta \ R_n^2 \ d\eta \tag{6.84}$$

The values of λ_n^2 and R_n can be found from the computations for the single phase solution [Sleicher and Tribus 1957] for given velocity function $f(\eta)$ and eddy diffusivity function $g(\eta)$.

For many practical gas-solid suspensions, β_3 is of the order of 10^5 or higher. The approximation leading to Eq. (6.81) is readily shown to be valid in physical systems with $N_{Pr} < 2$ and $N_{Re} <$ 500,000. The increase in heat transfer due to the presence of the solid particles is seen as a consequence of flattening of temperature profile by the suspended particles.

The Nusselt number of the system is given by, for mean temperature T of the mixture:

$$N_{Nu} = h(2R)/\bar{\kappa} = J_q(2R)/\bar{\kappa} \ (T_w - T_{mm}) \tag{6.85}$$

The Nusselt number for the case of uniform wall temperature is readily shown to be:

$$N_{Nu}(\xi) = \frac{\sum_n A_n \exp(-\lambda_n^2 \xi)}{\sum_n \frac{A_n \exp(-\lambda_n^2 \xi)}{\lambda_n^2 [1 + (\beta_4/\beta_3)]}} \tag{6.86}$$

Far downstream from the thermal entrance, all terms but the first become small so that the asymptotic Nusselt number is

$$(N_{Nu})_a = (1/2) \ [1 + (\beta_4/\beta_3)] \ \lambda_0^2 = (1/2) \ \bar{\lambda}_0^2 \tag{6.87}$$

Because of the assumption of similar velocities of the phases, the heat capacity ratio of β_4/β_3 in Eq. (6.80) is related to the ratio of specific heats times the mass flow ratio of solid to gas. This provides a basis of comparison of the above analytical results of Tien [1961] to the experimental results of Farbar and Morley [1957] as shown in Fig. 6.19. The latter illustrates the cases of N_{Re} of 27,000 and 13,500 for c_p/c of 1.2, at 1 bar and 15 C at low mass flow ratio, over a pipe length of 50 pipe diameters. Many other experimental results were reported by for instance, Pfeffer et al. [1966], and corresponding to drag reduction (Sec. 6.3), a decrease in Nusselt number were noticed as the mass flow ratio of solid to gas is increased toward 1 [Tien and Quan 1962].

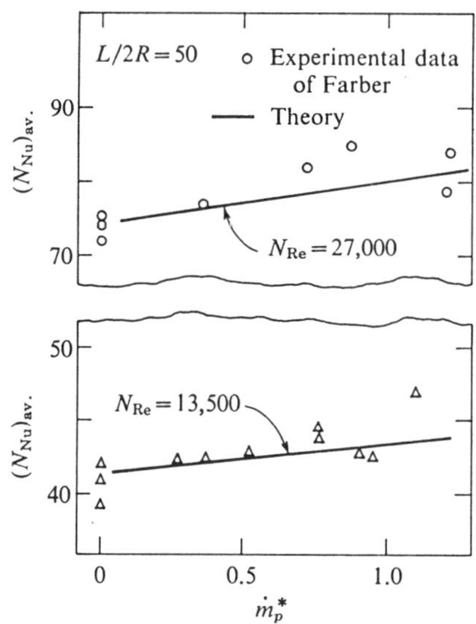

Fig. 6.19 Comparison with experimental results of Farbar and Morley [1957]

6.7 Cyclone Separation

Cyclone separators are among the most widely used collectors of
particulate matter from a gaseous or an acqueous suspension. It is
treated here as an extension of pipe flow because the method of
analysis belong to a similar class of differential equations and
boundary conditions, especially when considering a cyclone separator
with a steep cone. Circulatory motion in pipe flow has already been
identified as a possible mode (Fig. 6.12); an idealized swirling pipe
flow of a suspension was studied by Tung and Soo [1973] toward
understanding the motions in cyclone separators.

<u>Fluid velocity distribution</u>. Measurements by Rietema [1962] on flow
configuration in a cyclone were made on tangential velocity distributions. Fig. 6.20 shows the coordinates of the flow system and
the velocity distribution calculated for flow volume per unit time Q,
radius R(z) of the inside surface of the cyclone separator, with

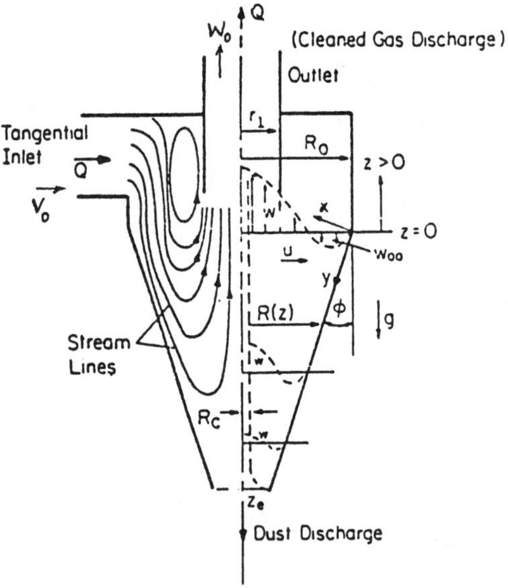

Fig. 6.20 Velocity distribution in a cyclone separator
         (coordinate system, streamlines, axial velocity
         distribution, and core radius)

components U, V, W in the radial, tangential, and axial directions.

The tangential velocity distributions was shown to consist of a core of solid body rotation at $r \leq R_c$, the core radius, and an outer free vortex motion. It was shown that outside of the boundary layer ($V = V_o$) at $R(z)$, we have [Soo 1970]

$$r \, V_o/C = 1 - \exp(-r^2/R_c^2) \tag{6.88}$$

where the core radius is given by

$$R_c^2 = 4 \, W_o \, \nu/[-(\partial P/\partial z)/\rho] \tag{6.89}$$

and C is the maximum vorticity in the system, ν is the kinematic viscosity of the fluid, W_o is the axial velocity in the core, (dP/dz) is the axial pressure drop, and ρ is the density of the fluid. Further study applying the momentum integral method to both the cases of laminar and the turbulent vortex motion over a plane gave [Soo 1973]:

$$W_o R_c/C = 1.37 \, (\nu/C)^{1/2} \tag{6.90}$$

for the laminar range, and

$$W_o R_c/C = 0.0643 \, (\nu/C)^{1/5} \tag{6.91}$$

for the turbulent range, with transition at $C/\nu = 10^4$. The pressure drop through the cyclone of a characteristic height L_c is given by combining Eqs. (6.89) and (6.90) or (6.91):

$$\Delta P^\star = \frac{\Delta P}{\rho \, C^2/2 \, R_o^2} = 4.26 \, \frac{L_c}{R_o} \, (\frac{W_o \, R_o}{C})^3 \tag{6.92}$$

for the laminar range, and

$$\Delta P^\star = 1940 \, (\frac{L_c}{R_o}) \, (\frac{W_o \, R_o}{C})^3 \, (\frac{\nu}{C})^{3/5} \tag{6.93}$$

for the turbulent range, a relation giving $\Delta P^\star$ close to that of typical cyclones (around 20) measured by Ogawa [1975].

The simplest or primary mode of distribution of W of the fluid can be represented by a polynomial ($r* = r/R(z)$):

$$\frac{W}{(Q/\pi R_0^2)} = [1 - \frac{z}{z_e}] (a_0 + a_1 r*^2 + a_2 r*^4 + a_3 r*^6) \equiv W* \quad (6.94)$$

and the continuity equation gives U as:

$$\frac{U}{(Q/\pi R_0^2)} = [1 - \frac{z}{z_e}] (\tan \frac{\phi}{2}) r*^3 (a_1 + \frac{4}{3} a_2 r*^2 + \frac{3}{2} a_3 r*^4)$$

$$+ (\frac{R}{2z_e}) r* (a_0 + \frac{1}{2} a_1 r*^2 + \frac{1}{3} a_2 r*^4 + \frac{1}{4} a_3 r*^6) \equiv U*$$

$$(6.95)$$

which satisfy the condition: $r = 0$, $U = 0$, $\partial U/\partial r = 0$, $\partial W/\partial r = 0$, with a_0, a_1, a_2, and a_3 determined by the following boundary conditions: $r = R(z)$, $W = 0$; $z < 0$, $\int_R^R Wrdr = 0$; at the exit, $r* = r_1*$, $2\pi \int_0^1 Wrdr = Q$; at the inlet plane, $2\pi \int_{r_1}^R Wrdr = -Q$; to give

$$a_0 = \alpha_1 (1 - \frac{1}{2} r_1*^2), \quad a_1 = \alpha_1 (-3 + 2r_1*^2)$$

$$a_2 = \alpha_1 (\frac{3}{2}) r_1*^2, \quad a_3 = 2\alpha_1 \quad (6.96)$$

where $1/\alpha_1 = r_1*^2 (1-r_1*^2)^2$, $R = R_0 + z \tan\phi$, and $\tan\phi = (R_0-R_e)/(-z_e)$. Additional effects such as dissipation at the wall can be accounted for by introducing a coefficient a_4, and so on. Here we have an adequate description of U and W as shown in Fig. 6.21 for the geometry of $r_1/R_0 = 1/3 = r_1*$. Negative value of U means radial inflow and U is one order of magnitude smaller than W and is approximately proportional to $-r$ in the case of pipe flow with superposed vortex motion (Eq. 6.88) [Soo 1970]. These results are also shown in Fig. 6.20. Toward the bottom of the cone of the cyclone separator, the velocity of upflow becomes small; reentrainment of the dust does not occur at the bottom. [Soo 1973]. The above relations, with Eq. (6.88), give the core radius R_c as, for $U \sim 0$ and $W = W_0$

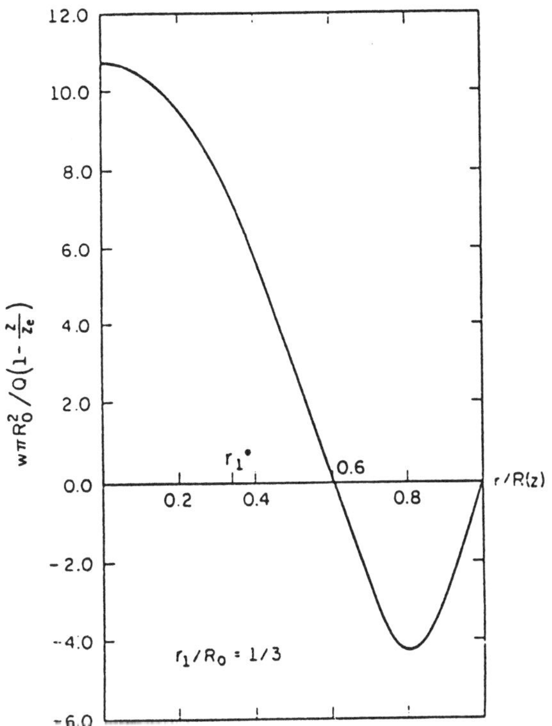

(a) Dimensionless axial velocity distribution

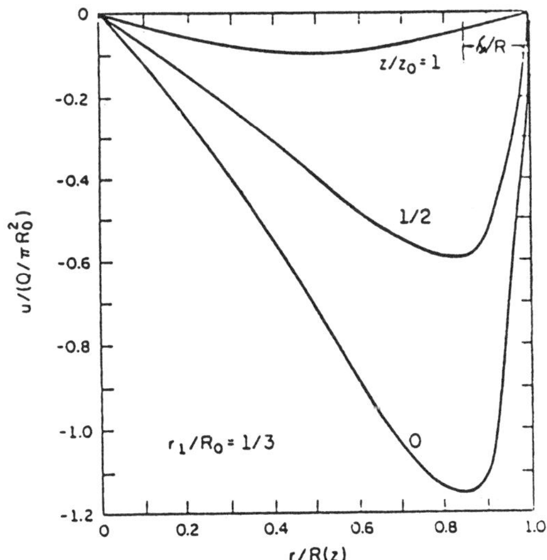

(b) Dimensionless radial velocity distribution

Fig. 6.21 Dimensionless velocity distributions in a cyclone
($r^* - r_1/R_0 = 1/3$) [Soo 1973]

$$\frac{R_c^2}{R_o^2} = \frac{4\pi D z_e}{Q a_o} \ln\left[\frac{1}{1 - \dfrac{z}{z_e}}\right] + \frac{R_{co}^2}{R_o^2} \tag{6.97}$$

and the radius R_{co} at $z = 0$, the entrance of the outlet pipe is:

$$R_{co}^2/R_o^2 = 4 W_o D \rho/(-\partial P/\partial z) R_o^2 \tag{6.98}$$

where D is the turbulent diffusivity of the fluid. Note that R_c actually decreases toward the outlet pipe (Prob. 6.10) as shown in Fig. 6.20 because of increasing W. Contrary to the popular con- jecture, the funnel shape in a cyclone separator is actually inverted, the core radius is larger near the bottom of the cone than at the outlet pipe. (Prob. 6.8)

<u>Distribution of particle density.</u> The density distribution of the particle phase and its deposition is strongly influenced by a finite particle diffusivity D_p and the field forces, together with finite interaction length L and sticking probability of particles at the wall or with the deposited layer of particles (Sec. 6.2). The field forces include the centrifugal, gravity, and electrostatic forces. The effect of electrostatic charges is prominent in gaseous suspensions.

For the flow system in Fig. 6.20, neglecting the inertia forces in comparison to field forces and viscous forces, the equation of radial and axial components of particle momentum are:

$$\rho_p \frac{v_p^2}{r} = -F \rho_p(U_p - U) + \rho_p E_r(q/m) \tag{6.99}$$

$$\rho_p g = -F \rho_p(W_p - W) + \rho_p E_z(q/m) \tag{6.100}$$

where E_r and E_z are the radial and axial components of the electric field, which is given by the Poisson Equation:

$$\nabla^2 V_e = -\nabla \cdot E = -(\rho_p/\varepsilon_0)(q/m) \tag{6.101}$$

where V_e is the electric potential and q/m is the charge to mass ratio. Substitution of Eqs. (6.98) to (6.101) into Eq. (6.6) gives the diffusion equation of the particles:

$$(U + \frac{q}{m}\frac{E_r}{F} + \frac{V_p^2}{rF})\frac{\partial \rho_p}{\partial r} + (W + \frac{q}{m}\frac{E_z}{F} - \frac{g}{F})\frac{\partial \rho_p}{\partial z}$$

$$+ \frac{2\rho_p}{F}\frac{V_p}{r}\frac{\partial V_p}{\partial r} = D_p \nabla^2 \rho_p - \frac{1}{F}\frac{\rho_p^2}{\varepsilon_o}(\frac{q}{m})^2 \tag{6.102}$$

which shows that the electrostatic, centrifugal, and gravity forces give rise to drift components on particles in addition to U and W.

At the boundary, for small thickness of deposits or a clean wall, we have, from Eq. (6.13) (coordinate x as shown in Fig. 6.20),

$$-D_p \frac{\partial \rho_p}{\partial x}\Big|_R = (1-\sigma)\ (1 - \frac{\bar{\rho}}{\rho_p})\ [-\frac{g}{F}\sin\phi + \frac{V_{pw}^2}{RF}\cos\phi]\ \rho_{pR}$$

$$+ [(1-\sigma)\ \rho_{pR} + \sigma_w'\ \rho_{pb}]\ (f_L/F)$$

$$+ (1-\sigma)\ (q/m)\ (-F_x/R)\ \rho_{pR} - \sigma_w\ \rho_{pR}(f_w/F) \tag{6.103}$$

A result of immediate interest is that unless the right hand side of Eq. (6.103) is less than zero, no deposition along the cone will actually occur.

The diffusion equation (6.102) can be expressed in dimensionless form by introducing $W_o = Q/\pi R_o^2$, and

$$r^* = r/R,\ z^* = z/R,\ U^* = U/W_o,\ W^* = W/W_o,\ V^* = VR/C$$

$$V_e^* = V/[\frac{\rho_{p1}}{\varepsilon_o}(\frac{q}{m})\ R_o^2],\ E^* = E/[\frac{\rho_{p1}}{\varepsilon_o}(\frac{q}{m})\ R_o],\ \rho_p^* = \rho_p/\rho_{p1} \tag{6.104}$$

where C is the maximum vorticity of the system, giving

$$(N_{Pe}\ U^* + \alpha_e E_r^* + \Omega\ \frac{V_p^{*2}}{r^*})\frac{\partial \rho_p^*}{\partial r^*} + (N_{Pe}\ W^* + \alpha E_z^* - \gamma)\frac{\partial \rho_p^*}{\partial z^*}$$

$$+ 2\ \Omega\ \rho_p^*\ \frac{V_p^*}{r^*}\frac{\partial V_p^*}{\partial r^*} = \nabla^{*2}\ \rho_p^* - \alpha_e \rho_p^{*2} \tag{6.105}$$

and the dimensionless parameters:

$$N_{Pe} = \frac{W_o R_o}{D_p}, \quad \alpha = \frac{\rho_{p1}}{\varepsilon_o}\left(\frac{q}{m}\right)^2 \frac{R_o^2}{D_p F}, \quad \Omega = \frac{C^2}{R_o^2 D_p F}, \quad \gamma = \frac{g R_o}{D_p F}$$

correlating convection, diffusion, and transport by various forces in relation to diffusion and relaxation phenomena. The centrifugal force parameter Ω plays an important part at the wall because of large relative velocity V_{pw} given by Eq. (6.15).

When applied to a steep cone (small angle ϕ in Fig. 6.20), the local condition can be determined by integrating Eq. (6.105) with the simplification of small axial electrical field when compared to the radial field, nearly uniform axial velocity, small gravitational effect on turbulence ($\gamma \sim 0$); and small radial velocity ($U^* \ll W^*$). Integration from zero to r^* gives:

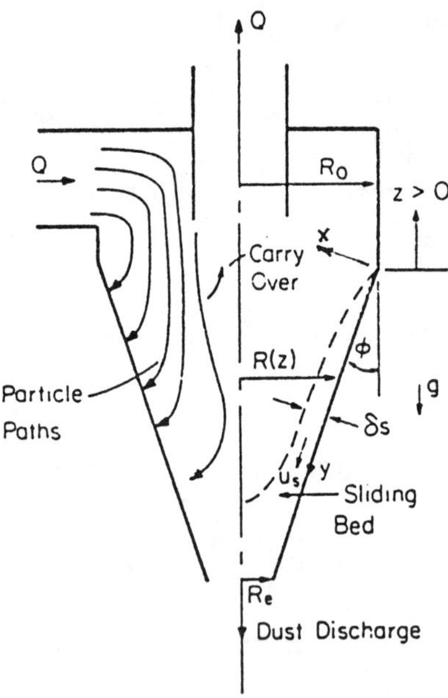

Fig. 6.22 Particle paths and sliding bed of collected particles

$$r^* \frac{\partial \rho_p^*}{\partial r^*} - \Omega \, \rho_p^* \, V_p^{*2} - \alpha \, \rho_p^* \int \rho_p^* \, r^* \, dr^* - \alpha_E \, \rho_p^* = N_{Pe} \frac{\partial}{\partial z^*} \int \rho_p^* \, W_p^* \, r^* \, dr^*$$

$$(6.106)$$

where

$$\alpha_E = \frac{r_E}{F} \frac{E_E}{D_p} \frac{q}{m} \tag{6.107}$$

E_E is the external field due to potential or surface charge density of the walls, and r_E is a characteristic dimension for the applied field field depending on its geometry. The right hand side of Eq. (6.106) is the dimensionless rate of deposition of particles.

With these simplifications, the efficiency of collection is given by: ($W_p^* \sim 1$)

$$\frac{2(\partial/\partial z) \int \rho_p^* \, W_p^* \, r^* \, dr^*}{\rho_p^*} \sim \frac{\partial \, \ell n(1 - \eta_c)}{\partial z}$$

Substitution of Eq. (6.103), and integrating once more with the approximation for density distribution due to centrifugal force alone:

$$\rho_p^* \simeq \rho_{pr}^* \, \exp[- \frac{\Omega}{2} (\frac{1}{r^{*2}} - 1)] \tag{6.108}$$

we get the collection efficiency η_c over a length L of the cyclone:

$$\eta_c \simeq 1 - \exp\{-2\sigma[\Omega + 4(2-k^*)\alpha + \alpha_E] \frac{L_c}{R} N_{Pe}^{-1}$$

$$+ 2[\sigma + \sigma_w' \frac{\rho_{pb}}{\rho_{pw}}] \frac{f_L}{F} \frac{L_c}{D_p} N_{Pe}^{-1} - 2\sigma_w \frac{f_w}{F} \frac{L_c}{D_p} N_{Pe}^{-1}\} \tag{6.109}$$

where

$$k^* = (\Omega/2) \exp(\Omega/2) \int_{\infty}^{\Omega/2} x^{-1} e^{-x} \, dx \leq 0[1] \tag{6.110}$$

for $\Omega > 0[10]$ and $\rho_{pR}^* = \rho_{PR}/\rho_{p1}$. Eq. (6.109) shows the influence of in decreasing the efficiency. Neglecting f_L and f_w Eq. (6.109) can also be expressed as, for large Ω:

$$\eta_c \simeq 1 - \exp\{- \sigma \ [\frac{(2\pi C^2 \ L_c}{FQ \ R_o^2} + 8\pi \ \frac{\rho_{p1}}{\varepsilon_o} \ (\frac{q}{m})^2 \ \frac{R_o^2 \ L_c}{F \ Q}$$

$$+ 2\pi \ E \ (\frac{q}{m}) \ \frac{R_o \ L_c}{FQ}]\} \tag{6.111}$$

for length L_c and volume flow rate Q. Note that $C^2 L_c/18FQR_o^2$ is just the empirical cyclone number of Rietema and Verver [1961]. Other terms in Eq. (9.111) show that by a combination of centrifugal force and electrostatic forces, a large diameter ($2R_o$) cyclone separator may have the same efficiency as a small one, for similar pressure drop of the gaseous suspension. (Prob. 6.9) This is not the case for hydrocyclones.

In the absence of surface adhesion, sedimentation occurs when ρ_{pb} is reached at the wall. For volume fraction solid, $\alpha_s = \rho_{pb}/\rho_p$ for sedimentation, and $\alpha_1 = \rho_{p1}/\overline{\rho_p}$ at the inlet, sedimentation begins at the condition given by Eqs. (6.58) and (6.59). The ideal condition would be a dense bed flowing down the cone toward the dust discharge as illustrated in Fig. 6.22.

<u>Motion of collected particles</u>. The ideal situation would be a dense bed collected by various forces flowing down the cone toward the discharge as shown in Fig. 6.22. Unloading of the collected particles at the bottom of the cyclone is similar to the condition of unloading of bins and flow of solids through an orifice. While the case of dense beds of suspensions will be treated in Chapter 8, here we have a limiting case of sliding bed that can be treated according to elementary considerations of the layer of deposits.

For the sliding bed as shown in Fig. 6.22, mass balance gives, for $dy = -dz/\cos\phi$, $R = R(z)$,

$$2\pi \int \sigma \ \rho_{pw} \ U_{pw} \ dy \ R \, dt = 2\pi \ \delta_s \ dy \ R \ \rho_{ps} \tag{6.112}$$

where δ_s is the thickness of the layer and ρ_{ps} is the mean density in this sediment layer. The deposition velocity U_{pw} is given by

$$U_{pw} = \frac{E_E}{F} \left(\frac{q}{m}\right) + \frac{v^2}{RF} + \frac{\rho_p}{C_o F} \left(\frac{q}{m}\right)^2 \pi R \tag{6.113}$$

The force balance is gilven by:

$$\tau_s \, Rdy + f\delta_s \, Rdy \, \rho_{ps} \, g \, \sin\phi = \delta_s \, Rdy \, \rho_{ps} \, g \, \cos\phi + \tau_f \, Rdy \tag{6.114}$$

where τ_s is the shear stress in the bed, f is the Coulomb friction coefficient, τ_f is the fluid shear stress which is negligible when compared to τ_s. Eq. (6.114) gives U_s via the approximation:

$$\tau_s = \mu_s \frac{\partial U_s}{\partial y} \tag{6.115}$$

for a bed viscosity μ_s and bed velocity U_s. The mean bed velocity is given by:

$$\frac{dy}{dt} = \overline{U}_s = \frac{1}{\delta_s} \int_o^{\delta_s} U_s \, dy = \frac{g \, \rho_{ps} (\cos\phi - f \, \sin\phi)}{\mu_s} \frac{\delta_s^2}{2} \tag{6.116}$$

Since

$$R = R_o + \frac{R_o - R_e}{Z_e} Z = R_o - y \sin\phi \tag{6.117}$$

Eq. (6.116) can be integrated to give, for a steep cone and E = 0, the bed thickness in the form:

$$\left(\frac{\delta_s}{R_o}\right)^3 \approx \left(\frac{\sigma \, \mu_s \, D_p}{g \, \rho_{ps} \, R_o^3}\right) \left(\frac{\rho_{p1}}{\rho_{ps}}\right) \left(\frac{y}{R_o}\right) \frac{(\Omega + \pi\alpha)}{(\cos\phi - f\sin\phi)} \tag{6.118}$$

or a 1/3 power relation to y or z direction starting from z = 0. For a thin layer of deposit, at a steady rate of removal, the weight is balanced by a shear stress,

$$\tau_s = \delta_s \, \rho_{ps} \, g(\cos\phi - f\sin\phi) \qquad\qquad (6.119)$$

To unload successfully, τ_s must be greater than the yield stress of the bed. Larger yield stress leads to a thick bed. The shear stress due to fluid flow is, in general, negligible. (Prob. 6.10)

Exercise Problems

6.1 Derive Eq. (6.13).

6.2 For flow of a dust cloud through a two-dimensional channel with a transverse electric field and mobility of dust particles K, show that the boundary condition at the wall with sticking probability σ is given by:

$$D_p \, \partial\rho_p/\partial y\big|_w = (1-\sigma) \, KE \, \rho_p \qquad\qquad (6.120)$$

where D_p is the particle diffusivity, ρ_p is the cloud density, and E is the electric field.

6.3 In a two-dimensional flow channel where only diffusivity and van der Waals force with sticking probability σ_w affect the particle distribution, determine the boundary condition at the wall. What would be the dust concentration at the wall?

6.4 Show the density distribution in Eq. (6.48).

6.5 Derive Eq. (6.55).

6.6 Derive Eq. (6.62) via a transformation of coordinates.

6.7 Determine the velocity of water produced by rising gas bubbles of radius a of constant density and volume fraction in fully developed laminar flow starting from the bottom of a vertical pipe.

6.8 Derive Eq. (6.97) from the tangential component of momentum of the fluid and Eq. (6.88).

6.9 A cyclone separator has $2R_0$ = 0.25 m, Q = 0.187 m^3/s. C = 4.17 m^2/s, L_c = 0.875 m, 2a = 10 μm, $\bar{\mu}$ = 4 x 10^{-6} kg/m s, $\bar{\rho}$ = 5 kg/m^3, F = 500 s^{-1}, ρ_{p1} = 2.98 kg/m^3, q/m = 10^{-4} C/kg, σ = 0.05. Compute the collection efficiency. A second cyclone has $2R_0$ = 1.5 m, Q = 6.43 m^3/s, C = 31.1 m^2/s, L_c = 5.25 m, and similar pressure drop and other parameters. Compute its collection efficiency. Compute the acceleration at R_0 due to centrifugal force and that due to electrostatic force for each. Discuss the contributing factors to collection efficiency. Ans. 97.3%, 98%; 13100, 2180m/s^2; 1330, 7950 m/s^2.

6.10 Compute the thickness of layer of deposit for a cyclone of $2R_0$ = 1.50 m, L_c = 5.25 m, ρ_{ps} = 822 kg/m^3 (0.6 volume fraction solid) and a yield stress of 476 N/m^2. Ans. 0.0672 m

Chapter 7

GENERAL MOTION OF DILUTE SUSPENSIONS

In the previous chapters we have accounted for the basic aspects of multiphase flow through the steps of recognition of primary interactions of phases, formulation of conservation relations, determination of transport properties, experimental evidence, and validations in the one-dimensional systems and pipe flow systems. Now we want to take up the general case of two- and three-dimensional flows based on the continuum approximations.

We are concerned with motion and deposition of particles, bubbles, or droplets suspended in a fluid. They are treated as spherical in the ideal sense and isometric particles in actual applications. The general flow regimes of liquid-vapor mixtures (Sec. 4.2, Fig. 4.10 and general liquid-gas flow, Sec. 5.4, for instance) call for extensive empiricism and are treated in a special volume on pipe flow [Govier & Aziz 1972] and in various handbooks (see, for instance, Hetsroni [1982], Cheremisinoff & Gupta [1983]). In this chapter, we shall treat cases of interaction of a suspension with a boundary, under the influence of hydrodynamic, gravity, and electrostatic forces. We shall take up at first the cases where rigorous computations can be made, followed by semiempirical methods. Similar solutions of fluid dynamics of suspensions will be discussed.

In this chapter we shall deal with cases in the following sequence:

Pseudo-potential motions

Laminar boundary layer motion

Turbulent boundary layer motion
Jets, sprays, and separated flow
Diffusion and fallout
Heat transfer and multiple phase interactions

7.1 Vortex Motion

This section belongs to the methodology which we define as pseudo-potential motion, that is, potential motion except that particle-fluid interactions are taken into account. Boundary friction is neglected.

Earlier work on motion of particles in a fluid in free vortex motion and uniform axial-flow velocity of both phases [Hirschkron & Ehrich 1964] computed particle trajectories as a basic study on cyclone separators (Sec. 6.7). Lagrangian coordinate system was used, which in itself is a unique case in fluid mechanics. For particle trajectories in swirling flow in the Lagrangian frame of reference, the radial (r), tangential (ϕ), and axial (z) components of the momentum equation take the forms:

$$(dU_p/dt) - (V_p^2/r) = F(U - U_p) \tag{7.1}$$

$$(dV_p/dt) + (U_pV_p/r) = F(V - V_p) \tag{7.2}$$

$$dW_p/dt = F(W - W_p) \tag{7.3}$$

where t is the time, U, V, W for the fluid, and U_p, V_p, W_p for the particle are the conjugate components of velocities. With the fluid-flow field given by:

$$U = 0, \quad W = W_o, \quad \text{and} \quad V = C/r$$

the last equality specifies free vortex motion with vorticity C. For the position of a particle given by coordinate r, ϕ, z,

$$U_p = dr/dt$$

Eqs. (7.1) to (7.3) become

$$(d^2r/dt^2) - (V_p^2/r) = -F(dr/dt) \qquad (7.4)$$

$$(d\ rV_p/dt) + F\ rV_p = FC \qquad (7.5)$$

$$(dW_p/dt) + FW_p = F\ W_o \qquad (7.6)$$

with inlet condition at z_o for a particle given by

$$U_p = U_{po}, \ V_p(r_o) = V_{po} \ \text{and} \ W_p = W_{po} \qquad (7.7)$$

Eqs. (7.4) to (7.6) now integrate to give

$$V_p = (C/r) + r^{-1} (r_p\ V_{po} - C)\ e^{-Ft} = r\ d\phi/dt \qquad (7.8)$$

$$W_p = W_o + (W_{po} - W_o)\ e^{-Ft} = dz/dt \qquad (7.9)$$

with U_p given by

$$\frac{d^2r}{dt^2} + F\ \frac{dr}{dt} - \frac{[(r_o V_{po} - C)\ e^{-Ft} + C]^2}{r^3} = 0 \qquad (7.10)$$

Eq. (7.9) is readily integrated to give $z(t)$, but Eqs. (7.8) and (7.10), being nonlinear, was integrated numerically to determine the deviation of particle trajectory from stream lines. Computation was carried out with the simplification of $r_o V_{po} = C$, and $W_{po} = W_o$.

Introducing the dimensionless time coordinate $t^* = tC^2/r_o^2$, and radial coordinate $r^* = r/r_o$, the drag effect is represented by the parameter Fr_o^2/C. Fig. 7.1 gives the particle paths in a z-plane and the time trace at various radii r/r_o, and Fig. 7.2 correlates the position to t^* for various drag parameters. This form of result does not determine directly the behavior of a particle cloud. The latter is facilitated by solution in the Eulerian system, which does not call for the determination of individual particle path.

<u>Distribution of particles in vortex motion.</u> In order to determine the distribution of particles, and, conceptually to account for cases where random motion of particles exists, the following solution was proposed. With similar simplifications as in the above, the continuity and momentum equations of particles in the Eulerian system of coordinates from Eqs. (2.111) and (2.113) are:

$$\frac{\partial r\, \rho_p\, U_p}{\partial r} + \frac{\partial r\, \rho_p\, W_o}{\partial z} = 0 \qquad\qquad (7.11)$$

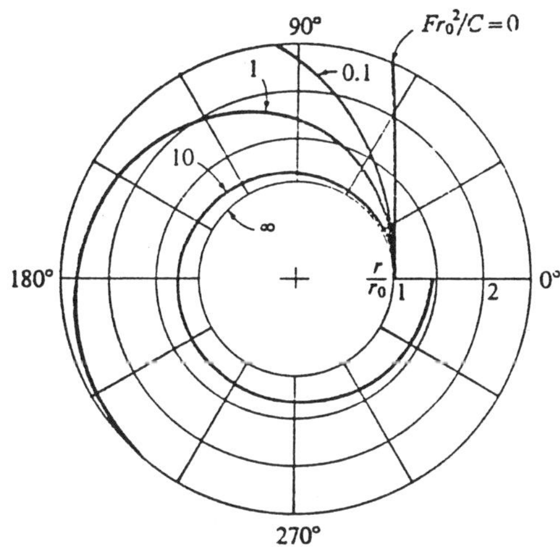

Figure 7.1   Spiral path, zero deviation of injection velocity.

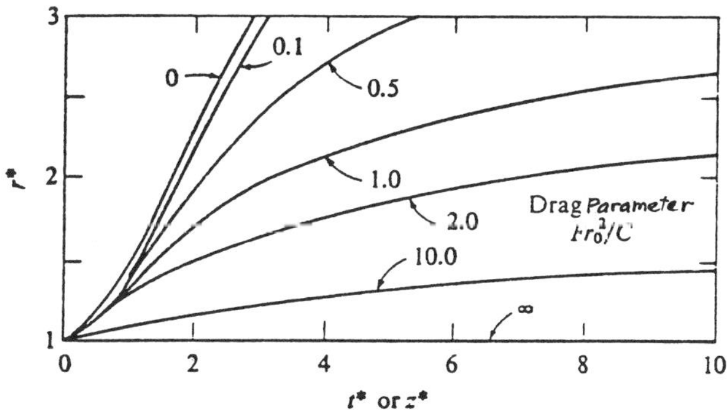

Figure 7.2   Displacement at zero deviation of injection velocity.

and

$$U_p \frac{\partial U_p}{\partial r} + W_o \frac{\partial U_p}{\partial z} - \frac{v_p^2}{r} = - F U_p \tag{7.12}$$

while the momentum equations in the ϕ and z directions vanish. These are first-order differential equations and solution is feasible for given conditions at z_o.

Introducing stream function ψ_p for the particle phase such that

$$r \, \rho_p W_o = (\frac{\partial \psi_p}{\partial r})_z, \quad r \, \rho_p U_p = -(\frac{\partial \psi_p}{\partial z})_r \tag{7.13}$$

which satisfies Eq. (7.11), and changing the independent variables from r and z to ψ_p and z, we get, from Eq. (7.12)

$$(\frac{\partial U_p}{\partial z})_{\psi_p} + \frac{F}{W_o} U_p = \frac{c^2}{W_o r^3} \tag{7.14}$$

r is now a dependent variable. Integration gives, for $U_p = 0$ at z = 0, along a streamline ψ_p,

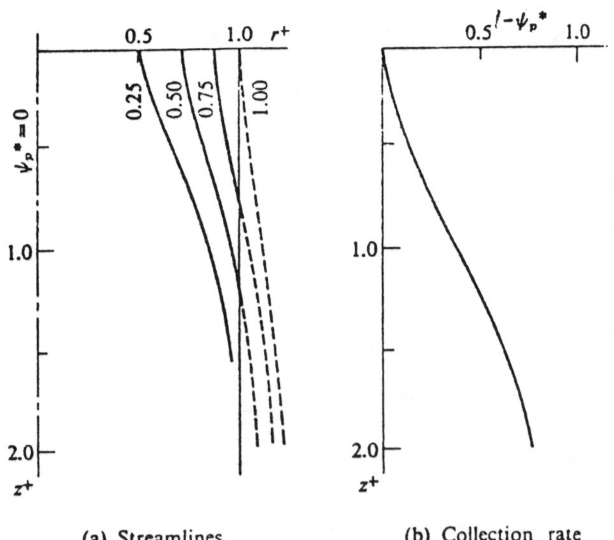

(a) Streamlines   (b) Collection rate

Figure 7.3   Flow and collection rate of particles in vortex flow
$(Fr_1^2/C = 10)$.

$$U_p = e^{-(Fz/W_o)} \int_o^z \frac{C^2}{W_o r^3} e^{(Fz/W_o)} \, dz \tag{7.15}$$

Also along a stream line, Eqs. (7.13) give

$$\frac{U_p}{W_o} = - \frac{(\partial \psi_p / \partial z)_r}{(\partial \psi_p / \partial r)_z} = \left(\frac{\partial r}{\partial z}\right)_{\psi_p} \tag{7.16}$$

Substitution of Eq. (7.14) and differentiating gives

$$\frac{d^2 r}{dz^2} + \frac{F}{W_o} \frac{dr}{dz} = \frac{C^2}{W_o^2 r^3} \tag{7.17}$$

With $z^* = (z/W_o)(C/r_1^2)$, Eq. (7.17) can be expressed as

$$\frac{d^2 r^*}{dz^{*2}} + \frac{F r_1^2}{C} \frac{dr^*}{dz^*} = \frac{1}{r^{*3}} \tag{7.18}$$

whose solution is also given in Fig. 7.2. The usefulness of this result is seen in Fig. 7.3(a) for the case of a uniform inlet mixture
with $Fr_1^2/C = 10$ for the streamline starting at $r = r_1$, $r^+ = r/r_1 =$
r^*, and $\psi_p^* = r^2 \rho_{po} W_o / r_1^2 \rho_{po} W_o$ at $z^* = 0$. For other streamlines
starting at r_1', $r^+ = (r_1'/r_1)r^*$, $z^+ = z^*(r_1'/r_1)^2$, $Fr_1'^2/C = (Fr_1^2/C)$
$(r_1'/r_1)^2$. The dashed parts of streamlines indicate particles coming
into contact with the wall. If the particles hitting the wall do not
become reentrained, the rate of collection is now obtained as shown
in Fig. 7.3(b), or the flux of particles hitting the wall is given by

$$2\pi \, \psi_p = 2\pi \, r_1 \int_o^z \rho_p U_p \, dz \tag{7.19}$$

that is, the intercept of particle streamlines at the wall. This is
an idealized view of collection of particles in a cyclone separator
(Sec. 6.7).

An alternate case is that of a centrifuge for fractionating
particles. Here the lag in the tangential velocity of particles is
significant [see, for instance, Schachman 1959]. (Prob. 7.1)

7.2   Electrohydrodynamic Flow

Extensive details of the performance of electrostatic precip-
itators are given in the treatise by White [1963].   We shall now
explore some of the fundamental aspects of electrohydrodynamic flow
and to identify similarity relations.   Here again only the electrostatic effects and particle-fluid interaction are taken into account;
boundary friction is neglected.   The nature of space charge effect is
illustrated by the system in Fig. 7.4, of slug flow of a cloud of
charged solid particles into a two-dimensional channel of width 2b
with grounded conducting walls.

At the inlet into the channel, the cloud density of identical
solid particles is uniform (ρ_{po}), the longitudinal velocity of fluid
and particles are uniform and identical $(U = U_p = U_o)$.   This case is
an extension of Case (2), Sec. 6.4 and has a similar solution to Sec.
5.8.   The collision of particles with the wall is taken to be inelastic.   Denoting $x' = x/U$, the continuity and momentum equations
become:

$$\frac{\partial \rho_p}{\partial x'} + \frac{\partial \rho_p V_p}{\partial y} = 0 \tag{7.20}$$

$$\frac{\partial V_p}{\partial x'} + V_p \frac{\partial V_p}{\partial y} = \left(\frac{q}{m}\right) E_y + F(-V_p) \tag{7.21}$$

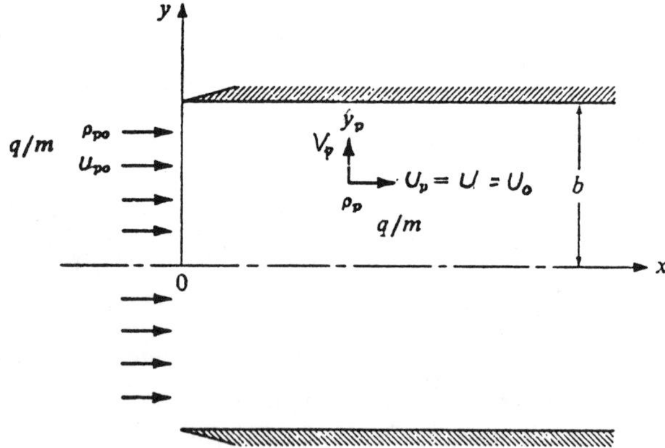

Figure 7.4   Flow of a uniformly charged cloud of solid particles into
a channel.   (The walls are grounded conductors.)

in the y-direction, V_p is the velocity of particles normal to the wall. Neglecting the change in the electric field in the x direction, the transverse electric field E_y is given by:

$$\partial E_y / \partial y = (q/m)\, \rho_p / \varepsilon_o \tag{7.22}$$

The inlet conditions are: $x = 0$, $V_p = 0$, and $x = 0$, $y = 0$, $\partial V_p / \partial y = 0$, with the leading edge effect neglected. Solution of Eqs. (7.20) to (7.22) gives the variation of particle concentration due to collection at the walls, and variation in V_p and E_y due to space charge. Eq. (7.20) is satisfied by introducing a stream function ψ such that:

$$\rho_p = -\partial \psi / \partial y, \qquad \rho_p V_p = \partial \psi / \partial x' \tag{7.23}$$

Replacing the independent variables (x', y) with (x', ψ), Eq. (7.21) is reduced to one with the derivative with respect to x' only and can be rewritten as:

$$\frac{\partial}{\partial x'} (V_p e^{Fx'}) = \left(\frac{q}{m}\right) E_y e^{Fx'} \tag{7.24}$$

Eq. (7.22) is reduced to:

$$-\frac{\partial E_y}{\partial \psi} = \varepsilon_o^{-1} \left(\frac{q}{m}\right) \tag{7.25}$$

Elimination of E_y between the last two equations gives

$$\frac{\partial^2}{\partial x' \, \partial \psi} (V_p e^{Fx'}) = -\left(\frac{q}{m}\right)^2 \varepsilon_o^{-1} e^{Fx'} \tag{7.26}$$

which has the particular solution

$$V_p = -\psi \left(\frac{q}{m}\right)^2 (1 - e^{-Fx'})/\varepsilon_o F \tag{7.27}$$

Substitution of Eq. (7.27) with Eq. (7.23) into Eq. (7.20) gives:

$$\frac{\partial \rho^\star}{\partial (x^{\star 2})} + \frac{\partial}{\partial y^\star} \left(\rho^\star \int \rho^\star \, dy^\star\right) = 0 \tag{7.28}$$

where $\rho^* = \rho_p/\rho_{po}$, $y^* = y/b$, and

$$x^{*2} = \frac{1}{2}\left[1 + \frac{U_o}{Fx}\,e^{-Fx/U_o}\right]\left[\left(\frac{q}{m}\right)^2 \frac{\rho_{po}b^2}{\epsilon_o U_o^2}\right]\frac{U_o}{Fb}\frac{x}{b} \qquad (7.29)$$

Similarity of Eq. (7.28) to Eq. (5.124) is readily seen, and the present coordinate system gives the distribution of cloud density by series expansion as:

$$\rho^* = (1 + x^{*2})^{-1} + y^{*2}(1 + x^{*2})^{-4} + \ldots \qquad (7.30)$$

The stream function or mass distribution is given by:

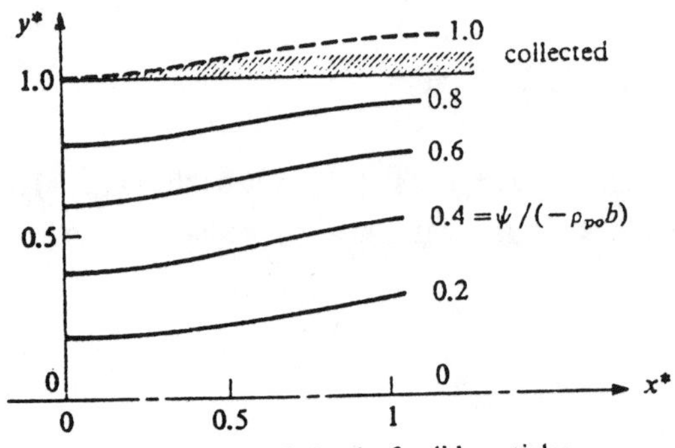

(a) Streamlines of cloud of solid particles

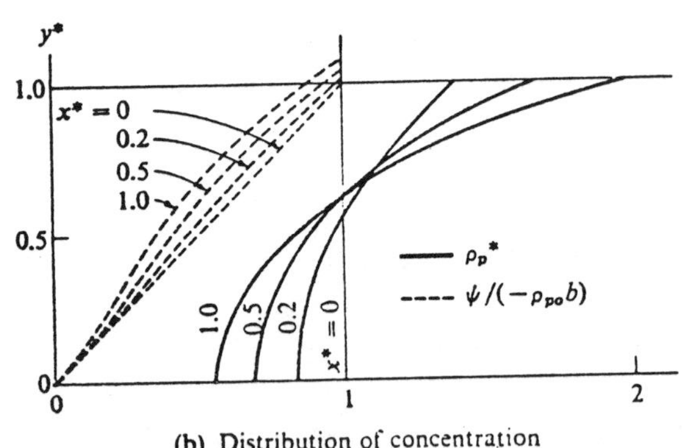

(b) Distribution of concentration
and stream function

Figure 7.5   Behavior of slug flow of a charged cloud of solid particles into a two-dimensional channel.

$$-\psi = \int_0^y \rho_p \, dy$$

$$= \rho_{po} b [y^*(1+x^{*2})^{-1} + \frac{y^{*3}}{3} (1+x^{*2})^{-4} + \ldots] \tag{7.31}$$

and the drift velocity distribution is:

$$\frac{V_p}{U_o} = (\frac{U_o}{Fb}) [1 - \exp(- \frac{Fb}{U_o} \frac{x}{b})] (\frac{q}{m})^2 (\frac{\rho_{po} b}{\varepsilon_o U_o^2})$$

$$\times [y^*(1+x^{*2})^{-1} + \frac{y^{*3}}{3} [1+x^{*2}]^{-4} + \ldots] \tag{7.32}$$

These relations are shown in Fig. 7.5. The amount collected is shown between the streamline marked 1.0 and the wall in Fig. 7.5(a). It is noted that

$$(\frac{q}{m})^2 \frac{\rho_{po} \, b^2}{\varepsilon_o \, U_o^2} = (N_{Re})_{eb}^{-1} \tag{7.33}$$

is the electric Reynolds number defined by Stuetzer [1962] and is the ratio of inertia force of the solid particle to the electrostatic force and the momentum transfer number $N_m = U_o/Fb$ is based on the dimension b. They are further related to the electro-viscous number N_{ev} according to, $N_{ev} = (\rho_{po}/4\varepsilon_o)(q/m)^2 (b^4/\bar{\nu}^2)$ (compare to Eq. 6.27):

$$N_m (N_{Re})_{eb}^{-1} = (N_{ev}) (\frac{8}{9} \frac{a^2}{b^2} \frac{\bar{\rho}_p}{\rho}) (N_{Re})_b^{-1} \tag{7.34}$$

where $(N_{Re})_b$ is the Reynolds number of the channel. These relations permit scale-up of test systems.

The collision rate per unit area of the channel is given by the product $[\rho_p V_p]_w$ at $y^* = 1$. Eq. (7.23) gives

$$[\rho_p V_p]_w = [\partial\psi/\partial x']_{y^*=1} = \frac{1}{2} \rho_{po} U_o (N_{Re})_{eb}^{-1} N_m (1 - e^{-Fx/U_o})$$

$$\times [\frac{1}{(1 + x^{*2})^2} - \frac{4}{3(1 + x^{*2})^5} + \ldots] \tag{7.35}$$

It is also seen that $\rho_p V_p = 0$ at $x = 0$, the inlet, and the initial rate of increase of $[\rho_p V_p]_w$ is given by

$$\left.\frac{\partial\,\rho_p\,V_p}{\partial x'}\right|_{x=0} = \rho_{po}\,U_o\,(N_{Re})^{-1}_{eb} \qquad\qquad (7.36)$$

regardless of the value of N_m. These results are plotted as shown in Fig. 7.6 for $(N_{Re})_{eb} = 10^{-2}$ and $N_m = 10^{-2}$ and 10^{-3}; the magnitudes of quantities for a physical system are shown in Prob. 7.2. Figure 7.6 shows that with higher flow velocity (turbulent flow), to a limit to be determined, the collection rate per unit area could be higher than proportionate to the low velocity (laminar) flow.

For a particle cloud of a range of size distribution, the above should be modified according to the basic equations, but the above transformation will not be applicable because of multiplicity of streamlines and mutually self-consistent field. Numerical solution, however, is feasible. A curve such as those shown in Fig. 7.6 will be obtained for each narrow size range; summation gives the total accumulation. This will give proportionately greater amounts of large particles near the entrance than in the entering suspension, a fact that was experienced [White 1963] but can now be explained.

The above electric Reynolds number and electroviscous number are in the form for self-field of a cloud of charged particles. For given characteristic electric field E_o, they take the form, for

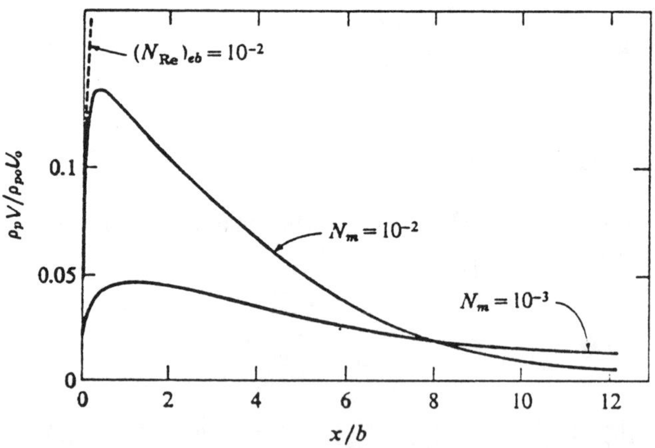

Figure 7.6   Dimensionless impact rate at wall of channel in Fig. 7.4.

example:

$$(N_{Re})_{eb} = U_o^2/E_o \; (q/m)b \tag{7.37}$$

and

$$(N_{ev})_b = [E_o \; (q/m) \; b^3/\bar{\nu}^2]^{1/2} \tag{7.38}$$

<u>Application to electrostatic precipitators</u> [Soo 1973c].   Typical
field and collector plate configurations for electrostatic precip-
itators are illustrated in Fig. 7.7.   The system consists of corona
wires and collector plates as shown in Fig. 7.7 (a) and a simplified
representation is given in Fig. 7.7 (b).   The suspension is flowing
with velocity U in the x direction, and a drift velocity KE in the y
direction.   K is the mobility of the particles, K = (q/m)/F, and E is
a uniform electric field.   Fig. 7.7 (b) also represents the situation
in a collector passage such as in a two-stage precipitator as shown
in Fig. 7.7 (c).   The slug flow model is a good approximation,
especially for cases of large interaction length giving a large
relative velocity from Eq. (3.122).

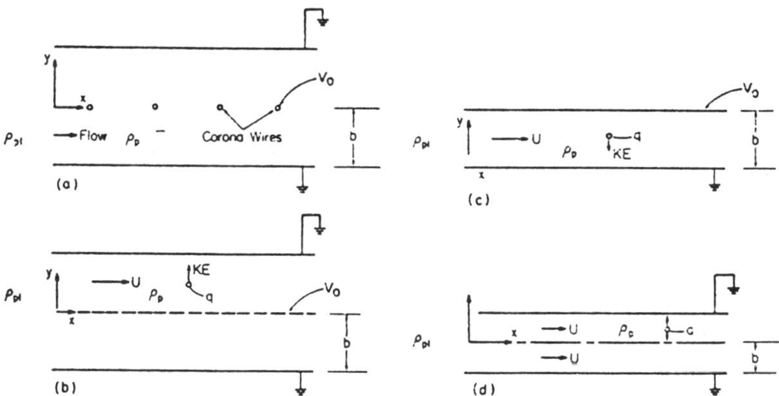

Figure 7.7   Field and boundary configurations of electro static
              precipitator system.   (a) Conventional precipitator
              passage; (b) Idealized equivalent passage of (a);
              (c) Collector passage formed by opposite charged plates;
              (d) Passage for collection by image force.

The diffusion equation of the particles of a given species takes the form:

$$N_{Pe} \frac{\partial \rho_p}{\partial x^*} + \alpha_e \frac{\partial \rho_p}{\partial y^*} = \frac{\partial^2 \rho_p}{\partial y^{*2}} \qquad (7.39)$$

for $x^* = x/b$ and $y^* = y/b$. N_{Pe} and α_e are the Peclet number and a parameter correlating drift and diffusion, respectively, and

$$N_{Pe} = \frac{bU}{D_p}, \qquad \alpha_e = \frac{KEb}{D_p} = \left(\frac{q}{m}\right) \frac{Eb}{FD_p} \qquad (7.40)$$

Using these definitions, the boundary conditions are (Eq. 4.90):

$$y^* = 0, \qquad \frac{\partial \rho_p}{\partial y^*} = \alpha_e \, \rho_p \qquad (7.41)$$

at the wires, and at

$$y^* = 1, \qquad \frac{\partial \rho_p}{\partial y^*} = (1-\sigma) \, \alpha_e \, \rho_p \qquad (7.42)$$

at the collector plate where deposition is desired. σ ranges from 0.1 to 0.5 at a clean surface in an electrostatic precipitator. The influence of other fluxes at the boundary is neglected at present. A solution of Eq. (7.39) by separation of variables was given by Soo & Rodgers [1971a]. It is interesting to note that:

(a) At small drift force or small α_e and small σ, the density distribution is approximated by:

$$\rho_p/\rho_{p1} = [1 + \alpha_e \, y^*] \, \exp[-\sigma \, \alpha_e \, N_{Pe}^{-1} \, x^*] \qquad (7.43)$$

for particle cloud density ρ_{p1} at the inlet. The collection efficiency η_c is obtained from the change in the total mass rate of particles according to:

$$1 - \eta_c \Big|_x = \int_0^b \rho_p U \, dy \Big|_x \Big/ \int_0^b \rho_p U \, dy \Big|_0$$

or

$$\eta_c \simeq 1 - \exp[-\sigma \, \alpha_e \, N_{Pe}^{-1} \, L_e/b]$$

$$= 1 - \exp[-\sigma \, (\tfrac{q}{m}) \, \frac{E}{F \, U} \, \frac{L_e}{b}] \qquad (7.44)$$

for passage length L_e. This is the Deutsch equation [White 1963] modified by the introduction of the sticking probability of particles.

(b) At large α_e, we eliminate the $\partial \rho_p/\partial y^\ast$ term in Eq. (7.39) by transforming with an exponential form to get:

$$\rho_p \propto (1 - y^\ast) \, \exp[\tfrac{1}{2} \, \alpha_e \, y^\ast - \tfrac{1}{4} \, \alpha_e^2 \, N_{Pe}^{-1} \, x^\ast] \qquad (7.45)$$

where $(1 - y^\ast)$ was obtained for small particle diffusivity, and

$$\eta_c \cong 1 - \exp[- \tfrac{1}{4} \, \alpha_e^2 \, N_{Pe}^{-1} \, L_e/b]$$

$$= 1 - \exp[- (\tfrac{q}{m})^2 \, \frac{E^2 b}{4F^2 \, D_p} \, \frac{L_e}{Ub}] \qquad (7.46)$$

Hence, for large α_e, the collection is not strongly influenced by the sticking probability; the influence of particle diffusivity is readily seen. (Prob. 7.3) Space charge is neglected in the above, which is appropriate for a dilute suspension in a strong applied electric field.

(c) Collection in the passage such as is shown in Fig. 7.7 (d) depends on space charge effect when there is substantial concentration of charged particles. It depends on image force when the suspension is dilute. Because of the presence of particle diffusivity, even in laminar fluid flow in the passage, a particle is not expected to remain at the mid-plane indefinitely, and collection is produced by the interaction of diffusion and the image force per unit mass f_y which gives a drift velocity

$$\frac{f_y}{F} \simeq \frac{q^2}{\pi \, \varepsilon_o \, m \, F \, b^2} \, \frac{y^\ast}{(1 - y^{\ast 2})^2} \qquad (7.47)$$

For a given sticking probability σ, as a particle reaches $y = b - a$, a being the particle radius:

$$\rho_p \simeq \rho_{p1} \exp[-\sigma \, \bar{\alpha} \, N_{Pe}^{-1} \, (a^2/b^2)x*] \tag{7.48}$$

where $\bar{\alpha} = (\bar{\rho}_{po}/\varepsilon_o) \, (q/m)^2(b^2/FD)$ based on the space charge effect of the material. The efficiency is given by:

$$\eta_c = 1 - \exp[\, - \, \sigma \, \frac{\bar{\rho}_p}{\varepsilon_o} \, (\frac{q}{m})^2 \, \frac{a}{FU_o} \, \frac{a}{b} \, \frac{L_e}{b}] \tag{7.49}$$

where ρ_p is assumed to be uniform at each section of x because of high particle diffusivity. (Prob. 7.4, 7.5)

Significant modification of the above relations is seen in the effects of back corona, nonuniform electric field, and electric wind. Computation of velocity of electric wind was given by Ramadan and Soo [1969] with experimental verification.

7.3 Laminar Boundary Layer Motion over a Flat Plate

Although most gas-solid systems encountered in practice are turbulent, the case of laminar boundary layer motion of a gas-solid suspension over a flat plate was treated [Soo 1968] to develop some basic understanding, via mathematical procedures, of interaction of a suspension with a boundary. This was done for the same reason laminar boundary layer motion of a gas was studied in spite of the fact that almost all flow machineries involve turbulent flow. We leave out the electrostatic effects for the present.

A simplest nontrivial case of two-dimensional motion of a gas-solid suspension over a flat plate consists of an incompressible gas phase with uniform solid concentration of particles of one size in the free stream only.

For viscosity μ and kinematic viscosity ν of the predominating component in the mixture ($\mu \sim \mu_{mixture}$), the Schmidt number of the particle phase is $N_{Sc} = \nu/D_p$, D_p is the diffusivity of the particles in the mixture. Similarly, the thermal conductivity of the mixture $\kappa_{mixture}$ is contributed mainly by that of the predominating component in the mixture $\kappa(\sim\kappa_{mixture})$ while that of the particles is related to the diffusivity D_p according to:

$$\kappa_{pm}/c_p\,\rho_p \equiv \alpha_p \sim D_p$$

according to Eq. (3.100), where α_p is the thermal diffusivity of the particle cloud.   For thermal diffusivity α of the predominating component ($\alpha = (\kappa/c_\rho)$, the Lewis number of component p is thus:

$$N_{Le} = \frac{\alpha}{D_p} = \frac{\alpha}{\nu}\frac{\nu}{D_p} = N_{Pr}^{-1}\,N_{Sc} \sim \frac{\alpha}{\alpha_p}$$

where N_{Pr} denotes the Prandtl number (ν/α).

Generality of results is achieved by the following simplifications:

(1)   The motion is subsonic and shear dissipation can be neglected in the energy equations.

(2)   Since $\rho \gg \rho_p$, the main motion can be treated as incompressible, or ρ = constant, allowing for the variation of ρ_p, however.

(3)   The range of temperature is such that thermal diffusion can be neglected.

(4)   The ranges of temperature and density variations are such that μ, μ_p, κ, κ_p, ν, D_p, α, α_p, F, and G can be taken as constants.

The solution obtained here is therefore in the nature of a simplest nontrivial solution in non-equilibrium flow.

We now consider the case of laminar flow over a semi-infinite flat plate as shown in Fig. 7.8.   The velocity components U, U_p, V, V_p, are in the x, y direction along and normal to the flat plate.

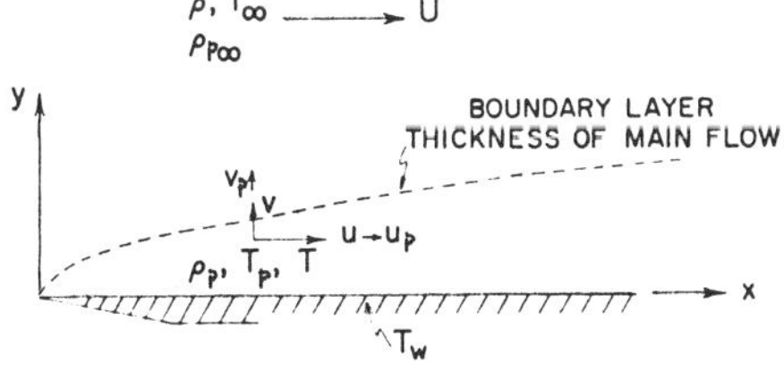

Figure 7.8   Coordinates and boundary layer thickness of predominating component in non-equilibrium flow over a flat plate.

The flow velocity $U_0 = U_{po}$, density ρ_{po}, and temperature $T_0 = T_{po}$ are uniform at infinity, ρ being a constant. The wall temperature T_w is constant and the pressure is constant throughout. With boundary layer approximations [Schlichting 1979] and neglecting conduction of heat in the x direction, the basic equations take the form:

$$\frac{\partial U}{\partial x} + \frac{\partial V}{\partial y} = 0 \tag{7.50}$$

$$U \frac{\partial U}{\partial x} + V \frac{\partial U}{\partial y} = \nu \frac{\partial^2 U}{\partial y^2} + F \frac{\rho_p}{\rho} (U_p - U) \tag{7.51}$$

$$U \frac{\partial T}{\partial x} + V \frac{\partial T}{\partial y} = \alpha \frac{\partial^2 T}{\partial y^2} + G \frac{C_p}{C} \frac{\rho_p}{\rho} (T_p - T) \tag{7.52}$$

for the fluid phase; and

$$\frac{\partial \, \rho_p U_p}{\partial x} + \frac{\partial \, \rho_p V_p}{\partial y} = 0 \tag{7.53}$$

$$U_p \frac{\partial U_p}{\partial x} + V_p \frac{\partial U_p}{\partial y} = F (U - U_p) + D_p \frac{\partial^2 U}{\partial y^2} \tag{7.54}$$

$$U_p \frac{\partial V_p}{\partial x} + V_p \frac{\partial V_p}{\partial y} = F(V - V_p) \tag{7.55}$$

$$U_p \frac{\partial T_p}{\partial x} + V_p \frac{\partial T_p}{\partial y} = G(T - T_p) + \alpha_p \frac{\partial^2 T_p}{\partial y^2} \tag{7.56}$$

for the particle phase. Note that the boundary layer approximation applies to the fluid phase only, the y-component of momentum equation of the particles is retained. The terms including D_p and α_p originate from Eqs. (3.97) and (3.98). The boundary conditions are:

$$y = 0, \; x > 0, \; U = V = V_p = 0, \; T = T_w;$$
$$\tag{7.57}$$
$$y = \infty, \; U = U_p = U_{po}, \; T = T_p = T_0, \; \rho_p = \rho_{po}$$

In addition, at $y = 0$, Eq. (7.54) gives, for $D_p = 0$,

$$U_{pw} = U_0 \left(1 - \frac{Fx}{U_0}\right) \tag{7.58}$$

over the range $0 < xF/U_0 < 1$, and $U_{pw} = 0$ for whatever value of D_p, for $xF/U_0 > 1$, assuming small interaction length (Eq. 1.72). With ξ = xF/U_0 as a perturbation parameter, additional boundary conditions can be specified over two ranges:

(1) $0 < \xi < 1$, $U_{pw} = U(1 - \xi)$ for $D_p = 0$; for the case $D_p \neq 0$, we have the situation of combined slip and diffusion of component p. The case $D_p = 0$ is treated here.

(2) $\xi > 1$, $U_{pw} = 0$ and accumulation of phase p is expected if diffusion is absent.

Because of our simplifying assumptions of $\rho_p \ll \rho$, the last terms of both Eqs. (7.51) and (7.52) can be neglected and the similar solution given by Blasius [Schlichting 1979] is valid. Thus, for $\eta = y(U_0/\nu x)^{1/2}$,

$$U = U_0 f'(\eta), \quad V = \frac{1}{2}\left(\frac{\nu U_0}{x}\right)^{1/2} (\eta f' - f) \text{ and } 2f''' + ff'' = 0 \qquad (7.59)$$

and the Blasius function f is well known. The energy Equation (7.52) gives

$$T^{\star} = \frac{T - T_0}{T_w - T_0} = t(\eta) = \int_0^{\eta} (f'')^{N_{Pr}} \, d\eta / \int_0^{\infty} (f'')^{N_{Pr}} \, d\eta \qquad (7.60)$$

Solution of Eqs. (7.53) to (7.55) to obtain U_p, V_p, and ρ_p has to be treated separately over the above two ranges. It is seen that as long as F is greater than zero, range (1) exists as a special character of non-equilibrium flow. It is recognized, however, that for small particles, the actual distance for the range for the range $0 < \xi < 1$ or $0 < x < U_0/F$ can be extremely short from the leading edge. Yet when applied to a suspension with $F \sim 100 \text{ s}^{-1}$, $U_0 \sim 10$ m/s, this length is about 10 cm from the leading edge.

Again because of the simplifying assumptions, Eqs. (7.50) to (7.55) are independent of the energy equations (7.58). Analogous effect of relaxation in momentum exchange, the variation of T is expected to be governed by the similarity parameter $\xi_t = xG/U_0 = (G/F)\xi$. From comparison of Eqs. (7.54) and (7.56), we know that complete similarity of the distributions of T_p and U_p is not expected except for $N_{Pr} = 1$, and $G = F$. The temperature distribution T_p will be treated independently.

We note that a compressible fluid phase can be readily treated, even though we choose the present simple case. The fluid phase stream function ψ can be determined by well known methods such as the Howarth transformation [1936] with the motion of the particle phase determined with $f(\eta) \propto \psi/x^{1/2}$. A separate set of ψ_p for the particle phase can be determined from the solution.

<u>Solution over the range $0 < \xi < 1$</u>. Since this range is extremely short in cases where diffusion is significant and since the initial period of disturbance from equilibrium is relaxation-controlled, our solution is carried out by neglecting diffusion, or $N_{Sc} \to \infty$ and $N_{Le} \to \infty$. Pertinent solutions of Eqs. (7.54) and (7.55) were obtained in the form:

$$U_p = U_o[g_o'(\eta) + \xi\, g_1'(\eta) + \xi^2\, g_2'(\eta) + \ldots\,] \tag{7.61}$$

$$V_p = \frac{1}{2}\, (\nu\, U_o/x)^{1/2}[m_o(\eta) + \xi m_1(\eta) + \xi^2 m_2(\eta) + \ldots\,] \tag{7.62}$$

Eq. (7.54) gives, for various orders of ξ:

$$g_o'' = 0, \; g_o' = 1, \; \frac{\eta}{2}\, g_1'' - g_o'g_1' = g_o' - f'$$

$$\frac{\eta}{2}\, g_2'' - 2g_o'g_2' = g_1'(1 + \frac{\eta}{2}\, g_1'' - g_1') - \frac{1}{2}\, g_1''\, m_1, \tag{7.63}$$

Eq. (7.55) gives,

$$m_o = 0, \; \frac{\eta}{2}\, m_1' + \frac{1}{2}\, m_1 = -(\eta\, f' - f)$$

$$\frac{\eta}{2}\, m_2' - \frac{5}{2}\, m_2 = -g_1'\, (\frac{\eta}{2}\, m_1' - \frac{1}{2}\, m_1) + \frac{1}{2}\, m_1 m_1' \tag{7.64}$$

Simultaneous solution of Eqs. (7.63) and (7.64) gives

$$g_o' = 1, \; g_1' = f' - 1 + \eta\, f'' + \eta^2 \int_\eta^\infty \frac{f'''}{\eta}\, d\eta$$

$$\frac{\left(\dfrac{g_2'}{4}\right)}{\eta} = -\frac{g_1'}{2}\left(\dfrac{g_1'}{2}\right)_\eta + \frac{2}{5}(\eta-f)\,g_1'' + 2\,\frac{g_1'}{5}_\eta \qquad (7.65)$$

$$m_o = 0, \quad m_1 = 2(\eta-f), \quad \left(\dfrac{m_2}{3}\right)_\eta{}' = -\frac{g_1'}{2}\left(\dfrac{m_1}{\eta}\right) + \frac{1}{4}(m_1{}^2) + \frac{2m_1}{4!}_\eta \qquad (7.66)$$

with the following limits as $\eta \to 0$,

$$f = \frac{A}{2!}\,\eta^2, \quad f' = A\eta, \quad f'' = A, \quad f''' = -\frac{A^2}{4}\,\eta^2, \quad \dots \qquad (7.67)$$

where $A = 0.332$ was given by Blasius, further

$$g_1' = -1, \quad g_2' = \frac{1}{2}, \quad m_1 = 2\eta, \quad m_2 = -6\eta \dots \qquad (7.68)$$

Eq. (7.53) is satisfied by :

$$\rho_p = \rho_{po}[h_o(\eta) + \xi h_1(\eta) + \xi^2 h_2(\eta) + \dots] \qquad (7.69)$$

the functions h_o, h_1, etc. can be determined in a similar manner as in the above. (Prob. 7.6)

The functions g, m, and h are readily computed, with g and m as shown in Fig. 7.9. Fig. 7.10 illustrates the changing nature of U_p, V_p, and ρ_p with ξ. Note that the coordinate η changes with ξ also.

For finite but large N_{Sc}, the above solution is modified to:

$$g_o' = 1, \quad g_1' = f' - 1 + N_{Sc}^{-1}f''' + N_{Sc}^{-2}f' + \dots \qquad (7.70)$$

Since $f'''(0) = 0$ but $f^V(0) = -A^2/2$, the boundary condition in Eq. (7.67) is modified to:

$$U_{pw}/U_o = 1 - \xi(1 + N_{Sc}^{-2}\frac{A^2}{2} + \dots) + \xi^2(\dots \qquad (7.71)$$

or a rapid slowing down of particles at the wall at small N_{Sc}. However, such an effect is small for $N_{Sc} \sim 10$ in this relaxation-controlled range.

<u>Solution for the range $\xi > 1$</u>.   Beyond $\xi = 1$, at $N_{Sc} = 0[1]$ the process is mainly diffusion-controlled which will lead to $U_p = U$, $V_p = V$, $T_p = T$, $\rho_p = \rho_{po}$.   Non-equilibrium in the range $\xi > 1$ will be experienced only in the cases of long relaxation times and large N_{Sc}.   We shall exclude the case of $N_{Sc} = \infty$ when deposition beyond $\xi = 1$ occurs, and the present basic formulation will be inadequate in describing the interactions.   For finite but large N_{Sc}, this range is satisfied by a solution in the form:

$$U_p = U_o[g_o' + \xi^{-1} g_{1-}' + \xi^{-2} g_{2-}' + \ldots] \tag{7.72}$$

$$V_p = \tfrac{1}{2}\,(\nu\,U_o/x)\,[m_{0-} + \xi^{-1}m_{1-} + \xi^{-2}m_{2-} + \ldots] \tag{7.73}$$

$$\rho_p = \rho_{po}[h_{0-} + \xi^{-1}h_{1-} + \xi^{-2}h_{2-} + \ldots] \tag{7.74}$$

Solutions of various orders of g_-, h_- and m_- in terms of the Blasius function are left as an exercise (Prob. 7.7).   Fig. 7.11 shows the

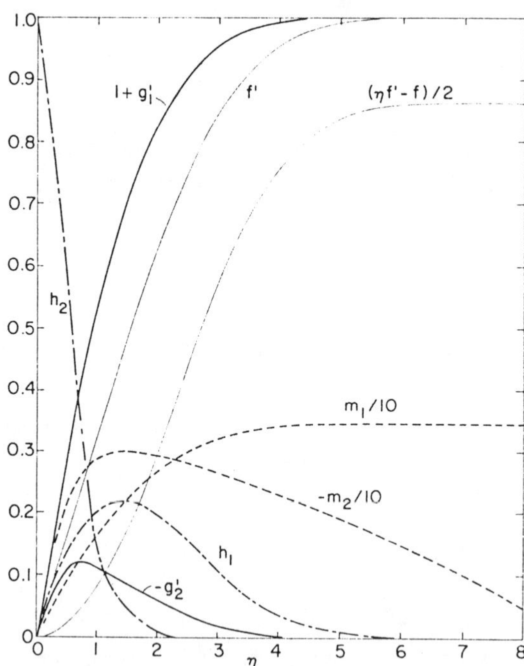

Figure 7.9   Functions for non-equilibrium velocity and density
distributions, $0 < \xi < 1$, $N_{Sc} \to \infty$.

distributions of velocities and density of component p for ξ = 5, 10, and ∞ for N_{Sc} = 10. It is seen that after acceleration in the range $0 < \xi < 1$ of component p to the velocity of the predominating component, as V decreases, V_p is being slowed down by relaxation and stays ahead of V, until $\xi \to \infty$, $V_p \to V \to 0$.

<u>Temperature distribution</u>. The temperature distribution of component p can be determined from Eq. (7.56) for T_p, U_p, and V_p given for the above ranges. The boundary condition for T can be determined form Eq. (7.56) at y = 0, which gives, for $N_{Sc} = \infty$,

$$T^{*}_{pw} = \frac{T_{pw} - T_o}{T_w - T_o} = 1 - (1-\xi)^{G/F} \qquad (7.75)$$

in the range $0 < \xi < 1$. For small ξ, and ξ_t = Gx/U

$$T^{*}_{pw} \sim \left(\frac{G}{F}\right)\xi - \frac{1}{2!}\left(\frac{G}{F}\right)\left(\frac{G}{F} - 1\right)\xi^2 + \ldots \sim \xi_t - \frac{1}{2}\left(1 - \frac{F}{G}\right)\xi_t^2 + \ldots \qquad (7.76)$$

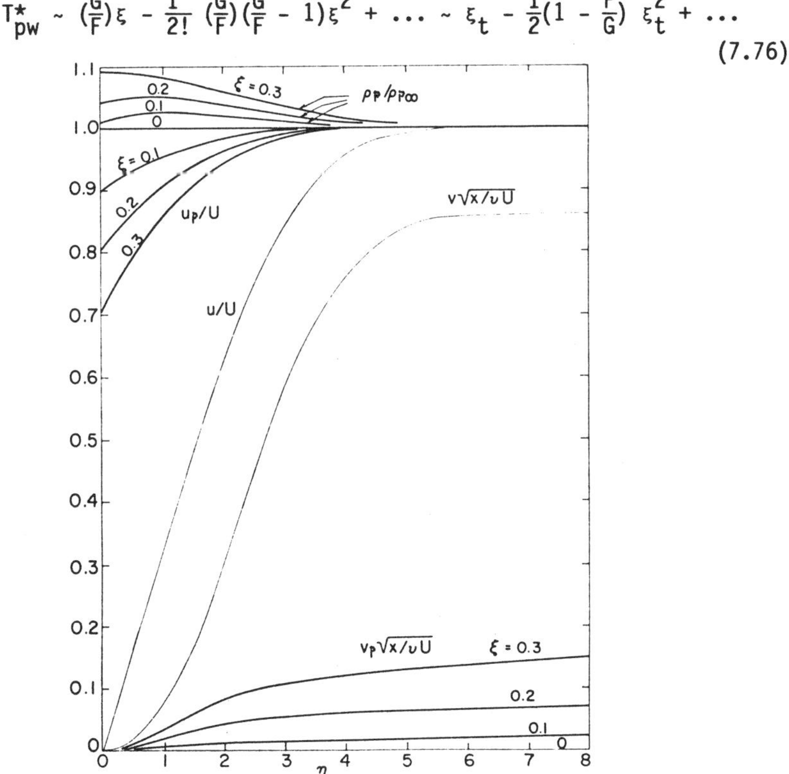

Figure 7.10 Non-equilibrium velocities and density at various ξ, $0 < \xi < 1$, $N_{Sc} = \infty$.

Therefore, the temperature distribution is influenced by both F and G. Solution in terms of various orders of ξ is feasible as in the above calculations. Solutions were again obtained for the two ranges of $0 < \xi < 1$ and $\xi > 1$ with the following series expansions:

$$T_p^* = \frac{T_p - T_o}{T_w - T_o} = t_o(\eta) + \xi\, t_1(\eta) + \xi^2 t_2(\eta) + \cdots \tag{7.77}$$

and

$$T_p^* = t_{o-} + \xi^{-1} t_{1-} + \xi^{-2} t_{2-} + \cdots \tag{7.78}$$

Determination of equations of different orders of t and t_ and their boundary conditions are left for an exercise (Prob. 7.8).

In both ranges, when $G/F = 1$, $N_{Pr} = 1$, the temperature distributions T* of phase p are similar to those of $1 - (U_p/U_o)$ in

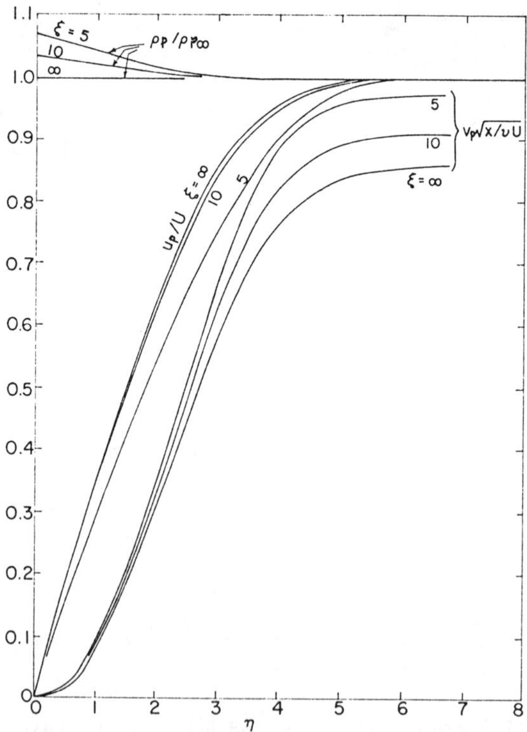

Figure 7.11 Non-equilibrium velocities at various $\xi, \xi > 1$, $N_{Sc} = 10$.

Figs. 7.10 and 7.11. When $G/F > 1$, $T_{pw} = T_w$ beyond $\xi = 1$. The well-known conclusion that smaller N_{Pr} gives a greater thermal boundary layer thickness than that of the fluid prevails here.

The nature of non-equilibrium flow is seen from various quantities in the physical coordinates as shown in Fig. 7.12. Using dimensionless quantities, we denote the coordinate

$$y^* = y/(\nu/F)^{1/2} \tag{7.79}$$

and the velocities of particle V_p in the y direction as

$$V_p^* = [\frac{V_p}{U_o} (\frac{x\ U}{\nu})^{1/2}]\ \xi^{-1/2} \tag{7.80}$$

and V^* is that for the fluid phase defined in a similar manner. The boundary layer thickness δ and δ_p are given by y^* at $U/U_o = 0.99$. The transition at $\xi = 1$ is clearly seen. Fig. 7.12 gives the distributions in ρ_p, U_p, U, V_p, V, δ_p, δ in a linear scale of ξ over the whole range of interest for $N_{Sc} = \nu/D = 10$ and negligible L_p. It is seen that:

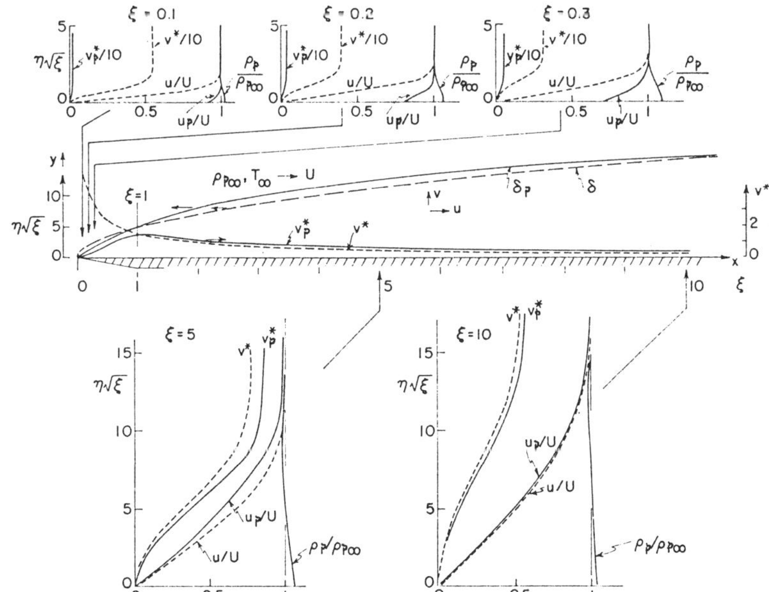

Figure 7.12 Coordinates, boundary layer thicknesses, and
distributions in laminar flow over a flat plate.

(1) Over the range $0 < \xi < 1$, even for very small L_p, U_p is finite at the wall (slip) for large N_{Sc} and is slowed down by the fact that $U = 0$ at the wall and $N_{Kp} \sim 0$. Although the singularity in the Blasius solution gives $V^* \to \infty$ at $x = 0$, $V_p^* = 0$ initially for finite values of N_{Sc}, V^* is accelerated from zero in this range while V^* is decreasing until the point $\xi = 1$ is reached where $V_p^* = V^*$, $U_p = U$. $\rho_{pw}/\rho_{po} > 1$ prevails near the wall due to slowing down of the mixture. In this range $\delta_p < \delta$.

(2) Equalization of velocities of phases at $\xi = 1$ does not mean that the disparity of the phases has died out. This is because V^* continues to decrease and in the range $\xi > 1$, $V_p^* > V^*$ due to the relaxation mechanism because of the inertia of the particles. $U > U_p$ at a given y^* in the range $\xi > 1$, and δ_p/δ. For the same reason, ρ_p/ρ_{po} decreases below 1 in the boundary layer although $\rho_{pw}/\rho_{po} > 1$ at the wall. At $\xi = \infty$, $U_p = U$, $V_p = V = 0$, $\rho_p = \rho_{po}$, $\delta = \delta_p$. The decrease in ρ_{pw} shows that this range is diffusion-controlled.

(3) It is well known that $(N_{Re})_x = xU_0/\nu \sim 10^5$ for the transition from laminar to turbulent boundary layer [Schlichting 1979]. This relation gives the transition at

$$\xi < 10^5 \, (U_0^2/\nu F)^{-1} = \xi_{tr} \tag{7.81}$$

for the laminar range. If we take, for example, $U_0 = 100$ m/s, $\nu = 10^{-4}$ m^2/s, $\xi_{tr} = 0.1$ for $F = 100$ s^{-1} (such as 100 µm sand in air); and $\xi_{tr} = 100$ for $F = 10^5$ s^{-1} (such as 1 µm aerosols in air). Therefore, in the former case, slip of p would not have considerably diminished and $\xi < 1$ before turbulence sets in at 10 cm from the leading edge, while in the latter case, the observable range include 1 mm for $\xi < 1$, and 10 cm for the range $1 < \xi < 100$.

(4) The previous items shows that, when transition to turbulence occurs within the range $0 < \xi < 1$, presence of phase p in sufficient quantity would delay the transition since $V_p < V$. Yet, when transition to turbulence occurs beyond $\xi = 1$, phase p enhances transition at lower (N_{Re}) than that in a classical fluid because

$V_p > V$. Therefore, in the above physical examples, transition to turbulence will be suppressed by the 100 μm sand in air, while the 1 μm aerosols would aid the transition.

7.4 Particulate Suspension in a Turbulent Fluid

The case of an semi-infinite homogeneous charged suspension in a system represented by Fig. 7.8 is physically unattainable and mathematically divergent. Stukel and Soo [1969] chose the system of flow into a channel formed by two parallel flat plates as shown in Fig. 7.13 to study the boundary layer interaction of a charged suspension of particles.

With the two parallel flat plates having sharp leading edges situated at a distance 2b apart, the coordinate x is aligned with the flat plates and the coordinate y normal to it. The corresponding components of velocities are U, V for the fluid and U_p and V_p for the particles. Each of the particles has radius a with a material density $\bar{\rho}_p$, and is suspended in a fluid of material density $\bar{\rho}$. The suspension is dilute.

For the system defined above, with flow in the subsonic range, the carrier fluid may be treated as incompressible. However, the dependence of ρ_p on the transport processes must be taken into account. We therefore have the following continuity equations:

$$(\partial U/\partial x) + (\partial V/\partial y) = 0 \tag{7.82}$$

$$(\partial/\partial x)\,(\rho_p U_p) + (\partial/\partial y)(\rho_p V_p) = 0 \tag{7.83}$$

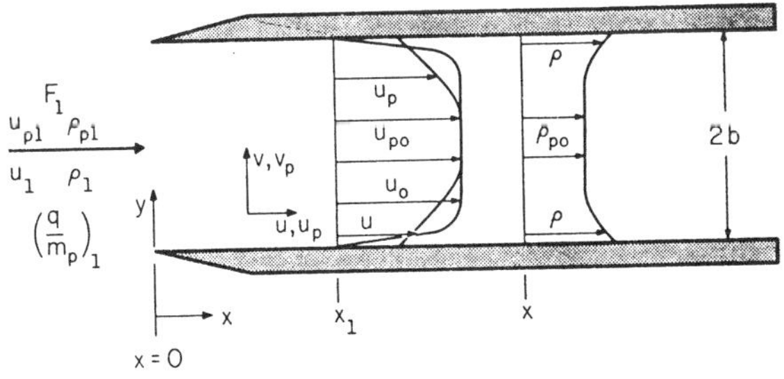

Figure 7.13 Flow system and coordinates.

The equations for the x-components of momenta of the phases with boundary layer simplifications are:

$$\rho U(\partial U/\partial x) + \rho V(\partial U/\partial y) = -(\partial P/\partial x) - KF\,\rho_p(U-U_p)+(\partial\tau/\partial y), \qquad (7.84)$$

$$U_p(\partial U_p/\partial x) + V_p(\partial U_p/\partial y) = F(U-U_p) + E_x(q/m) + \rho_p^{-1}(\partial\tau_p/\partial y) \qquad (7.85)$$

where τ, τ_p are the shear stresses, P is the static pressure, and E is the electric field in the x-direction. The y-component of particle motion is given by:

$$U_p(\partial V_p/\partial x) + V_p(\partial V_p/\partial y) = F(V-V_p) + E_y(q/m). \qquad (7.86)$$

The E-field is given by the equation:

$$(\partial E_x/\partial x) + (\partial E_y/\partial y) = \rho_p(q/m)/\varepsilon_o \qquad (7.87)$$

where ε_o is the permitivity.

Because of symmetry, we can outline the boundary conditions at one of the plates, such that, at $x = 0$, $U = U_p = U_1$, $V = V_p = 0$, $\rho_p = \rho_{p1}$; and at $x > 0$: $y = b$, $U = U_o(x)$, $U_p = U_{po}(x)$, $V = V_p = 0$, $\rho_p = \rho_{po}(x)$, $E_y = 0$. At $y = 0$: $U = 0$, $U_p = U_{pw}(x)$, $V = 0$, $\rho_p = \rho_{pw}(x)$. Eq. (7.85) and the slip condition further give:

$$U_{pw} = U_1(1 - \xi) + L_p(\partial U_p/\partial y)_{y=0} \qquad (7.88)$$

where the first term on the right-hand side accounts for slip at the inlet, with $\xi = Fx/U_1$ at $0 < x < U_1/F$, and $\xi = 1$ at $x > U_1/F$, and the second term account for finite interaction length defined by Eq. (1.72). Inside the channel, it is assumed that $(\partial E_y/\partial y)>>(\partial E_x/\partial x)$. Hence for given τ for the turbulent fluid and τ_p for the dilute suspension given by Eq. (3.97) to be:

$$\tau_p \sim \rho_p\,D_p(\partial U_p/\partial y) \qquad (7.89)$$

Eqs. (7.82) to (7.87) are solved for U, V, U_p, V_p, ρ_p, and E for given ρ, ρ_{p1}, U_1, subject to the above boundary conditions and stated assumptions.

Furthermore, for a dilute suspension, Eq. (7.83) can be replaced by the equation of diffusion under field forces in the form

$$U \frac{\partial \rho_p}{\partial x} + V \frac{\partial \rho_p}{\partial y} = - \frac{\partial}{\partial y} \left[\left(\frac{q}{m}\right) \frac{E_y \rho_p}{F} \right] + \frac{\partial}{\partial y} \left(D_p \frac{\partial \rho_p}{\partial y}\right) \tag{7.90}$$

according to Eqs. (2.123) and (2.124) and $|E_y| >> |E_x|$ in Eq. (7.87).

Reduction of the above equations to those of momentum integrals was carried out by introducing, from Eqs. (7.82) and (7.83),

$$V = -\int (\partial U/\partial x) \, dy \tag{7.91}$$

$$\rho_p V_p = \int (\partial \rho_p U_p/\partial x) \, dy + \rho_{pw} V_{pw} \tag{7.92}$$

where V_{pw} accounts for deposition of particles (or erosion from the wall). Further, the following momentum integrals for boundary layer thicknesses δ and δ_p of the fluid and particle phases were introduced:

$$\delta^* = \int_0^\delta \left(1 - \frac{U}{U_o}\right) dy, \qquad \theta = \int_0^\delta \frac{U}{U_o} \left(1 - \frac{U}{U_o}\right) dy,$$

$$\delta_p^* = \int_0^{\delta_p} \left(1 - \frac{\rho_p U_p}{\rho_{po} U_{po}}\right) dy, \qquad \theta_p = \int_0^{\delta_p} \frac{\rho_p U_p}{\rho_{po} U_{po}} \left(1 - \frac{U_p}{U_{po}}\right) dy,$$

and a gas-particle interaction thickness to correlate particle-fluid interactions,

$$\delta_{pg} = \int_0^\delta \left(1 - \frac{\rho_p U}{\rho_{po} U_o}\right) dy$$

Eqs. (7.84) to (7.86) were reduced to

$$(d/dx)(U_o^2 \theta) + \delta^* U_o (dU_o/dx) + KF(\rho_{po}/\rho_p)(U_o \delta_{pg} - U_{po} \delta_p^*) = \tau/\rho \tag{7.93}$$

$$(d/dx)(\rho_{po} U_{po}^2 \theta_p) + \rho_{po} \delta_p^* U_{po}(dU_{po}/dx) + \rho_{pw} V_{pw}(U_{po} - U_{pw})$$

$$= F \rho_{po}(U_o \delta_{pg} - U_{po} \delta_p^*) + \rho_{pw} D_{pw}(\partial U_p/\partial y)_w, \qquad (7.94)$$

$$U_o(\delta_{pg} - \delta_p)(d\rho_{po}/dx) + \rho_{po}(d/dx)[U_o(\delta_{pg} - \delta^*)]$$

$$= \left(\frac{\rho_{pw}-\rho_{po}}{\epsilon_o F}\right)\left(\frac{q}{m}\right)^2 \rho_{po}(b-\delta_p)+\left(\frac{q}{m}\right)^2 \left(\frac{\rho_{pw}}{F\epsilon_o}\right) \int_0^{\delta_p} \rho_p dy + D_{pw}\left(\frac{\partial \rho_p}{\partial y}\right)_w$$

$$(7.95)$$

for the transfer of momentum of the phases and the distribution of particle density.

As in the case of single phase turbulent boundary layers, evaluation of the integrals has to be made semi-empirically. Experiments with a suspension of 10 μm magnesia in air flowing over the inlet of a channel formed by two flat plates was made for various flow

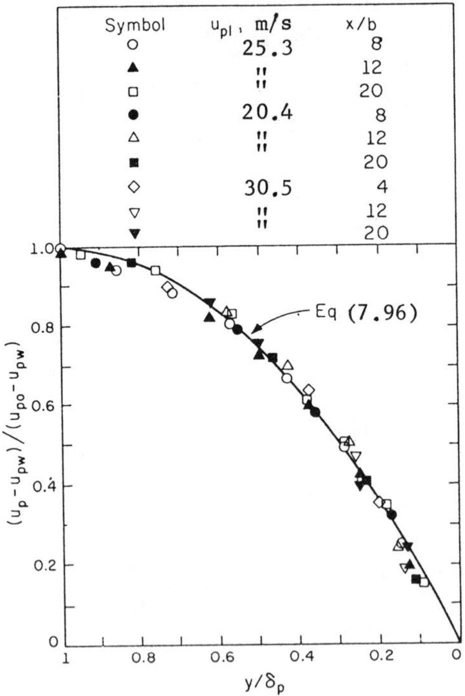

Figure 7.14 Particle velocity distribution.

velocities, plate gap widths, and mass flow ratios of solids to air. Experiments were carried out in a 30 cm by 30 cm section wind tunnel with flow velocities up to 36.6 m/s, plate gap widths of 6.4, 25.4, and 51 mm, and mass flow ratios up to 0.1 kg particles per kg air. Primary measurements included the determination of the particle and air velocities, the particulate mass flow and density distributions, and the particle size distributions as influenced by the flow response; probes were described in Sec. 6.1.

Measurements showed that the motion of the gaseous phase satisfies the turbulent boundary layer theory [Schlichting 1979]. The displacement thickness of the fluid was shown to follow the 1/7-th velocity profile.

Over the range of this investigation, namely for m_p^* varying between 0.01 to 0.125, with U_1 varying from 20.4 to 30.4 m/s, the velocity profile of the particle phase follows:

$$\frac{U_p - U_{pw}}{U_{po} - U_{pw}} = 2\left(\frac{y}{\delta_p}\right) - \left(\frac{y}{\delta_p}\right)^2 \tag{7.96}$$

Eq. (7.96) is shown in Fig. 7.14 along with the experimentally

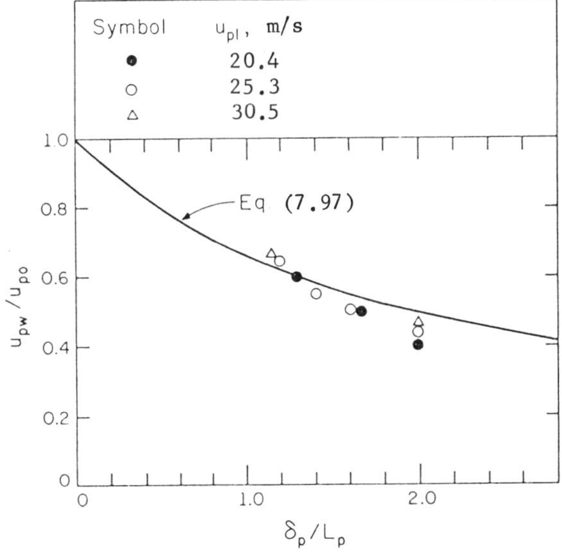

Figure 7.15 The particle slip velocity as a function of δ_p/L_p.

determined profiles. The agreement between the power law profile and the data is seen. The particle phase is in the regime of laminar slip motion (Sec. 3.5). One obtains from Eq. (7.88), neglecting $(1 - \xi)$, the slip velocity at the wall in relation to the core velocity:

$$\frac{U_{pw}}{U_{po}} = [\frac{1}{2} \frac{\delta_p}{L_p} + 1]^{-1} \tag{7.97}$$

Figure 7.15 gives a comparison of U_{pw}/U_{po} calculated from Eq. (7.88) ($\xi = 1$) with experimentally determined values of δ_p, L_p, and U_{pw}/U_{po}.

The particle density distribution was shown to follow Eq. (7.90). From the velocity in the core ($U = U_o$, $V = 0$), the density in the core is given by:

$$\frac{\rho_{p1}}{\rho_{po}} = 1 + (\frac{q}{m})^2 \; (\frac{b \; \rho_{p1}}{U_1 \; F\varepsilon_o}) \; (\frac{x}{b}) \; [1 - 0.0257 \; \varepsilon + 0.0001649 \; \varepsilon^2 - \; ...] \tag{7.98}$$

The charge to mass ratio was at the level of 10^{-3} C/kg. Similar to the particle velocity, a core region is seen to exist for the concentration distributions; $\varepsilon = (x/b)^{4/5} \; (U_1 b/\nu)^{-1/5}$. (Prob. 7.9, 7.10)

At the steady state condition, the drift flux of particles toward the wall $\rho_p V_p$ is balanced by the diffusion from the wall.

Nondimensionalizing of Eqs. (7.93) to (7.95) shows that in addition to the parameter $N_{Re} = U_1 b/\nu$ normally used for scale-up, the above equations give:

$$N_m \equiv \frac{U_1}{bF} \; , \; N_{ED}^2 \equiv (\frac{q}{m})^2 \; \frac{\rho_{p1} \; b^4}{\varepsilon_o \; D_p^2} , \; N_{Pe} \equiv \frac{U_1 \; b}{D_p}$$

Similarity of these correlation parameters to those in Sec. 6.2 is readily noted.

The effect of radiation on heat transfer in boundary layer motion can be treated by combining with the procedures given in Sec. 4.1.

7.5 Jets and Sprays

Jets of multiphase systems exist in air-cleaning systems such as venturi scrubbers, air-atomizing systems, and exhaust of metallized propellant rockets, to name a few. In all these cases, distribution of particulate matter in the jet is a quantity of interest. We illustrate with the case of a circular jet of charged dilute suspension; the principal fluid is neutral and incompressible and the background fluid is identical to the principal fluid of the jet [Soo 1970b].

The principal fluid jet is as defined by Schlichting and Tollmien for motions in the laminar range and the turbulent range [Schlichting 1979]. External field is absent. In their analysis, the coordinate x is measured from a virtual origin or source. Its relation to the location of the physical orifice is seen diagrammatically in Fig. 7.16. A potential flow core regime exists when there is a substantial pressure difference between the plane of the orifice and the flow field. For given mean velocity at the exit U_1, the velocity profile at the exit of the orifice depends on the length of the orifice which gives, in the limit, fully developed pipe flow. The velocity distribution at the exit will follow a parabolic profile for fully developed laminar motion or a nearly 1/7th velocity law prior to leaving the orifice for fully developed turbulent

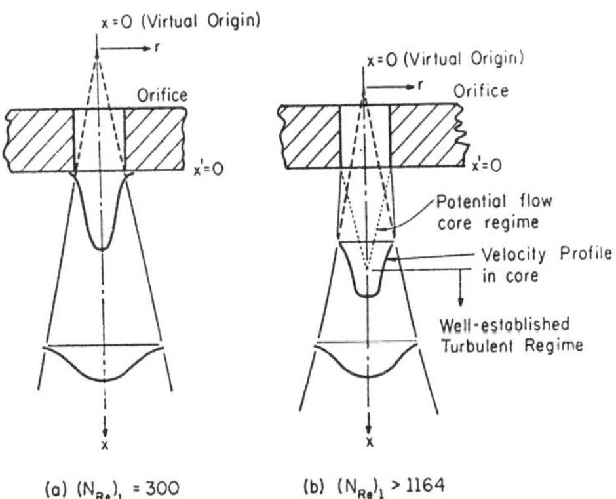

Figure 7.16 Flow regimes of gaseous phase in free jets.

motion. Slug flow is expected for an ideal knife-edged orifice.

Since the relaxation controlled range of even a laminar jet occurs over a short distance from the orifice, only the diffusion controlled range is considered here. The forces acting on the particles are such that:

Fluid (drag) force > electrostatic force > inertia force.

For a dilute suspension, the velocity of the principal fluid is unaffected by the presence of the particles. Schlichting and Tollmien gave the following axial (x-direction) and radial (r-direction) components (U, V) of velocity distribution:

$$U = D \, f'/x\eta, \quad V = D(\eta f' - f)/x\eta \tag{7.99}$$

where D denotes kinematic viscosity ν of the fluid in laminar motion, and apparent kinematic viscosity or eddy diffusivity in turbulent flow, where $D \sim r_1 U_1/39$ as given by Tollmien and others. In Eq. (7.99), r_1 is the orifice radius and U_1 is the uniform velocity at the exit of the orifice; other quantities are $\eta = r/x$, $\xi = \gamma\eta$, γ $(3^{1/2}/4)(r_1 U_1/D)$, $f = \xi^2/X$, $f' = 2\gamma^2 \eta/X^2$, and $X = 1 + (\xi^2/4)$. The above velocity distribution is valid for $x > (3^{1/2}/2)\gamma r_1$. (Prob. 7.11)

The jet boundary r at a given x is given by

$$(2/3^{1/2}) \; \gamma \; r_1/x = \xi_s^2/X_s$$

or

$$X_s^{-1} = 1 - [\gamma \; r_1/2(3)^{1/2} \; x] \tag{7.100}$$

Under the specifications of the present system and the boundary layer assumptions, diffusion and self-field effects are small in the x-direction when compared to those in the r-direction. The flux of the particles consists of that due to diffusion and that due to electric self-field J_{pE}. For densities of principal and dilute phases ρ, ρ_p, in the suspension, and $\rho \gg \rho_p$, the diffusion takes the form:

$$U \frac{\partial \rho_p}{\partial x} + V \frac{\partial \rho_p}{\partial r} = \frac{1}{r} \frac{\partial}{\partial r} r [D_p \frac{\partial \rho_p}{\partial r} - J_{pE}] \tag{7.101}$$

where D_p is the diffusivity of component p in the mixture. When the inertial force of the particle is small, the radial momentum equation gives

$$\rho_p U_p \frac{\partial V_p}{\partial x} + \rho_p V_p \frac{\partial V_p}{\partial x} = 0 = \rho_p F(V-V_p) + E_r \rho_p(q/m)$$

$$= - F J_{pE} + E_r \rho_p(q/m) \tag{7.102}$$

where q is the charge and m is the mass of each particle and E_r is the electric field. E_r $(\gg E_x)$ is given from the Poisson equation to take the form:

$$r^{-1}(\partial/\partial r) (r E_r) = \rho_p(q/m)/\epsilon_o \tag{7.103}$$

Combining the above three equations, we get, for D_p = constant,

$$\frac{1}{D_p}(U \frac{\partial \rho_p}{\partial x} + V \frac{\partial \rho_p}{\partial r}) = \frac{1}{r} \frac{\partial}{\partial r} r \frac{\partial \rho_p}{\partial r} - \frac{(q/m)^2}{D_p \epsilon_o F} \frac{1}{r} \frac{\partial}{\partial r} (\rho_p \int_o^r \rho_p r \, dr). \tag{7.104}$$

Eq. (7.104) is solved by changing the independent variables from x, r to x, η with $(\partial/\partial x)_r = (\partial/\partial x)_\eta - (\eta/x)(\partial/\partial \eta)_x$, and $(\partial/\partial r)_x = (1/x)(\partial/\partial \eta)_x$, U, V, given by Eq. (7.99). ρ_p is then given by:

$$\rho_p = (\beta/x) [h_o(\eta) + x^* h_1(\eta) + x^{*2}(\eta) + \ldots] \tag{7.105}$$

where the perturbation parameter $x^* = [(q/m)^2/F \gamma^2 D_o \epsilon_o]\beta x$, β being a constant to be determined from the conservation of total flow of particles. For $D/D_p = N_{SC} = N$, the Schmidt number, different orders of x^* are given for boundary conditions $h'(0) = 0$, $h(\infty) = 0$. Solution of equations of various orders of h_n give for $N_{SC} > 1/2$:

$$h_o = X^{-2N}$$

$$h_1 = (1 - 2N)^{-1} X^{-4N+1}$$

$$h_2 = -(1 - 2N)^{-2} X^{-2} [\int \frac{dX}{(X-1)^{6N-4}} - \int \frac{dX}{X-1}] . \qquad (7.106)$$

which has integral values for $6N - 4 = \alpha$, $\alpha \geq 2$,

$$h_2 = -(1-2N)^2 [\sum_{(\lambda-1)}^{(\alpha-1)} \frac{1}{\lambda X^{\lambda+2}} - \frac{\ell n\ X}{X^2}] \qquad (7.107)$$

Conservation of overall particle flow rate gives the jet boundary at r_{ps}, and

$$\beta = (\rho_{p1}\ U_1\ r_1^2/8D)\ (1 + 2N)\ (1 - X_{ps}^{-2N-1})^{-1} \qquad (7.108)$$

The perturbation parameter x* is now given by

$$x^* = (1-X_s^{-2N-1})^{-1} [\frac{\rho_{p1}}{\varepsilon_o} (\frac{q}{m})^2 \frac{r_1^4}{D_p^2}][\frac{D_p}{F\ r_1^2}] \frac{4}{3}(\frac{r_1 U_1}{\nu})^{-1} (\frac{D}{\nu})(1 + 2N_{Sc})(\frac{x}{r_1})$$

$$= (1 - X_s^{-2N-1})^{-1}\ N_{ED}\ N_{DF}\ N_{Re}^{-1}\ (\frac{D}{\nu})\ (1 + 2N_{Sc})\ \frac{x}{r_1} \qquad (7.109)$$

where N_{ED} and N_{DF} are as given in Sec. 6.2. N_{Re} is the Reynolds number based on orifice radius. $D/\nu = 1$ for laminar motion, and $D/\nu = N_{Re}/39$ for a turbulent jet. In the latter case, x* is unaffected by N_{Re}.

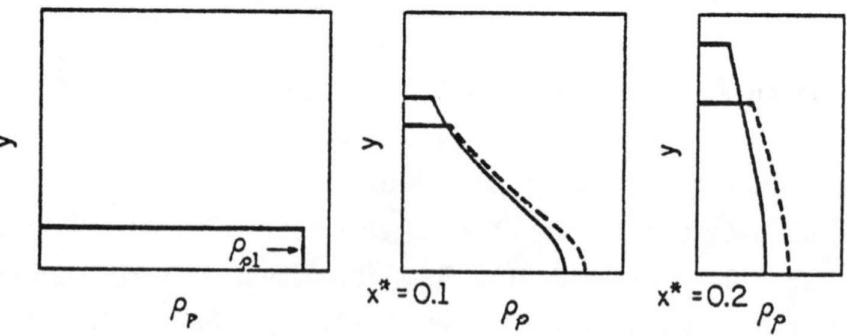

Figure 7.17   Comparison of density distribution of particles with and without space charge effect ($N_{Sc} = 1$).

The spreading of a jet of a particulate suspension is illustrated in Fig. 7.17, comparing the cases with and without electric charges on particles. It is seen that $x^* = 0[10^{-1}]$ is obtained at $x = 20$ cm for $r_1 = 1$ cm, $\rho_p = 0.1$ kg/m^3, $(q/m) = 10^{-3}$ C/kg, $D_p = 10^{-3}$ m^2/s, $F = 10^4$ s^{-1} (1 μm aerosol in air at STP), in a turbulent jet. Experimental verification of the above correlation was given by Lee and Soo [1970]. Using approximate forms of turbulent-kinetic-energy and dissipation-rate equations, the influence of particles on the turbulence in a jet was computed by Elghobashi et al. [1984].

<u>Free jet boundary</u>. Near the exit of a jet nozzle, or in the general case of separated flow, the mechanism via which the discontinuity of velocity is smoothed out must be taken into consideration. Even in the case of small characteristic Reynolds numbers, say, based on the length of the nozzle, the laminar flow velocity profile in the wake of the nozzle possesses a point of inflexion, and is extremely unstable. Therefore, it is appropriate to treat separated flow as involving turbulent mixing in general. The method here in dealing with separated flow of a dilute suspension extends from that due to Goertler [Schlichting 1979], who gave for a meeting of two fluid streams of velocities U_1 and U_2 at $x = 0$, with $U_1 > U_2$,

$$U = \bar{U} (1 + \Lambda \text{ erf } \xi) = \bar{U} f'(\xi) \qquad (7.110)$$

where $\bar{U} = (U_1 + U_2)/2$ and $\Lambda = (U_1 - U_2)/(U_1 + U_2)$, and $\xi = \sigma y/x$, $\sigma \sim 13.5$ is an empirical constant, x is along the direction of $\bar{U}$, and y is normal to x (Fig. 7.18). It is readily shown that

$$V = -\frac{\bar{U}}{\sigma} \Lambda [\int_{\xi}^{\infty} (1 - \text{erf } x) \, dx - \xi(1 - \text{erf } \xi)]$$

$$= -\frac{\bar{U}}{\sigma} [f - \xi f'] \qquad (7.111)$$

with $f''' + 2 ff'' = 0$.

Now we want to consider the case with the particulate phase suspended in U_1 and at $x = 0$, $U_p = U_1$, $\rho_p = \rho_{p1}$. The continuity and momentum equations of the particulate phase are:

$$\frac{\partial(\rho_p U_p)}{\partial x} + \frac{\partial(\rho_p V_p)}{\partial y} = 0 \tag{7.112}$$

$$U_p \frac{\partial U_p}{\partial x} + V_p \frac{\partial U_p}{\partial y} = F(U - U_p) \tag{7.113}$$

with additional boundary conditions $y = \pm \infty$, $U_p = \overline{U}(1 \pm \Lambda)$, $V_p = 0$; $y = \infty$, $\rho_p = \rho_{p1}$.

As when dealing with a circular jet, for a suspension, it is desirable to identify the jet boundary of stream 1, which is given by

$$\xi_s (1 - \Lambda) = \Lambda \int_{-\infty}^{\xi_s} (1 - \text{erf } x) \, dx \tag{7.114}$$

for example, at $\Lambda = 1/2$, $\xi_s = -0.314$, or $y_s = \xi_s x/\sigma$ in the physical coordinate (Fig. 7.18, dotted line). From similar considerations as when dealing with flat-plate flow, it is seen that for $x^* = Fx/U_1 \gtrsim O[1]$, $y_{sp} \sim x^* y_s = (\xi_s/\sigma)(F/U_1)x^2$, y_s and y_{sp} merge at $x^* \gtrsim O[1]$ and the jet boundary of the particle phase is shown by solid line in Fig. 7.18. It is further seen that

$$f(\xi) = \xi - \xi_s + \Lambda \int_{\xi_s}^{\xi} \text{erf } \xi' \, d\xi'$$

$$= (1 + \Lambda) \xi - \Lambda \int_{-\infty}^{\xi} (1 - \text{erf } x) \, dx - 2\xi_s \tag{7.115}$$

Pertinent solutions of the motion and density of the particulate phase are given by

$$U_p = \overline{U} \, [g_0' + x^{*-1} g_1' + x^{*-2} g_2' + \dots] \tag{7.116}$$

$$V_p = -\frac{\overline{U}}{\sigma} \, [m_0 + x^{*-1} m_1 + x^{*-2} m_2 + \dots] \tag{7.117}$$

$$\rho_p = \rho_{p1} [1 + \frac{1+\Lambda}{1-\Lambda} x^* (h_0 + x^{*-1} h_1 + \dots)] \tag{7.118}$$

Substitution into Eqs. (7.112) and (7.113) and that of the y-component gives

$$g_0' = f', \quad g_0 = f$$

$$g_1' = ff'', \quad \ldots$$

$$h_0 = (\text{const.})f, \quad m_0 = f - \xi f', \quad \ldots \tag{7.119}$$

following similar procedures as when dealing with the case of a circular jet. $g_0' = f'$, and g_1' are shown in Fig. 7.18. In this inertia controlled range, the layer of high particle density will be thrown along the curve y_{sp} versus x, and the peak concentration profile decreases with greater x.

When significant effect of diffusion is present and $\rho_{pz} = 0$, the density distribution of particles is again given by

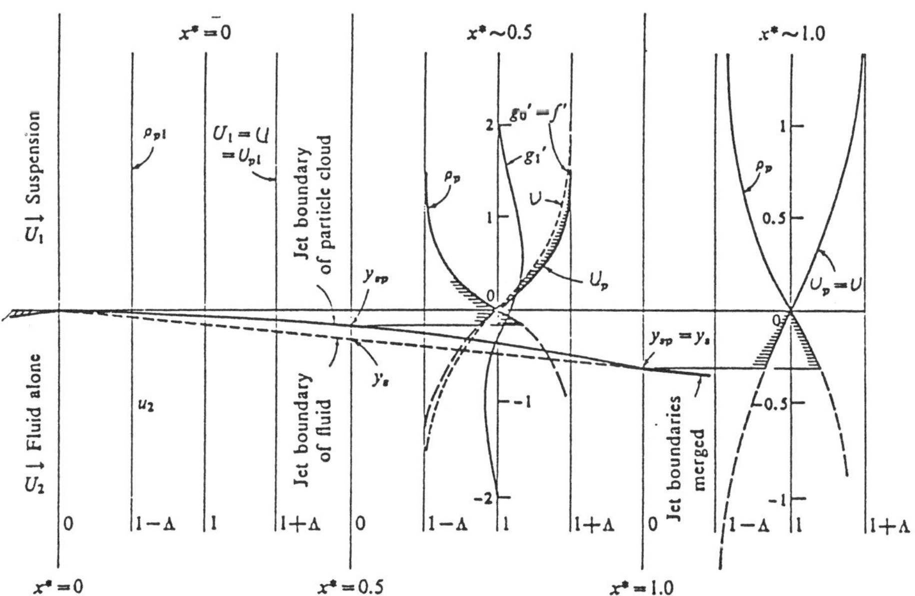

Figure 7.18 Jet boundary, velocity, and density distribution of a suspension and a clean fluid.

$$D \frac{\partial^2 \rho_p}{\partial y^2} = U \frac{\partial \rho_p}{\partial x} + V \frac{\partial \rho_p}{\partial y} \qquad (7.120)$$

at $x^* > 1$, $U_p \sim U$, $V_p \sim V$, for eddy diffusivity given by $\epsilon = Ux/2\sigma^2$, and $N_{Sc} = \epsilon/D = \bar{U}x/2\sigma^2 D$ following Goertler. One readily shows that Eq. (7.120) integrates to

$$\rho_p = \rho_{p1}[erf\ (N_{Sc}^{1/2}\ \xi) + 1]/2 \qquad (7.121)$$

[Soo 1967].(Prob. 7.12)

<u>Sprays</u>. A jet of a suspension of liquid droplets in a gas constitutes a spray. Study of generation and transport processes of sprays is a field by itself [see, for instance, Faeth 1983]. Distributions in a spray affect evaporation and combustion as seen in the single droplet behaviors in Sec. 1.4. The dynamics of a spray of droplets formed by gas injection is in the diffusion controlled range of a jet. The situation is quite different in the case of atomization by injection of a liquid jet into an initially stagnant gas at a high velocity. Oscillation of a liquid filament, breakup, and penetration into the gas have been extensively studied [Levich 1962, Bogy 1979]. Some of the interrelations are seen in the results of Robert [1946], based on a distribution in droplet size given by the Rosin-Rammler relation:

$$f = exp\ [- (a/\bar{a})^n] \qquad (7.122)$$

where f is the volume fraction of the spray composed of drops greater than radius a, $\bar{a}$ is the size constant, and n is the distribution constant. The rate of evaporation $\dot{m}_v$ was approximated by:

$$\dot{m}_v = \frac{4\pi\ a\ D'}{K(P_g - P_o)} \qquad (7.123)$$

where D is the diffusion constant, P_g the saturation pressure, P_o the vapor pressure of the surrounding air, and K is a constant depending on air temperature and density. The evaporation rate of a drop at time t was given as

$$(d/dt) \ a^3 = -Ka \qquad (7.124)$$

Deviations from relations for evaporation and combustion of single droplets (Sec. 1.4) may be attributed to group behaviors of droplets in a spray.

7.6 Diffusion and Fall-out from Point Sources

Atmospheric dispersion of stack effluents has been the subject of many studies such as those presented in Stern [1968]. The present section deals with cases which account for the simultaneous effects of diffusion and fall-out. The aim is to provide primary parameters necessary for modeling, simulation, and correlation of experimental results, as well as suggesting a method for making logical estimations.

Following Stern, a stack plume is assumed to originate from an equivalent point source at a height h. This "horizontal straight plume" has a fully accommodated horizontal wind velocity U aligned with the x-axis. The coordinate z is the height measured from the ground and y is measured along the ground normal to the plane of x and z. The plume height is obtained by computing the plume path from the stack exit at a given velocity W_1, the Grashof number which is a measure of the buoyancy of the gas, the Richardson number of the atmosphere, and the Reynolds number of the stack exhaust gases [Stern 1968]. The y and z components of the diffusivity, D, in the atmosphere given by Sutton [1953] can be represented by the general form:

$$D_y = a \ f'(x), \quad D_z = b \ f'(x) \qquad (7.125)$$

where a and b are constants. The diffusion in the x-direction is assumed to be negligible in comparison with the mass transport associated with the velocity U. It is seen that each contaminant species, k, may have a diffusivity in the form:

$$D_{ky} = a_k \ f'(x), \quad D_{kz} = b_k \ f'(x) \qquad (7.126)$$

depending on the relation of D_k/D; D_k could be different from D because of the difference in inertia of species k from that of the atmosphere. Any coupling between various D_k's can be neglected for $\rho_p \ll \rho$, that is, the density of each phase k is much smaller than that of the atmosphere.

When fall-out due to the gravitational field is included, the diffusion equation for the k-th species having density ρ_k take the form, under the above specifications, and for constant ρ, Eq. (2.123) becomes:

$$U \frac{\partial \rho_k}{\partial x} = D_{kz} \frac{\partial^2 \rho_p}{\partial z^2} + D_{ky} \frac{\partial^2 \rho_k}{\partial y^2} + \frac{g}{F} \frac{\partial \rho_k}{\partial z} \qquad (7.127)$$

where F_k is the inverse relaxation time for momentum transfer from the atmosphere to species k and g/F_k is the terminal velocity of free fall of species k [Soo 1971].

Eq. (7.127) can be solved analytically with the following transformation of coordinates:

$$\xi = b_k \, f(x)/U \qquad (7.128)$$

$$\eta = (b_k/a_k)^{1/2} \, y \qquad (7.129)$$

where $b_k f(x) = b_k \int_0^x f'(x)dx$. We further introduce

$$\beta_k = g/2F_k b_k f'(x) \qquad (7.130)$$

$$\rho_k = F \exp(-\beta_k z - \beta_k^2 \xi) \qquad (7.131)$$

one notes that, for cases in which D_{kz} and D_{ky} are constants, $\xi = D_{kz}x/U$, $\eta = (D_{kz}/D_{ky})^{1/2}y$, $b_k f(x) = D_{kz}x$, and $\beta_k = g/2F_k D_{kz}$. It is seen that regardless of the form of $f'(x)$, as long as $\beta_k =$ constant, Eq. (7.127) reduces to:

$$\frac{\partial^2 F}{\partial \eta^2} + \frac{\partial^2 F}{\partial z^2} = \frac{\partial F}{\partial \xi} \qquad (7.132)$$

which has the general solution

$$F = F_0 \exp [- (\eta^2 + z^2)/4\xi] \tag{7.133}$$

where F_0 is the constant of integration.

The distribution of ρ_0 can be influenced both by the extent of the "ground reflection", and by the absorption of the species k by the ground (or body of water). Introducing a reflectivity A_k defined such that $0 < A_k < 1$, it is noted that for $A_k = 0$, there is complete absorption of species k reaching $z = 0$, and for $A_k = 1$ complete reflection occurs which is the condition of the original Sutton diffusion equation. (Note the difference from the relation in Eq. 4.90). Further, when $A_k = 1$ and $1/F_k = 0$, the total flux is equal to the original source strength of species k, or

$$U \int_{-\infty}^{\infty} \int_{0}^{\infty} \rho_k \, dy \, dz = \rho_{k1} W_1 \pi r_1^2 \tag{7.134}$$

where ρ_{k1} is then the density of species k at the stack of radius r_1 at velocity W_1.

When these conditions are applied to Eq. (7.133), one obtains, after some rearranging, an equation for the density distribution of k:

$$\rho_k = \rho_{k1} \left(\frac{1}{4x_k^*}\right) \left(\frac{r_1}{h}\right)^2 \left(\frac{W_1}{h}\right) \left(\frac{D_{ky}}{D_{kz}}\right)^{1/2} \exp\left[- \frac{y^2}{4x_k^* h^2} \frac{D_{kz}}{D_{ky}}\right]$$

$$\times \left\{\exp\left[- \frac{1}{4x_k^*} \left(\frac{z}{h} + \tau_k^* x_k^* - 1\right)^2\right] + A_k \left[- \frac{1}{4x_k^*} \left(\frac{z}{h} + \tau_k^* x_k^* + 1\right)^2\right]\right\} \tag{7.135}$$

where $\tau_k^* = 2\beta_k h$ is a diffusion-fall-out interacting parameter (ratio of terminal velocity to diffusion velocity), and

$$\tau_k^* x_k^* = 2\beta_k b_k f(x)/Uh = gx/F_k Uh \tag{7.136}$$

$$x_k^* = b_k f(x)/Uh^2 \tag{7.137}$$

$(\tau_k^* = gh/D_{kz} F_k, \; x^* = D_{kz} x/Uh^2$ for constant D_k).   Note that when compared to Eq. (4.90), the boundary condition for absorption or reflection only starts to become applicable when phase k start to reach the ground. (Prob. 7.13)

Eq. (7.135) reverts to the Sutton diffusion equation for $1/F_k = 0$, and

$$D_{ky} = D_y = \frac{(2-n)}{4} \, c_y^2 \, U \, x^{(1-n)}$$

$$D_{kz} = D_z = \frac{(2-n)}{4} \, c_z^2 \, U \, x^{(1-n)} \tag{7.138}$$

where C_y and C_z are the Sutton diffusion coefficients and n is a turbulent index. The C's have the unit in $(m)^{n/2}$ for elevation m in meters.   In SI units, $0.05 < C_z < 0.58$, $0.19 < n < 0.58$; $C_y = C_z$ for $m > 25$ meters; for a 100 m stack, $U = 10$ m/s, $n = 0.20$, $C_y = C_z = 0.21 m^{1/10}$. We now have

$$x_k^* = (1/4) \, c_z^2 \, x^{(2-n)}/h^2 \tag{7.139}$$

It is also noted that $h \, \tau_k^* x_k^* = gx/UF_k$ is the change in plume height due to free settling suggested in the U. S. Weather Bureau report [1955] as a modification of the Sutton equation. It is seen that in Eq. (7.135) the correction to the reflected contaminant is also in the direction of gravity.

Because of fall-out, either with or without ground reflection, the total amount of contaminant transported, J_{kT}, can be calculated using Eq. (7.134). The result is:

$$J_{kT}^* = J_{kT}/\rho_{k1} \, \pi \, r_1^2 \, W_1$$

$$= \frac{1}{2} \left\{ \mathrm{erfc} \left[\frac{\tau_k^* x_k^* - 1}{2x_k^{*1/2}} \right] + A_k \, \mathrm{erfc} \left[\frac{\tau_k^* x_k^* + 1}{2x_k^{*1/2}} \right] \right\}$$

$$= J_{kTF}^* + A_k \, J_{kTR}^* \tag{7.140}$$

where erfc is the complementary error function with $\mathrm{erfc}(-\infty) = 2$, $\mathrm{erfc}(0) = 1$, and $\mathrm{erfc}(\infty) = 0$. The subscript F denotes fall-out and R denotes reflection. Fig. 7.19 depicts $J^\star_{kTF}$ as a solid line and $J^\star_{kTR}$ as a dashed line for each value of $\tau^\star_k$. It shows that the ground reflection is negligible for large values of $\tau^\star_k$ when $J^\star_{TR} \ll J^\star_{TF}$. This is readily seen when D_{kz} is taken to be a constant. It also can be seen that $J^\star_{kT}$ decreases due to fall-out even for $A_k = 1$. Moreover, $1 - J^\star_{kT}$ gives the total fall-out fraction deposited along x.

The ground level flux is seen to be given by drift as well as diffusion:

$$-J_{kG} = -(g/F_k)\,\rho_k\big|_{z=0} - D_{kz}\,\partial\rho_k/\partial z\big|_{z=0} \tag{7.141}$$

which gives, after rearranging:

$$
\begin{aligned}
J^\star_{kG} &= \left[J_{kG}\,\frac{r_1^2\,\rho_{k1}\,W_1}{4(a_k b_k)^{1/2}} \right] \exp\left[\frac{y^2}{4x^\star_k h^2}\,\frac{D_{hz}}{D_{ky}} \right] \\
&= \frac{1}{2}\left(\frac{\tau^\star_k x^\star_k + 1}{x^{\star 2}_k} \right) \exp\left[-\frac{(\tau^\star_k x^\star_k - 1)^2}{4x^\star_k} \right] \\
&\quad + \frac{A_k}{2}\left(\frac{\tau^\star_k x^\star_k - 1}{x^{\star 2}_k} \right) \exp\left[-\frac{(\tau^\star_k x^\star_k + 1)^2}{4x^{\star 2}_k} \right] \equiv J^\star_{kGF} + A_k\,J^\star_{kGR}
\end{aligned}
\tag{7.142}
$$

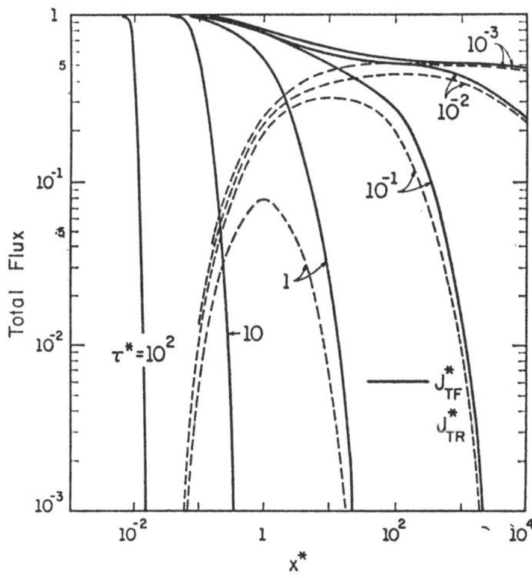

Figure 7.19 Components of total flux above the ground level.

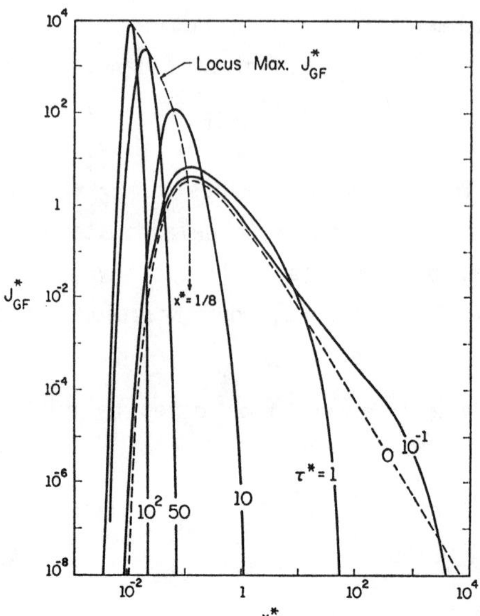

Figure 7.20   Flux of fall-out at the ground level with no reflection.

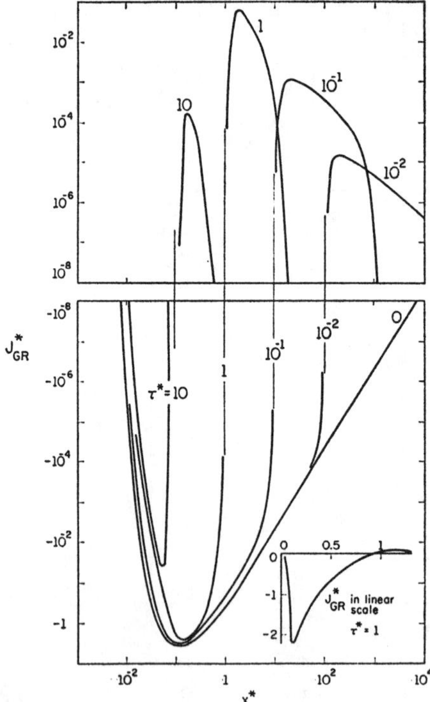

Figure 7.21   Flux at the ground level by reflection.

Fig. 7.20 gives the relation for the first term J^*_{kGF} whereas Fig. 7.21 gives the relation for the second term of J^*_{kGR} (reflected). One notes that as $\tau^*_k \to 0$, a maximum value for J^*_{kGF} occurs at $x^* = 1/8$. As the gravity effect increases or as the value of τ^*_k becomes large, the maximum of J^*_{kGF} shifts toward lower values of x^*_k. In Fig. 7.21, the bottom part shows negative values of J^*_{kGR} while the upper part indicates positive values of J^*_{kGR}, and $J^*_{kGR} = 0$ when $\tau^*_k x^*_k = 1$. A negative value of J^*_{kGR} means a decrease in the fall-out due to reflection while a positive value means that the fall-out increases the ground level flux and thus deposition is assisted by diffusion. At large τ^*_k gravitational effects predominate with J^*_{kGR} negligible even for large values of A_k. Only the diffusion flux is reflected when the rate of absorption at the ground level is smaller than the incoming flux. As long as the species are dilute, summation over species k for ρ_k and J_k can be applied to get the total quantities. Limiting cases are seen in Prob. 7.14.

The present method can also be extended to the transient problem of a contaminant species k of mass m_k released at a point $x = y = 0$, $z = h$ at $t = 0$, t being the time. The solution is analogous to Eq. (7.135) with x/U replaced by t with diffusivities D_{kx}, D_{ky} and D_{kz} being constant or functions of t. In this case, the density distribution takes the form:

$$
\begin{aligned}
\rho_k = {} & \frac{m_k}{\pi^{3/2} h^3} \frac{D_{kz}}{(D_{kx} D_{ky})^{1/2}} \exp\left[-\frac{x^2}{4D_{kx} t} - \frac{y^2}{4K_{ky} t}\right] \\
& \times t_k^{*-3/2} \left\{ \exp\left[-\frac{1}{4t^*_k}\left(\frac{z}{h} + \tau^*_k t^*_k - 1\right)^2\right] \right. \\
& \left. + A_k \exp\left[-\frac{1}{4t^*_k}\left(\frac{z}{h} + \tau^*_k t^*_k + 1\right)^2\right]\right\}
\end{aligned}
\tag{7.143}
$$

where $t^*_k = D_{kz} t/h^2$, $\tau^*_k = gh/F_k D_{kz}$, and $\tau^*_k t^*_k = gt/F_k h$. This result is similar to that obtained by Banister and Davis [1962] except that they treated the one-dimensional case with variable density of the atmosphere. The present solution gives the dimensionless ground level flux as:

$$J'^{\star}_{kG} = J'_{kG}\,\frac{\pi^{3/2}\,h^4}{m_k\,D_{kz}}\,\frac{(D_{kz}\,D_{ky})^{1/2}}{D_{kx}}\;\exp\!\left[\frac{x^2}{4D_{kx}\,t} + \frac{y^2}{4D_{ky}\,t}\right]$$

$$= \frac{1}{2}\left(\frac{\tau^{\star}_k\,t^{\star}_k + 1}{t^{\star}_k{}^{5/2}}\right)\exp\left[-\frac{(\tau^{\star}_k\,t^{\star}_k - 1)^2}{4t^{\star}_k}\right]$$

$$+ \frac{A_k}{2}\left(\frac{\tau^{\star}_k\,t^{\star}_k - 1}{t^{\star}_k{}^{5/2}}\right)\exp\left[-\frac{(\tau^{\star}_k\,t_k + 1)^2}{4t^{\star}_k}\right. \equiv J'^{\star}_{kGF} + A_k\,J'^{\star}_{kGR}$$

$$(7.144)$$

The parametric relations are shown in Fig. 7.22. As $\tau^{\star}_k \to 0$, the maximum value of $J'^{\star}_{kGF}$ and minimum value of $J'^{\star}_{kGR}$ occur at $t^{\star} = 1/10$. Fig. 7.22 shows that for $\tau^{\star} = 100$, the fall-out begins to be feld at nearly $t^{\star} = 0.003$ after release, reaching a peak at $t^{\star} = 0.01$, and ending at $t^{\star} = 0.02$. As $\tau^{\star} \to 0$, for $A_k = 0$, the ground level flux reaches a maximum at $t^{\star} = 1/10$ after release due to diffusion alone. (Prob. 7.15)

The above procedure has been extended to treating transient dispersion by wind from nonpoint source by Chen and Soo [1983]

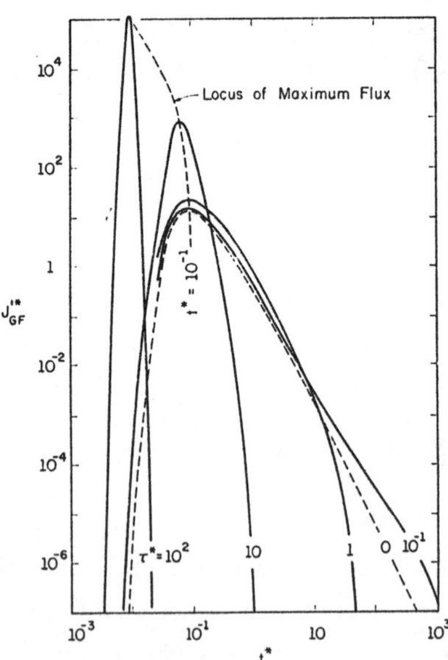

Figure 7.22 Flux of transient fall-out at ground level from release of a contaminant at an altitude h, $t^{\star} = D_z t/h^2$, $\tau^{\star} = gh/D_z F$

7.7 Flow over a Cylinder

Motivated by the possibility of improving heat transfer co-efficient of a gas, studies were made on the utilization of gas-liquid drop suspensions. Goldstein et al. [1967] presented an analytical study of laminar flow of a gas-liquid droplet suspension over a circular cylinder. The physical system is represented in Fig. 7.23. The undisturbed fluid, consisting of liquid droplets suspended in the gas, has velocity U_o, temperature T_o, and volume fraction of liquid α_{po}. These drops form a liquid film of thickness $\delta(x)$ along the surface upon impinging on the cylinder. It may be expected to alter its form near the point of separation due to boundary layer in-stability or gravity. This case illustrates the nature of interactions in combined dispersed flow and stratified flow.

At the liquid-gas interface, the temperature T_s and the surface velocity U_s are unknown and must be determined from interface balance (p. 44). All temperatures T_o, T_s, and T_w of the wall are assumed to

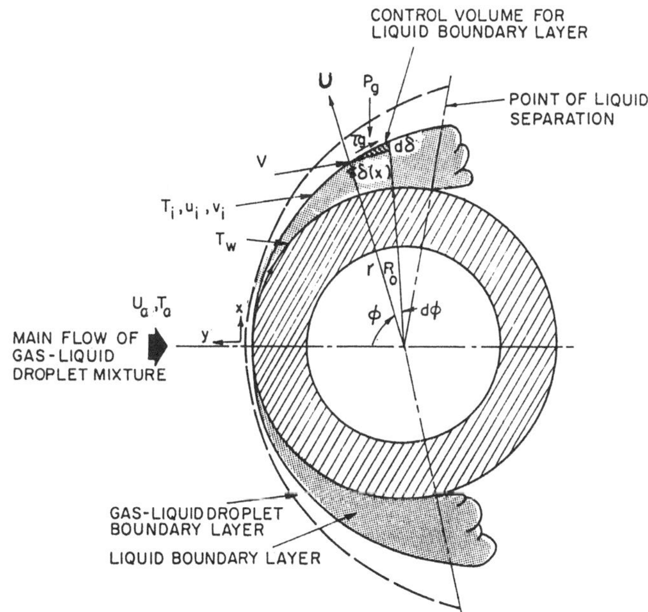

Figure 7.23 Liquid-gas, two-phase flow over a cylinder. [Goldstein et al. 1967]

be lower than the saturation temperature of both components of the mixture in order that no phase change occurs in either phase. T_w is taken to be uniform.

The system consists of a pair of interacting boundary layers; an inner one consisting of the liquid phase and an outer one consisting of the suspension. Several flow regimes may occur: laminar (inner layer)-laminar (outer layer), turbulent-laminar, or turbulent-turbulent. Laminar-laminar steady flow model was studied, with these assumptions:

(1) The liquid drops contained in the gas are uniformly distributed and follow a certain trajectory in the free stream and gas boundary layer. They travel with the velocity of the gas far upstream.

(2) Wavy motion of the liquid layer and splashing are neglected.

(3) The effects of surface tension (Eq. 2.20), compressibility, gravity, and heating due to frictional dissipation are negligible.

(4) All physical properties are constant and the phases are represented by air and water.

(5) The regions of wake and separated flow are not considered.

Liquid film and interface balance. With the foregoing assumptions, the liquid film is treated as a discrete phase. The equations of conservation, when applied to the control volume for the liquid film in Fig. 7.23, take the form, for cylindrical coordinates r, ϕ:

$$\partial rU/\partial r + \partial V/\partial\phi = 0 \tag{7.145}$$

$$U\frac{\partial U}{\partial r} + \frac{V}{r}\frac{\partial U}{\partial\phi} - \frac{V^2}{r} = -\frac{1}{\rho}\frac{\partial P}{\partial r}$$
$$+ \nu\left[\frac{\partial}{\partial r}\left(\frac{1}{r}\frac{\partial rU}{\partial r} + \frac{1}{r^2}\frac{\partial^2 U}{\partial\phi^2} - \frac{2}{r^2}\frac{\partial V}{\partial\phi}\right] \tag{7.146}$$

$$U\frac{\partial V}{\partial r} + \frac{V}{r}\frac{\partial V}{\partial\phi} + \frac{UV}{r} = -\frac{1}{\rho r}\frac{\partial P}{\partial\phi}$$

$$+ \nu\left[\frac{\partial}{\partial r}\left(\frac{1}{r}\frac{\partial rV}{\partial r}\right) + \frac{1}{r^2}\frac{\partial^2 V}{\partial \phi^2} + \frac{2}{r^2}\frac{\partial V}{\partial \phi}\right] \tag{7.147}$$

$$U\frac{\partial T}{\partial r} + \frac{V}{r}\frac{\partial T}{\partial \phi} = \alpha\left[\frac{1}{r}\frac{\partial}{\partial r}r\frac{\partial T}{\partial r} + \frac{1}{r^2}\frac{\partial^2 T}{\partial \phi^2}\right] \tag{7.148}$$

where α is the thermal diffusivity; according to the conservation of mass, components of momentum, and energy (unsubscripted quantities refer to the liquid film).

At the interface between the liquid film and the two-phase region, the balance relations (Sec. 2.1) yield (at $r = R_o + \delta$)

$$\rho\, V\frac{d\delta}{d\phi} - \rho\, U(R_o + \delta) = \Omega \tag{7.149}$$

$$-\tau_{r\phi}(R_o + \delta) + \tau_{\phi\phi}\frac{d\delta}{d\phi} + (R_o + \delta)\,\tau_g + \rho_g\frac{d\delta}{d\phi}$$

$$= \Omega\,(V - V_p) \tag{7.150}$$

$$-\tau_{rr}(R_o + \delta) + \tau_{r\phi}\frac{d\delta}{d\phi} - P_g(R_o + \delta) + \tau_g\frac{d\delta}{d\phi}$$

$$= \Omega(U - U_p) \tag{7.151}$$

$$\kappa\frac{\partial T}{\partial r}(R_o + \delta) - \frac{\kappa}{R_o+\delta}\frac{\partial T}{\partial \phi}\frac{d\delta}{d\phi}$$

$$+ J_{qg}\left[(R_o + \delta)^2 + \left(\frac{d\delta}{d\phi}\right)^2\right]^{1/2} = \Omega\, c_p(T_o - T) \tag{7.152}$$

where

$$\tau_{rr} = -P + 2\mu\frac{\partial U}{\partial r}$$

$$\tau_{\phi\phi} = -P + 2\mu\left[\frac{1}{r}\frac{\partial V}{\partial \phi} + \frac{U}{r}\right] \tag{7.153}$$

and

$$\tau_{r\phi} = \mu\left[\frac{1}{r}\frac{\partial U}{\partial \phi} + \frac{\partial V}{\partial r} - \frac{V}{r}\right] \qquad (7.154)$$

are the shear components in the liquid film and

$$\Omega = \rho \; \alpha_p \; \left[V_p \frac{d\delta}{d\phi} - (R_o + \delta) \; U_p\right] \qquad (7.155)$$

The boundary conditions imposed at the cylinder surface are $r = R_o$: $U = V = 0$, $T = T_w$. The shear stress exerted by the suspension is assumed to be that of the gas (τ_g) only. (Prob. 7.16)

The governing equations and the boundary conditions are then expressed in a curvilinear coordinate system of coordinates x and y, whose x axis is in the direction of the cylinder wall and the y axis normal to it. In dimensionless forms, $x^* = \phi$, and $y^* = (r/R_o) - 1$.

For sufficiently large Reynolds number $(N_{Re} = U_o R_o/\nu)$ and small α_p, the volume fraction of the liquid droplets, the dimensionless liquid-film thickness $\delta^* = \delta/R$, is of the order of $(N_{Re})^{-1/2}$ for small values of a film thickness parameter $N_E = \alpha_p (N_{Re})^{-1/2}$. For large values of N_E, however, δ^* becomes the order of $2\alpha_{po}$. The criterion for this change in the dependency of δ on N_{Re} or α_{po} was found to be $N_E = 1$. In order to perform boundary layer simplifications, it is desirable to reduce the governing equations and boundary conditions in the x, y coordinates into dimensionless forms. The dependency of δ on N_{Re}, the quantities y^* and δ^* must be defined in two different forms depending on values of N_E.

Introducing a stream function ψ such that

$$U = \partial\psi/\partial y, \quad V = -\partial\psi/\partial x \qquad (7.156)$$

which satisfies Eq. (7.145) in coordinates x and y. The basic relations are reduced to dimensionless forms by introducing:

$$\psi^* = \psi/R_o \; U_o \; \alpha_{po}, \qquad P^* = 2P/\rho U_o^2$$

$$T^* = (T-T_o)/(T_w-T_o), \qquad x^* = x/R_o$$

$$U^* = U/U_o, \qquad\qquad V^* = V/V_o$$

and

$$y^* = y/2\alpha_{po}, \qquad\qquad \delta^* = \delta/2\alpha_{po}$$

for

$$N_E \geq 1;$$

$$y^* = (y/2)(N_{Re})^{1/2}, \qquad \delta^* = (\delta/2)(N_{Re})^{1/2}$$

for

$$N_E \leq 1.$$

In addition, terms of the order of $(N_{Re})^{-1/2}$ and α_p are neglected. Eqs. (7.146) to (7.148) become:

$$\frac{\partial\psi^*}{\partial y^*}\frac{\partial^2\psi^*}{\partial x^*\partial y^*} - \frac{\partial\psi^*}{\partial x^*}\frac{\partial^2\psi^*}{\partial y^{*2}} = -\frac{2}{(N_E)^2}\frac{\partial P^*}{\partial x^*} + \frac{1}{N_E}\frac{\partial^3\psi^*}{\partial y^{*3}} \qquad (7.157)$$

$$\partial P^*/\partial y^* = 0 \qquad (7.158)$$

$$\frac{\partial\psi^*}{\partial y^*}\frac{\partial T^*}{\partial x^*} - \frac{\partial\psi^*}{\partial x^*}\frac{\partial T^*}{\partial y^*} = \frac{1}{N_{Pr}\,N_E}\frac{\partial^2 T^*}{\partial y^{*2}} \qquad (7.159)$$

for $N_E \leq 1$. For $N_E \geq 1$, the above equations are modified by replacing y^* with y^*/N_E, and δ^* with δ^*/N_E.

The boundary conditions in Eqs. (7.150) to (7.152) become, at $y^* = \delta^*$:

$$\frac{2\,\tau_g}{\rho\,U_o^2\,\alpha_{po}} - \frac{\partial^2\psi^*}{\partial y^*} = -J\left(2V^* - N_E\frac{\partial\psi^*}{\partial y^*}\right) \qquad (7.160)$$

$$P - P_g + \frac{2}{N_{RE}^{1/2}}\,\tau_g\,\frac{d\delta}{dx} = 0 \qquad (7.161)$$

$$\psi^* = -\int_0^{x^*}\frac{\alpha_p}{\alpha_{po}}\,U^*\,dx^* \equiv \int_0^{x^*} J^*\,dx^* \qquad (7.162)$$

$$\frac{\partial T^*}{\partial y^*} = - N_{Pr} \, N_E \, J^* \, T^* - \frac{2R_o}{\kappa N_{Re}^{1/2}} \, J_{qg} \qquad (7.163)$$

Heat transfer from the drops to the gas boundary layer is neglected. The boundary conditions at the cylinder surface become, at $y^* = 0$:

$$\psi^* = \frac{\partial \psi^*}{\partial y^*} = 0, \; T^* = 1 \qquad (7.164)$$

Interactions with the droplet suspension have to be accounted for to evaluate the terms τ_g, V^*, and P_g, T_g, and J_{qg} appearing in Eqs. (7.160) to (7.163).

Two-phase region. The two-phase region consists of shear motion and a boundary layer next to the liquid film. The presence of the liquid droplets may contribute to shear stress outside the gas boundary layer. The gas and the liquid phases interact at two interfaces, that of the liquid layer and that between the droplets and the gas. From a detailed order of magnitude comparison, Goldstein et al. [1968] identified two limiting cases:

(i) $N_E \geq 0.1$, where the effects of the gas boundary layer on the liquid film are negligible and the droplet trajectory is straight. This leads to the simplification of $\alpha_p = \alpha_{po}$, $U^* = -\cos x^*$, and $V^* = \sin x^*$. The effects of the gas boundary layer on the liquid film are negligible. This case represents the case of low velocity mist flow; the readers are referred to the original article for the results of the solution.

(ii) $N_E \leq 0.1$ corresponds to the case in which the presence of the liquid film does not influence the gas boundary layer for any drop trajectories. Furthermore, if the volume fraction is less or equal to 0.1 (ρ_g/ρ), the gas flow may be treated as potential flow over a cylinder (Sec. 3.2) [Langmuir and Blodgett 1948]. Under these conditions the effects of inertia forces in the liquid film are negligible compared with other terms in the governing equations as the liquid has a low velocity. However, in the energy equation, the convection term multiplied by the Prandtl number N_{Pr} becomes comparable with the conduction term for liquids such as water which has a large Prandtl number. The shear stress and pressure gradient along the surface may be expressed in dimensionless form given by Froessling [Schlichting 1979] as

$$\frac{2\,\tau_g}{\rho_g\,U_o^2}\,(N_{Re})^{1/2} = 8\,Z(x^\star) \tag{7.165}$$

and

$$-\frac{\partial}{\partial x^\star}\left[\frac{2\,P_g}{\rho_g\,U_o^2}\right] = Z\,\sin 2x^\star \tag{7.166}$$

where

$$Z(x^\star) = f_1'(0)\,x^\star - \sum_{n=2}^{\infty}\frac{(-1)^n(2n)}{(2n-1)!}\,f_{2n-1}''(0)\,x^{\star\,2n-1} \tag{7.166}$$

and the first six values $f''(0)$ are given by Froessling as $f_1'' = 1.2326$, $f_3'' = 0.7244$, $f_5'' = 1.0320$, $f_5'' = 2.036833$, $f_9'' = 0.280140$, $f_{11}'' = 67.637501$. For this case the physical system is depicted in Fig. 7.24. The coordinates x_p, y_p represent the distances of a droplet from the center of cylinder, parallel and normal to the direction of the free stream. The equation of motion of a spherical droplet at x_p, y_p and velocity U_p may be expressed as

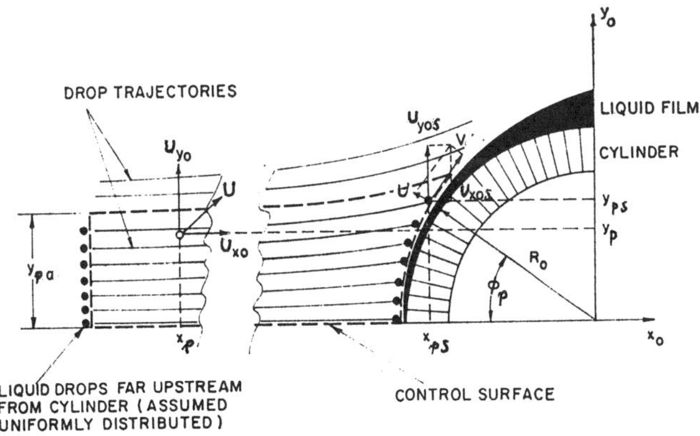

Figure 7.24 Control volume and coordinates for dynamic equations of liquid droplet and its trajectories. [Goldstein et al. 1967]

$$\frac{d\,U^*_{pxs}}{dt^*} = -\frac{3}{8}\frac{C_D}{(a/R_o)}\frac{\rho_g}{\rho}(U^*_{pxs}-U^*_{gx})[(U^*_{pxs}-U^*_{gx})^2 + (U^*_{pys}-U^*_{gy})^2]^{1/2}$$

$$= \frac{d^2X^*_p}{dt^{*2}} \tag{7.167}$$

where t is the time and $t^* = tU_o/R_o$, and

$$\frac{dU^*_{pys}}{dt^*} = -\frac{3}{8}\frac{C_D}{(a/R_o)}\frac{\rho_g}{\rho}(U^*_{pys}-U^*_{gy})[(U^*_{pxs}-U^*_{gx})^2 + (U^*_{pys}-U^*_{gy})^2]^{1/2}$$

$$= \frac{d^2Y^*_p}{dt^{*2}} \tag{7.168}$$

The gas velocity is given by potential motion around a cylinder:

$$U^*_{gx} = 1 - \frac{x^{*2}_p - y^{*2}_p}{(x^{*2}_p + y^{*2}_p)^2} \tag{7.169}$$

and

$$U^*_{gy} = -\frac{2x^*_p\,y^*_p}{(x^{*2}_p + y^{*2}_p)^2} \tag{7.170}$$

The initial conditions of these equations are similar to those of
Eq. (3.22), and similar solution techniques apply. The control
surface in Fig. 7.24, which is bounded on two sides by drop
trajectories, on the third side by the liquid film, and on the fourth
side by a line perpendicular to the drop trajectories at a distance
far upstream from the cylinder. The flux of liquid across the
control surface corresponding to angle ϕ_p from the forward stagnation
point $-\int_0^{\phi_p}\alpha_p U^*_p R_o\, d\phi$ must balance with the flux through the control
surface at the upsteam $U^*_p \alpha_{po} y^*_{po}$, one obtains

$$\int_0^{x^*} J^*\, dx^* = y^*_{po} \tag{7.171}$$

or

$$J^* = x^*_{ps}/(dy^*_{ps}/dy^*_{po}) \qquad (7.172)$$

The component of drop velocity at the edge of the liquid film (x^*_{ps}, y^*_{ps}) may be obtained from the geometry as

$$V^*_p = -U^*_{xos}\, x^*_{ps} + U^*_{yos}\, y^*_{ps} \qquad (7.173)$$

and

$$x^* = \tan^{-1} (y^*_{ps}/x^*_{ps}) \qquad (7.174)$$

The quantities on the left-hand side of Eqs. (7.171) to (7.174) are those given by Eqs. (7.160) to (7.163).

Solutions for both cases of N_E were given by Goldstein et al. [1968] via lengthy derivations and numerical integrations. With a

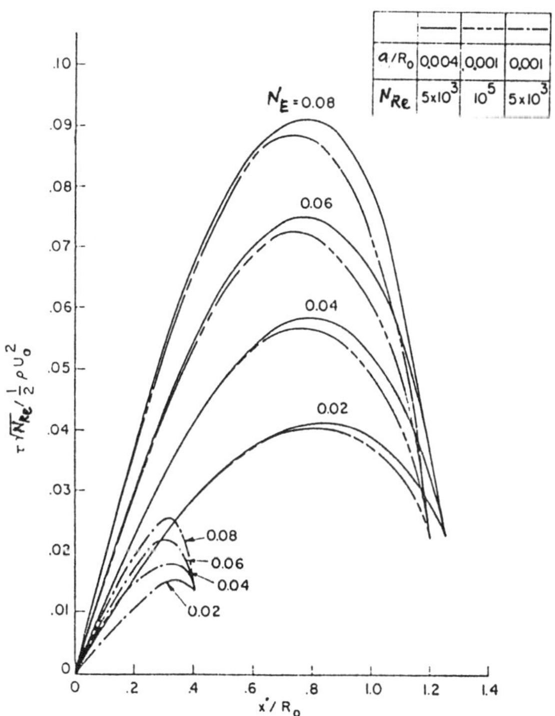

Figure 7.25   Local wall shear stress for $N_E \leq 0.1$.
[Goldstein et al. 1967]

new variable $\eta = y^*/\delta^*$, the local heat transfer and skin friction are given according to:

$$J_q = -\kappa \left.\frac{\partial T}{\partial y}\right|_{y=0} = \frac{\kappa(T_0 - T_w)}{2R_0 \alpha_p \delta^*} \left[T_w(\eta)\right]_{\eta=0} \tag{7.175}$$

and in terms of the local heat transfer coefficient $h = J_q/(T_w - T_0)$ the Nusselt number $N_{Nu} = hx/\kappa$ is given by:

$$N_{Nu}(N_{Re})^{-1/2} = (N_E \, \delta^*)^{-1} \left[T_\eta^*\right]_{\eta=0} \tag{7.176}$$

The skin friction due to shear stress exerted on the cylinder surface by the flowing liquid may by determined from $\tau = -\mu(\partial U_g/\partial y)_{y=0}$. In terms of the transformed variables, this becomes

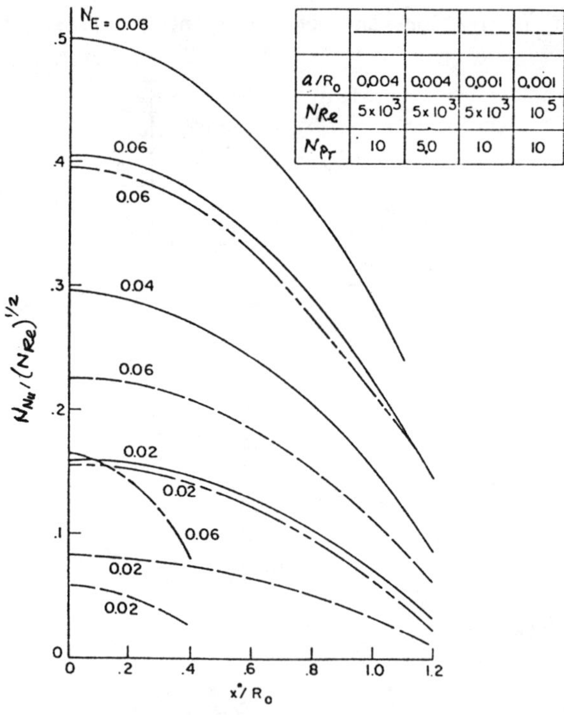

a/R_0	0,004	0,004	0,001	0,001
N_{Re}	5×10^3	5×10^3	5×10^3	10^5
N_{Pr}	10	5,0	10	10

Figure 7.26   Local Nusselt number for $N_E \leq 0.1$.
[Goldstein et al. 1967]

$$\frac{2\ \tau}{\rho\ U_0^2}\ (N_{Re})^{1/2} = (N_E\ \delta*^2)^{-1}\ [\frac{\partial^2 \psi*}{\partial \eta^2}]_{\eta=0} \tag{7.177}$$

These results for the case of $N_E < 0.1$ are shown in Figs. 7.25 and 7.26. It is seen that the local shear stress increases from zero at the forward stagnation point to a maximum at a certain location along the cylinder surface followed by a decrease in the downstream direction. In Fig. 7.26 the local Nusselt number decrease along the surface from a maximum at the forward stagna- tion point for all values of N_E. The local Nusselt number increase with an increase in Prandtl number of the liquid especially in the downstream direction. Termination of curves at different values of x/R indicates boundary layer separation via a sharp increase of boundary layer thickness in the computation.

A comparison to the experimental results of Acrivos et al. [Goldstein et al. 1967] is shown in Fig. 7.27.

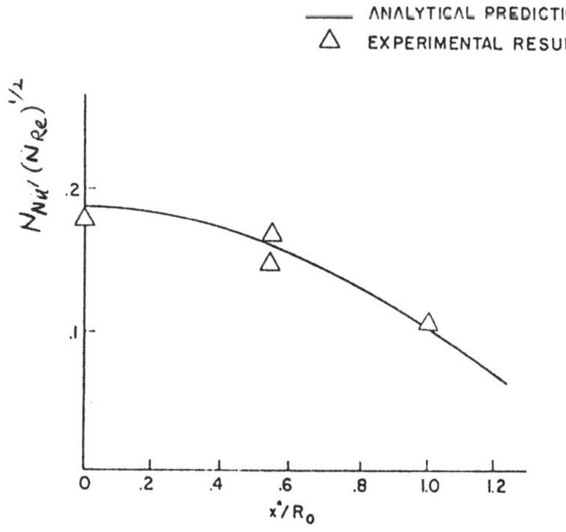

Figure 7.27 Comparison of analytical and experimental results of heat transfer from a circular cylinder, for $N_{Re} = 8 \times 10^4$, $a/R_0 = 0.004$, $N_{Pr} = 10$ and $N_E = 0.02$. [Goldstein et al. 1967]

Exercise Problems

7.1 Write the equations of particle trajectories for a centrifuge, where the fluid is in solid body rotation. Include the virtual mass force and lag in angular velocity of the particle having a radial velocity. Compute, for a steady state, the angular velocity and radial velocity of a particle at a radius of 0.1 m and angular velocity of 5000 rad./s for $F = 1.4 \times 10^4$.

Ans. 4500 rad./s, 154 m/s

7.2 Compute, for flow of a particle-gas suspension entering a channel with $\rho_{pq} = 0.5$ kg/m^3, b = 0.02 m (Fig. 7.4), $U_0 = 2$ m/s, q/m = 3×10^{-3} C/kg, $F = 10^5 s^{-1}$, $\bar{\rho}_p = 2500$ kg/m^3, $\bar{\rho} = 1.2$ kg/m^3, 2a = 1 μm: E_y at y = b, $(N_{Re})_e$, and N_{ev}.

Ans. 2.5 x 10 V/m, 0.02, 2900

7.3 Derive Eqs. (7.45) and (7.46).

7.4 Derive Eqs. (7.48) and (7.49).

7.5 Compute L_c for $\sigma = 0.5$, $\bar{\rho}_p = 10^3$ kg/m^3, q/m = 0.1 C/kg, a = 1 μm, $F = 10^5$ s^{-1}, U = 3 m/s, b = 4 mm, and η_c of 95% for Case (a) Sec. 7.2 with an applied voltage of 10^3 V, and for Case (c).

Ans. 48 mm, 6.36 m.

7.6 From Eq. (7.69), determine the equations for h up to h_2 in terms of f's, g's, and m's. Note that as $\eta \to 0$, $h \to A\eta$. $h_2 \to 1$.

7.7 From Eqs. (7.72) and (7.73), determine the different orders of g 's, m 's in terms of f and finite N_{Sc} up to order 2. Determine the condition at large ξ and N_{Sc}.

7.8 From Eqs. (7.77) and (7.78), determine t's and t_'s up to order 2, and discuss the influence of N_{Sc} and N_{Pr}.

7.9 For the core velocity to inlet velocity ratio of channel flow between two parallel plates as in Fig. 7.13 given by

$$\frac{U_o}{U_1} = 1 + K_1 \, \varepsilon + K_2 \, \varepsilon^2 + \ldots \qquad (7.178)$$

where $\varepsilon = (x/b)^{4/5} \, (U_1 b/\overline{\nu})^{-1/5}$, and $K_1 = 0.0463$, $K_2 = 0.001712$, .. determine the core velocity of the particle cloud.

7.10 Determine the density of particle cloud in the core from Eq. (7.90) and the result of Prob. 7.9.

7.11 Determine the position of the virtual origin of a circular jet for core velocity equal to U_1 at r_1.

7.12 When diffusion effect predominates, determine the density distribution in separated flow, postulating that $\rho_p = \rho_p(\xi)$.

7.13 Determine from Eq. (7.135), the relation of its boundary condition to the one given by Eq. (4.90), and the condition for commonality.

7.14 From Eq. (7.142), determine J^*_{kG} for constant D_k. Determine (1) J^*_{kG} for $A_k = 0$, $\tau^*_k \, y^*_k \gg 1$, (2) Very small τ^*_k. Note that the latter case gives $J^*_{kG} = 0$ for $A_k = 1$, identical to the Sutton equation.

7.15 The diffusion and fall-out from a jet engine at take-off can be represented by coordinate x along the ground level in the vertical plane of take-off at an angle of ascent with the ground and speed U. Here the problem becomes one of moving source. The correspondence to the case of stack plume is seen in that $x^* = D'_{kz} (Ut - x \cos\theta)/Ux^2\sin^2\theta,$, $\tau^* = g \, x \cos\theta \, \sin\theta/F_k \, D'_{kz}$, and the variation of deposition rate at a given x with time is analogous to the change of position along x in the case of the stack. D'_{kz} is the diffusivity in the direction normal to that of ascent. Here we again account for only the conditions after the effect of the jet velocity W itself has been diffused. Use the chart in Fig. 7.20, compute for the conditions: $U = 100$ m/s, $x = 2000$ m, $\theta = 30°$, $D_k = 10$ m^2/s, $F = 100$ s^{-1}, $g = 9.8$ m/s^2, total exhaust flow of 100 kg/s with 2% contaminant. Compute the time when fall-out began, the time it reaches a maximum, and the time it ends. What is the maximum fall-out rate? Compute for the time for the fall-out to end in the case of $F_k = 1000$ s^{-1}.

> Ans. 22.3 s, 42.5 s, 167.3 s, 0.16 kg/m^{2}s; 2 x 10^{4}s.

7.16 Derive the balance relations in Eqs. (7.149) to (7.155) from considering Eqs. (2.19), (2.20), and (2.22), neglecting the effects of surface tension.

Chapter 8

DENSE SYSTEMS

We have so far treated cases of dilute suspensions and have demonstrated principally the effects of particle-fluid interactions in multiphase flow. An exception is the significance of electrostatic effects in particle-particle interaction. In practice, we often encounter problems in systems of dense suspensions of particulates. Much remains to be developed in our knowledge on dense systems such as liquid slurries and fluidized beds. Packed bed and porous solid constitute the upper limit in the density of such systems. The basic equations in Chapter 2 still apply, except that we need additional constitutive relations.

Our understanding of dense systems is largely empirical at this stage. Extensive experimental data on specific systems are available in handbooks [for instance, Cheremisinoff and Gupta 1983] and are not included here. Topics treated in this chapter are intended as an introduction to further fundamental studies.

We further note that detailed treatment of flow through porous media, packed beds and moving beds, flow of granular solids, and fluidized beds are fields of studies by themselves. This is also the order of complexity from considering phase interactions; the two simple extremes appear to be dilute suspensions on the one end and fixed dense beds on the other. Fluidized beds are complicated by strong phase interactions and large scale motions. This sequence is also the order of presentation in this chapter.

8.1 Some Fundamental Nature of Dense Suspensions

Macroscopic treatment following the approach in rheology identi-
fies dense suspensions or slurries as non-Newtonian fluids. The non-
Newtonian nature is seen in that a finite value of shear stress must
be applied before the fluid deformation can begin. Once the finite
shear stress, τ_y, which is referred to as the yield stress [Bingham
1922], is reached and the initial movement has started, the fluid
behaves in the same manner as a Newtonian fluid, with the shear
stress increasing linearly with the rate of shear. This rheological
behavior is represented by the relation for flow through a circular
pipe of radius R:

$$\tau - \tau_y = \eta(dW/dr) \tag{8.1}$$

within $r_c \leq r \leq R$, where r_c is a characteristic radius, and η is
defined as the coefficient of rigidity or plastic viscosity and is
also referred to as the differential viscosity of a Bingham plastic
fluid. The apparent viscosity, μ_a, is defined as

$$\mu_a = \tau_y(dW/dr)^{-1} + \eta \tag{8.2}$$

which decreases from a large value at low shear rates to a limiting
value at high shear rates; therefore data of shear stress as a
function of shear rate is required.

The yield stress, τ_y, as given in Eq. (8.1), is such that at $r =
r_c$, $\tau = \tau_y$; and $dW/dr = 0$, for $0 \leq r \leq r_c$, and $\tau = 0$ at $r = 0$. The
shear stress at the wall is given by $\tau = \tau_w$ at $r = R$, and $\tau/\tau_w = r/R$
for $r > r_c$, according to Eq. (6.2) for $g = 0$. The mean velocity W_m
is given by integrating over the velocity profile to give:

$$\frac{8W_m}{(2R)} = \frac{1}{\eta}\left[\tau_w - \frac{4}{3}\tau_y + \frac{1}{3}\frac{\tau_y^4}{\tau_w^3}\right] \tag{8.3}$$

Eq. (8.3) shows the influence of τ_y and reduces to the case of laminar Newtonian flow when $\tau_y = 0$, but does not otherwise facilitate the determination of τ_y and τ_w. For $\tau_w \gg \tau_y$, Eq. (8.3) reduces to give an effective viscosity μ_e:

$$\mu_e = \eta \left[1 + \frac{\tau_y(2R)}{6\eta W_m} \right]. \tag{8.4}$$

η is thus the limiting viscosity at high shear rates.

Most non-Newtonian fluids fall into the class of pseudoplastic fluids, whose apparent viscosity and differential viscosity decreases with increasing rate of shear. One of the explicit expression for apparent viscosity is the power law of Ostwald-de Waele (see, for instance, Skelland [1967]):

$$\mu_a = K(dW/dr)^{n-1} \tag{8.5}$$

where K is called a consistency index, an example for its composition is given by Eq. (3.104), and n is called a flow behavior index. The shear stress is thus:

$$\tau = K(dW/dr)^n \tag{8.6}$$

where n = 1 is clearly the case of a Newtonian fluid; when n < 1, the fluid is said to have pseudoplastic or shear-thinning behavior, while n > 1 defines a dilatant or shear-thickening fluid. Pseudoplastic behavior is known to have been observed among suspensions such as paper pulp, fine coal in water, and other liquid suspensions of fine inert unsoluted solids. Dilatancy has been exhibited in titanium dioxide-water mixture and galena-water suspension with mean particle sizes of 20 to 30 μm [Cheremisinoff and Gupta 1983], and suspensions of spheres of high concentration, but only 12% by volume for iron oxide, which are highly non- spherical particles [Brodkey 1967]. Metzner and Whitlock [1958], however, showed that both ranges of n may occur depending on the shear rate. To understand these behaviors the nature of particle-particle and particle-fluid interactions needs be explored.

We have noted that the dilute suspensions (Sec. 2.7) are characterized by weak interactions and large relative motions between
phases. This nature simplifies the formulations and facilitates
rigorous solution of problems. The mean free path of particle-
particle collision is meaningful when the volume fraction of
particles is less than 10% (11.8% for spheres). Above that, relative
motions are provided by the availability of vacant spaces in the
suspension and strong interactions among particles and fluids have to
be accounted for. Some simplification is possible for a very dense
system when the volume fraction of particles is very high, say, 40 to
74% depending on the size distribution of the particulates. At these
high densities, strong interactions are accompanied by small relative
motions. In between these two extremes of concentrations, the
situation becomes complicated and motions tend to be unstable. A
comparison of dilute and dense suspensions is summarized in Table
8.1. Available details of transport properties are seen next.

Table 8.1

Comparison of Dilute and Dense Suspensions [Soo 1987]

	Dilute Suspension	Dense Suspension
Relative motion between particles	Large	Small
Particle-particle interactions	Weak	Strong
Particle diffusivity, D_p	Large	Small
Apparent viscosity of particle phase	Due to particle-fluid interaction $\mu_{pL} = \alpha_p \rho_p D_p$	Due to particle-particle interaction $\mu_{pb} \propto 1/D_p$
Flow regime in applications	Steady, turbulent	Pseudo-laminar to unsteady and stratified
Motion above minimum transport velocity	Stable	Stable in liquid slurry, unstable in gaseous suspension
Analogy in molecular systems	Rarefied gas flow	Molecular theory of liquids

When a suspension is so dense that the concentration is close to
that of a packed bed, the apparent viscosity or shear resistance of
the mixture is high while the diffusivity of the particles is low
because of reduced vacancy for random movement. The fluid turbulence
could be completely damped out. As a result, a form of laminar
motion could occur and is non-Newtonian in nature [Brodkey 1967]. In
this case, particle-particle interaction tends to overshadow the
effect of presence of the fluid [Savage 1983]. Motion of the
particle phase is influenced by the drag force exerted by the fluid
and shear force of the particles. The viscosity of particle-particle
interaction given by Eq. (3.104) can be represented in a simplified
form as:

$$\mu_{pp} \simeq C_\mu \alpha_p^2 \, a^2 \, \bar{\rho}_p [\partial W_p / \partial r] \qquad\qquad (8.7)$$

which conforms to the Ostwalde-de Waele model in Eq. (8.5). In Eq.
(8.7), C_μ is a constant of order 1 for spherical particles with
elastic collisions, α_p is the volume fraction of particles of radius
a and material density $\bar{\rho}_p$, W_p is the particle velocity, and r is the
radial coordinate of a pipe flow system for our illustration. C_μ
can, in general, be treated as an empirical coefficient to account
for non-sphericity, size distributions, inelastic collisions and
sliding frictions. This relation is only valid for isometric
particles and when the effect of collision overshadows friction due
to relative motion; namely, the particles cannot be too small.
Savage included $C_\mu \alpha_p^2$ in an empirical function; Schuegerl [1971]
followed a similar correlation for viscosity of gas fluidized beds of
glass and quartz particles corresponding to $C_\mu |\partial W / \partial r| \sim O[100]$.
Equation (8.7) represents the case of a shear thickening fluid such
as a pulverized coal slurry (Prob. 8.1). The general rheological
behavior of solid-liquid mixtures (see, for instance, Brodkey [1967])
are not treated here.

Diffusion of particles arises from displacement of particles
into vacant spaces in the suspension. Extension of kinetic theory of
liquids suggests the particle diffusivity be proportional to the
ratio of: (kinetic energy of fluid acting on each particle due to

shear motion)/(viscous resistance). This kinetic energy imparted by the fluid per particle is given by:

(mass of fluid/vol.)(vol./particle)(kinetic energy/mass)

$$= [\bar{\rho}(1-\alpha_p)][\frac{4\pi}{3}\frac{\alpha^3}{\alpha_p}]\ \frac{1}{2}\ [(\frac{4\pi}{3})^{1/3}\ \frac{a}{(1-\alpha_p)^{1/3}}\ |\frac{\partial W}{\partial r}|]^2$$

$$= (1-\alpha_p)\ \bar{\rho}\ (4\pi/3)^{2/3}\ a^2(1-\alpha_p)^{-2/3}|\partial W_p/\partial r|^2(4\pi/3)a^3/\alpha_p$$

while the viscous resistance is proportional to $6\pi a\mu_{pp}$, and the shear rate of the particle cloud is close to that of the fluid, $|\partial W/\partial r|$. A simplified form for particle diffusivity D_p is therefore:

$$D_p \simeq C_p(\bar{\rho}/\bar{\rho}_p)(a^2/\alpha_p^3)|\partial W/\partial r| \qquad\qquad (8.8)$$

where C_p is an empirical coefficient, $C_p \sim O[10]$, $\bar{\rho}$ and W are the material density and velocity of the fluid phase. Drag coefficient and inverse relaxation times of dense suspensions are given in Sec. 3.1.

When applied to relative motion of phases at low speed, the inertia forces are negligible, Eq. (2.92) is reducible to:

$$0 = (1 - \alpha_p)\ \rho\ F_{fp}(\mathbf{W}_p - \mathbf{W}) - \nabla P_f + (1 - \alpha_p)\rho_f\mathbf{f} \qquad (8.9)$$

according to Sec. 2.7 Eq. (8.9) is a form of Darcy's law ($W = 0$) for flow through porous media [Scheidegger 1957], with

$$\bar{\mu}(\mathbf{W}_p - \mathbf{W})/k_p = (1 - \alpha_p)\ \bar{\rho}F_{fp}(\mathbf{W}_p - \mathbf{W}) = \alpha_p\bar{\rho}_pF_{pf}(\mathbf{W}_p - \mathbf{W}) \quad (8.10)$$

where k_p is the permeability of the solid phase, and F_{pf} is the inverse relaxation time for momentum transfer from fluid to particle (Eq. 2.102 for the last equality). The limiting case of a dense suspension or slurry is slow relative motion with the solid carried forward by the drag force exerted by the fluid. For the following development, F_{pf} is given by:

$$F_{pf} = C_F \; 75a^2 \; \alpha_p \bar{u}/2\alpha \; [\bar{\rho}_p + (\bar{\rho}/2)] \qquad (8.11)$$

for small relative motion of particle cloud to fluid; the coefficient $C_F \sim (1 - \alpha_p)^2 \cdot (1 - \alpha_p)$.

8.2 Pipe Flow

We now apply the properties of a dense suspension to the case of pipe flow in Fig. 6.9 and treat the case of a horizontal pipe. For steady flow to exist, the gravity effect toward settling must be balanced by the shear lift force in laminar motion given by Eq. (6.25) and (6.26) with $W_p \sim 0$. Equations (6.42) and (6.43) are modified by changing to coordinates $x = r \sin \phi$, and $y = r \cos \phi$, and the density distribution in the vertical plane is given by, for $y^* = y/R$:

$$\alpha_p^{-3} \; (\partial\alpha_p/\partial y^*) = -\eta_c |\partial W^*/\partial y|^{-1} + \Lambda_c W^* |\partial W^*/\partial y^*|^{-1/2} \qquad (8.12)$$

where $W^* - W/W_o$, Λ_c and η_c are given by:

$$Rf_L/D_p F_{pf} = \Lambda_c \alpha_p^2 \Delta W^* |\partial W^*/\partial y^*|^{1/2} \qquad (8.13)$$

characterizing the shear lift force versus the effect of diffusion, with

$$\Lambda_c = 0.04113(W_o R/\nu)^{1/2}(R/a)[1 + (\rho^*/2)]\rho^* C_p C_F \qquad (8.14)$$

where $\rho^* = \bar{\rho}/\bar{\rho}_p$,

and

$$R(\bar{\rho}_p - \bar{\rho})g \; \sin\theta/D_p F_{pf}\bar{\rho}_p = \eta_c \alpha_p^2 |\partial W^*/\partial y^*|^{1/2} \qquad (8.15)$$

characterizing the gravity force versus the diffusion effect, with

$$\eta_c = \frac{2}{75}(1 - \rho^*)(1 + \frac{\rho^*}{2})\rho^{*-2}\left(\frac{Rg\sin\theta}{W_o^2}\right)\left(\frac{W_oR}{\nu}\right)/C_pD_F \qquad (8.16)$$

Eq. (8.12) integrates to

$$\alpha_{po}^{-2} - \alpha_{p\pi}^{-2} \simeq 2\left[\eta_c|\partial W^*/\partial y^*|_{av}^{-1} - \Lambda_c\Delta W_{av}^*|\partial W^*/\partial y^*|_{av}^{-1/2}\right] \qquad (8.17)$$

where α_{po} is the volume fraction of the particles at the top of the
pipe and $\alpha_{p\pi}$ is that at the bottom in relation to gravity. The subscript av specifies the average dimensionless velocity and velocity
gradient. We not that for steady nondepositing flow, the condition
of

$$\eta_c|\partial W^*/\partial y^*|_{av}^{-1} \sim \Lambda_c\Delta W_{av}^*|\partial W^*/\partial y^*|_{av}^{-1/2} \qquad (8.18)$$

must be met. Reverting to physical parameters, the above condition
translates to:

$$(ag/W_o^2)(W_oR/\nu)^{1/2} \sim 1.542\,\rho^*(1 - \rho^*)^{-1}|\frac{\partial W^*}{\partial y^*}|_{av}^{1/2} \qquad (8.19)$$

which correlates the gravity effect to the flow behavior (Prob.
8.2). Note that this does not mean small particles will always be
suspended, since agglomeration may occur. This trend for a liquid
slurry agrees with the results of Roco and Shook [1981], who also
used a two-dimensional system for the computations.

The non-Newtonian nature of a steady dense suspension is seen in
the simple case where we can neglect the gravity effect of the
velocity distribution and treating the density as nearly uniform.
Eq. (6.33) for the momentum of the fluid phase is reduced to:

$$-(1/2)(1 - \alpha_p)\rho^*N_m(\partial P^*/\partial z^*) - \alpha_p^2(1 - W_p^*) = 0 \qquad (8.20)$$

and the momentum equation of the mixture is given by:
$(-dP^*/dz^* = C)$

$$(C/2) + N_{Re}^{-1}\,\beta_c\alpha_p^2 r^{*-1}\frac{\partial}{\partial r^*}\left[r^*|\frac{\partial W_p^*}{\partial r^*}|\left(\frac{\partial W_p^*}{\partial r^*}\right)\right] = 0 \qquad (8.21)$$

where $\rho^* = \bar{\rho}/\bar{\rho}_p$, $N_m = W_o\alpha_p/F_{pf}R$, $z^* = z/R$, $r^* = r/R$, $P^* = P/(\bar{\rho}W_o^2/2)$, $N_{Re} = W_oR/\bar{\nu}$, and $\beta_c = C_\mu(a/R)^2 \rho^{*-1}$. This simplification is available to any realistic physical system where N_{Re} is large and the viscous stresses due to the particle phase is much greater than those due to the fluid phase. Eq. (8.21) integrates to, for $W_p < W$,

$$W_p = (7/3)(\dot{m}_p/\pi R^2 \alpha_p \bar{\rho}_p)[1 - (r/R)^{2/3}] \tag{8.22}$$

where $\dot{m}_p$ is the flow rate of particles (velocity profiles for general values of n in Eq. 8.5 are given in Brodkey [1967]). The pressure drop is given by

$$- \frac{dP}{dz} = 49\, C_\mu \left(\frac{a}{R}\right)^2 \bar{\rho}_p \left(\frac{\dot{m}_p}{\pi R^2 \bar{\rho}_p \alpha_p}\right)^2 / 2R \tag{8.23}$$

In terms of physical parameters, C_μ is of order 10 for a practical slurry pipe flow system with a mean particle size of 1 mm. Eq. (8.20) gives the core velocity of the particle phase W_{po} according to:

$$1 - \frac{W_{po}}{W_o} = \frac{7}{25\pi} \left(\frac{C_\mu}{C_F}\right)\left(\frac{1 - \alpha_p}{\alpha_p}\right)\left(\frac{a}{R}\right)^4 [1 + \frac{\bar{\rho}}{2\bar{\rho}_p}]\left(\frac{\dot{m}_p}{\mu R}\right) \tag{8.24}$$

This velocity lag provides the drag force to move the solid particles; this lag is small for small particles. (Prob. 8.3) The flow condition is close to that of flow through a porous body of solids with porosity given by Eq. (8.10).

Pipe flow at intermediate concentrations below a volume fraction of 30% demonstrates unsteadiness or wave motion evidenced by measurable correlations of velocity and density of particles. Sheen et al. [1984], using an ultrasonic mass flow instrument on pipe flow of a coal-oil slurry showed that at a mean velocity of 0.4 to 3.8 m/s, the flow regime is turbulent for volume fractions of 0–10.6 percent and laminar from 30 to 46.5 percent, with cross-correlation function in counts decreases with increase in concentration, a tendency toward steady laminar motion (Prob. 8.4). Cross-correlation due to unsteady motion in pipe flow of a gas solid suspension of 2 percent volume fraction solid (nearly 20 to 1 mass flow ratio) was used by Bobis et al. [1986] for measuring particle velocity and density by a capacitive flowmeter. In the case of a dense gas-solid suspension, plugs of solids were observed in horizontal pipe flow by Tsuji and Morikawa [1982] with fluidization assisted by a subpipe. Choking in vertical pipe flow of various particles occurred at reduced gas velocities was correlated by Chong & Leung [1986].

8.3 Porous Media and Moving Beds

We have noted that flow of a very dense suspension is similar to the limiting case of relative motion of fluid through a moving porous solid or packed bed. A basic macroscopic property of a packed bed is its permeability k_p. The Kozeny-Carman equation [Carman 1937] gave:

$$k_p = \varepsilon^3/S^2 a \tag{8.25}$$

where ε is the fraction void called porosity, S is the surface area per unit volume, and a is a constant including a tortuosity factor and a shape factor; a ~ 5. Marshall [1962] gave

$$k_p = \frac{1}{8}\,\varepsilon^2 n^{-2} \sum_{i=1}^{n} (2i - 1)\, r_i^2 \tag{8.26}$$

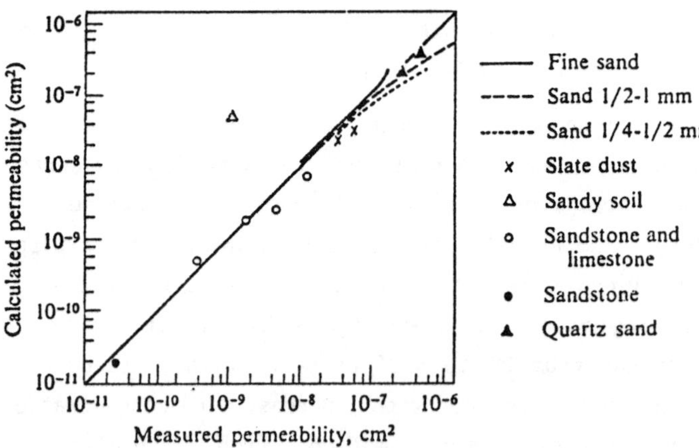

Figure 8.1   Comparison of measured and calculated permeability by Eq. (8.26) [Marshall 1962]

for $r_i > r_{r+1}$, and r_1, r_2,..., r_i, ..., r_n represent the mean radius of the pores in decreasing order of size in each of n equal fractions of the pore space. A comparison of the computed and measured values of k is shown in Fig. 8.1. A value of $k_p < 10^{-5}$ m^2 is usually considered to be large. Eq. (8.9) may be regarded as a derivation leading to the Darcy's law for the velocity of flow through a statistically homogeneous and isotropic packed bed given by:

$$U = -\frac{k_p}{\mu_d} (\nabla P - f) \tag{8.27}$$

where μ_d is defined as a dynamic viscosity. Solutions of cases are given in Scheidegger [1957] as a special field of study. Seepage of fluid through fine cracks of rocks may be treated according to Eq. (8.27) (Prob. 8.5).

A basic difference of a bed of solids from a suspension is the stress existing within the bed. Brandt and Johnson [1963] considered the resistance to flow of one particle past another or past the stationary wall of the container due to solid friction, and the pressure drop of the fluid as contributing an additional body force similar to gravity. In the representation in a cylindrical coordinate system (Fig. 8.2), they identified three normal compressive stresses σ_z, σ_θ, σ_r, normal to the planes and six shear

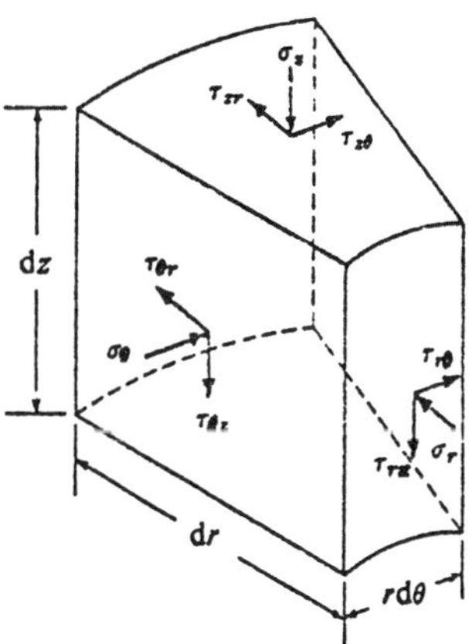

Figure 8.2   Forces acting in a moving bed.

stresses τ_{rz}, $\tau_{r\theta}$, etc., along the planes of stress. Stress is distributed by particle-particle friction, and no equilibrium shear stress can be higher than that determined by internal friction. For instance, if $\tau_{zr}/\sigma_z \geq f^*$, the coefficient of internal friction, slipping will occur along the z-plane and the stress pattern will change over the whole system. The forces acting on the boundaries of a bed are the radial and shear stresses at the wall and the average vertical stress over the cross-section of the containing vessel, as shown in Fig. 8.3. Delaplaine [1956] measured the stresses acting both at the boundaries and within downward flowing bed of sand, glass, and bead catalyst. He showed that the stresses are not proportional to bed depth but increase asymptotically to constant values at bed depths greater than 4 to 6 tube diameters. Force balance of the system in Fig. 8.3 gives, for average vertical stress $(\sigma_z)_a$ and wall shear stress $(\tau_{rz})_w$

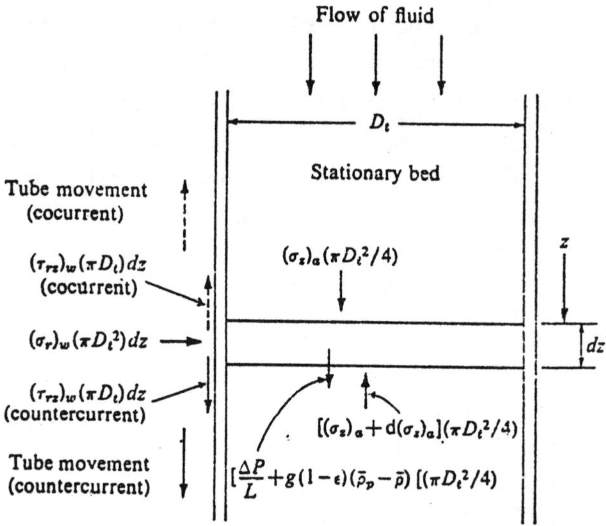

Figure 8.3   Force balance of boundary and average stresses in moving bed [Brandt and Johnson 1963]

$$\frac{d(\sigma_z)_a}{dz} = g\rho_p - \frac{4(\tau_{rz})_w}{D_t} \tag{8.28}$$

where D_t is the tube diameter. Substitution of the stress ratios $f^*_w = (\tau_{rz})_w/(\sigma_r)_w$, and $f^*_a = (\sigma_r)_w/(\sigma_z)_a$ gives:

$$\frac{d(\sigma_z)_a}{dz} = g\rho_p - \frac{4f^*_a f^*_w (\sigma_z)_a}{D_t} \tag{8.29}$$

Assuming f^*_a and f^*_w as constants, Eq. (8.29) has the solution:

$$(\sigma_z)_a = \frac{g\rho_p D_t}{4f^*_a f^*_w} [1 - \exp(-\frac{4f^*_a f^*_w z}{D_t})] \tag{8.30}$$

Thus $(\sigma_z)_a$ approaches $g\rho_p D_t/4f^*_a f^*_w$ in deep beds. When fluid flow occurs, Hancher and Jury [1959] gave

$$(\sigma_z)_a = \pm \frac{AD_t}{4f^*_a f^*_w} [1 - \exp(-\frac{4f^*_a f^*_w z}{D_t})] \tag{8.31}$$

where the (+) sign for cocurrent movement and (-) sign for counter-current movement, and

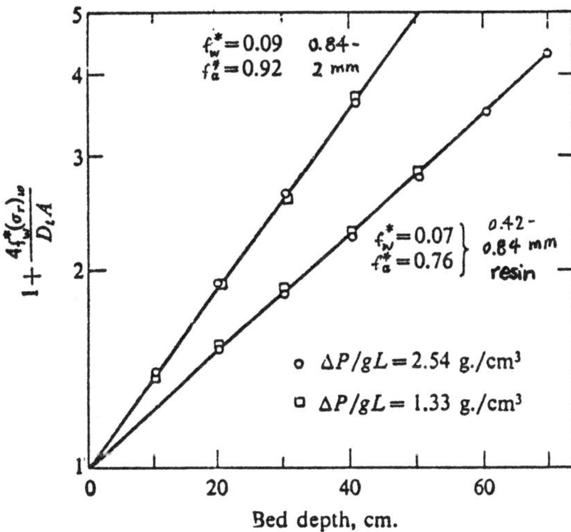

Figure 8.4   Comparison of Eq. (8.34) with experimental results (radial stress is moving bed in countercurrent) [Brandt and Johnson 1963]

$$A = \frac{\Delta P}{L} + (1 - \varepsilon)(\overline{\rho}_p - \overline{\rho})g \qquad (8.32)$$

$(\Delta P/L)$ being the pressure drop, $f_a^* \sim 1$ and $f_w^* =$ constant. The gradient of compressive stress is

$$\frac{d(\sigma_z)_a}{dz} = A \mp \frac{4f_a^* f_w^* (\sigma_z)_a}{D_t} \qquad (8.33)$$

Brandt and Johnson made measurements of the average vertical stress and radial stress at the wall of the tube containing the bed, with cocurrent or countercurrent motion (at 1 to 30 cm/min.) relative to the fluid (water), by a balance and pressure transducer at the tube wall. Experiments were carried out with particle sizes in various ranges between 0.15 to 2 mm. The coefficient of friction depends on solid velocity and particle size. Considerable internal friction was found in a bed of glass beads but not in that of resins. In counter-current flow, close correlation was made with integrated equation of force balance in the cross-section of the bed but in cocurrent flow, only 0.42 to 0.84 mm resin was represented by the same equation. Fraction void of the bed was not reported. Fig. 8.3 shows a typical comparison of Eq. (8.31) for countercurrent movement with experimental results, in which case Eq. (8.31) is expressed as

$$\ln \left[1 + \frac{(\sigma_r)_w}{A} \frac{4f_w^*}{D_t} \right] = \frac{4f_w^* f_a^*}{D_t} z \qquad (8.34)$$

It is possible that diffusion of particles may occur for small particles in cocurrent motion.

<u>Dry bulk materials.</u>

A simple model for the discharge of collected solids from the bottom of a cyclone was given in Sec. 6.7 for the movement of a thin layer of solids under gravity. The general case of flow of frictional, cohesive solids from bins or bunkers under gravity again constitutes a field by itself [Jenike and Johanson 1969]. A number of attempts have been made to introduce inertia terms into force balance models of flow of dry bulk materials under gravity, with or without the influence of the air present in the interstitial spaces. Savage [1965] assumed uniform flow through a straight-walled converging channel (Fig. 8.5). Polar coordinates with origin at 0 are used. The wall friction is set equal to zero to that there is no variation in the θ-direction. A steep mass flow channel is taken so that $\cos \theta_0 \simeq 1$. For a conical channel, the equation of motion in the r-direction is given by:

$$- \frac{d\sigma_r}{dr} + 2 \left(\frac{\sigma_\theta - \sigma_r}{r} \right) - \rho_p g = \rho_p U \frac{dU}{dr} \tag{8.35}$$

where σ_r and σ_θ are the stresses in the radial and tangential directions, and U is the velocity in the r-direction. The continuity equation gives $U = C / r^2$ in this case. The stress field in the hopper is assumed to be passive so that

$$\sigma_\theta = K \sigma_r \tag{8.36}$$

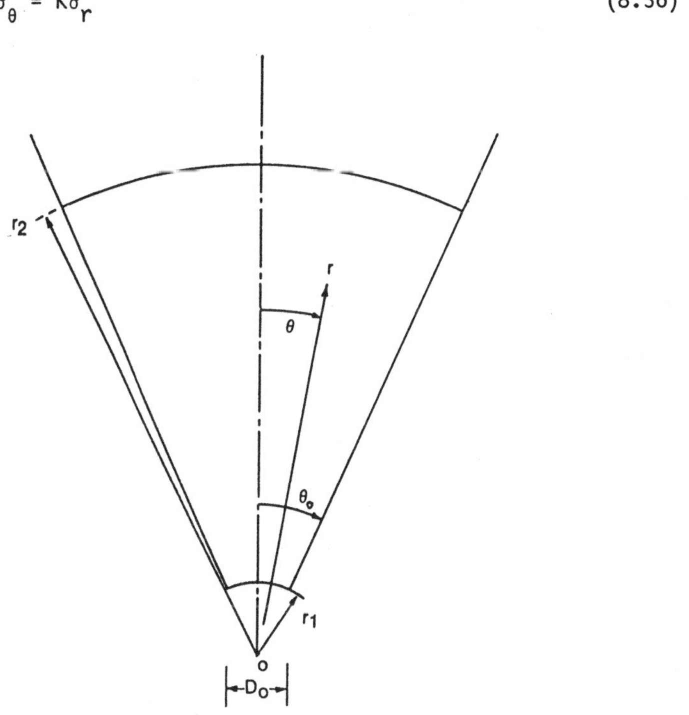

Figure 8.5   Coordinate system for flow in a converging channel.
             [Savage 1965]

where $K = (1 + \sin \delta)/(1 - \sin \delta)$; δ being an effective angle of internal friction determined experimentally. Substituting U and Eq. (8.35) and integrating with boundary conditions $\sigma_r(r_1) = \sigma_r(r_2) = 0$, gives an expression for U. At the outlet, we have

$$U_1 = - \left[\frac{(1 + K)gD_o}{2[2(K - 1) - 1]\sin \theta_o} \left(\frac{1 - (r_1/r_2)^{2(K-1)-1}}{1 - (r_1/r_2)^{2(K+1)}}\right)\right]^{1/2} \qquad (8.37)$$

For a deep bin, $r_2 >> r_1$, Eq. (8.37) reduces to

$$U_1 \approx \left[\frac{(K + 1)gD_o}{2(2K - 3) \sin \theta_o}\right]^{1/2} \qquad (8.38)$$

that is, the velocity is independent of the head. The solid flow rate is given by:

$$\dot{m}_p = \left[\frac{K + 1}{2(2K - 3) \sin \theta_o}\right]^{1/2} \frac{\pi}{4} \rho_p g^{1/2} D_o^{5/2} \qquad (8.39)$$

which has been experimentally validated [Franklin and Johanson 1955]. Actual flow rate tends to be lower because of wall friction and various degrees of cohesiveness (modification by Johanson [1964] by introducing a flow factor of hopper). Internal friction can be reduced by partial fluidization [Davidson and Nedderman 1973].

8.4 Observations on Fluidized Beds

Fluidized beds, due to their important applications as devices for mixing and heat and mass transfer, and as chemical reactors, particularly in the field of petroleum refining, have been the subject of many intensive studies. Still, among gas-solid and liquid-solid systems, the fluidized bed is least susceptible to rigorous analysis. With the availability of treatises by Davidson et al. [1985] and Kunii and Levenspiel [1968], to name a few, we shall cover only the basic concepts and methods related to the aspects of fluid dynamics.

Continuous transition from the regimes of packed bed, fluidized bed, to transport flow was studied by Wilhelm and Kwauk [1948]. McCune and Wilhelm [1949] treated the mass and momentum transfer in

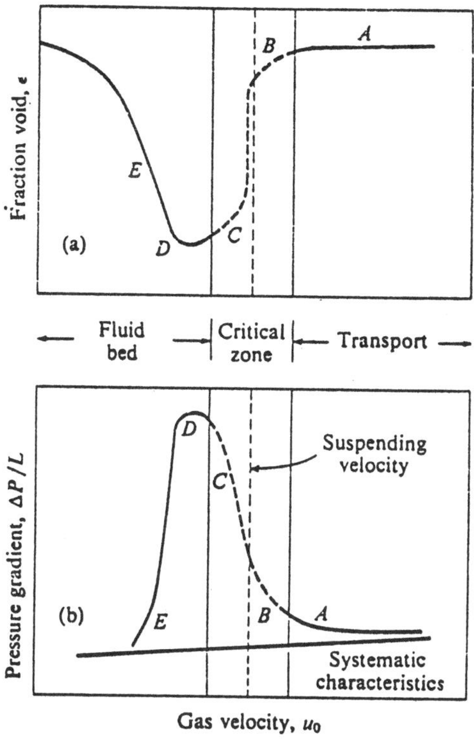

Figure 8.6   Typical trend of fraction void and pressure gradient at given feed rate [McCune and Wilhelm 1949]

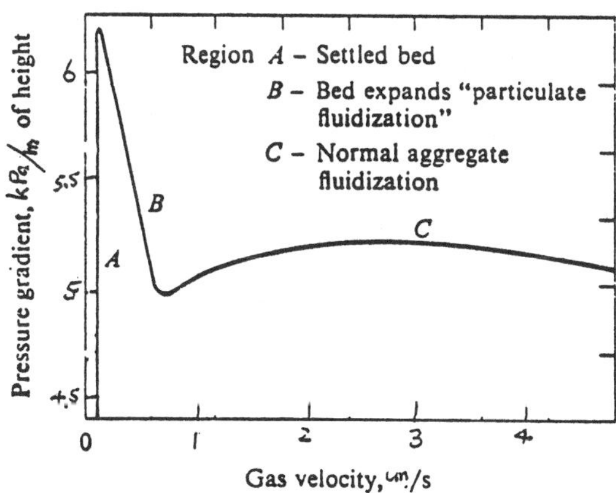

Figure 8.7   Fluidization of cracking catalyst [May and Rossell 1954]

fixed and fluidized beds, and gave the trend for transition from fluidization to transport as shown in Fig. 8.6 giving the relation between gas velocity, fraction void, and transport. In the representation of May and Rossell [1954], three steps of the fluidizing process were recognized, and are summarized in Fig. 8.7. In Range A, low velocity gas percolates through the bed without agitation of individual particles, the gas phase is in viscous flow, and the pressure drop increases linearly with velocity, but less than the weight of the bed. Sufficient increase in gas velocity (to 0.3 to 0.6 cm/s for cracking catalyst) leads to transition to Range B; the solids are suspended by the gas and there is a substantial increase in the bed volume; the pressure drop becomes equal to or slightly in excess of the weight of the solids present. Above 0.6 cm/s, the minimum fluidizing velocity in this case, we have Range C. Further increase in gas flow is accompanied by "bubble" flow.

The density of fluidized bed changes due to expansion of the bed. A large bed expansion in the "particulate" region is characteristic of smooth fluidization. The effects of particle characteristics and of gas velocity on bed density are shown by the typical relationship in Fig. 8.8; that is, for similar superficial velocity (based on the cross-section of the bed), larger particles give rise to higher bed density [Gohr, 1956]. The minimum fluidization velocity is defined as the minimum superficial fluid velocity at which the pressure drop through the bed reaches nearly a constant value. It is also given by equating the pressure drop given by Eq. (3.1) to that required to support the bed weight (Sec. 3.1).

The flow pattern in the bed has been a subject of many studies and conjectures. Leva [1962] showed that at a relatively low fluid velocity, a cylindrical bed amounts to two beds in series, with the bottom part acting as a packed bed with very little solid motion while the top bed is in intensely fluidized state (Fig. 8.9). In the upper bed, the solids move upward in the center and downward near the wall. At the interface between the top and bottom beds, the solids move radially inward. As the fluid velocity is increased, the bottom bed may become nearly depleted other than a stagnant zone at the corner. This flow pattern of ascending at the center and descending near the wall will be referred to as ACDW. Whitehead et al. [1976] through their observations, however, suggested that the reverse, AWDC, is the case.

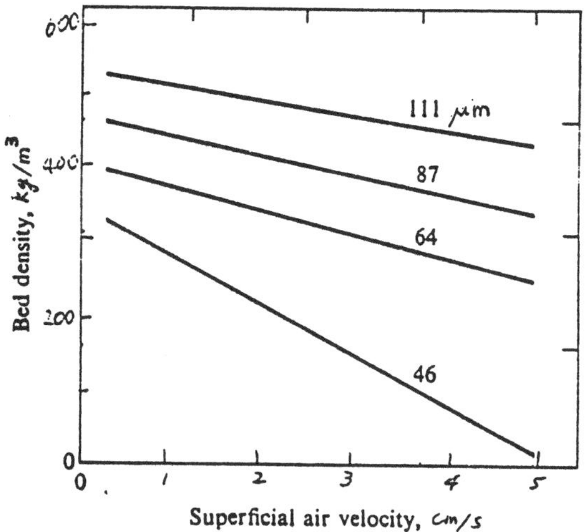

Figure 8.8   Effect of gas velocity on fluid-bed density of cracking catalyst [Gohr 1956]

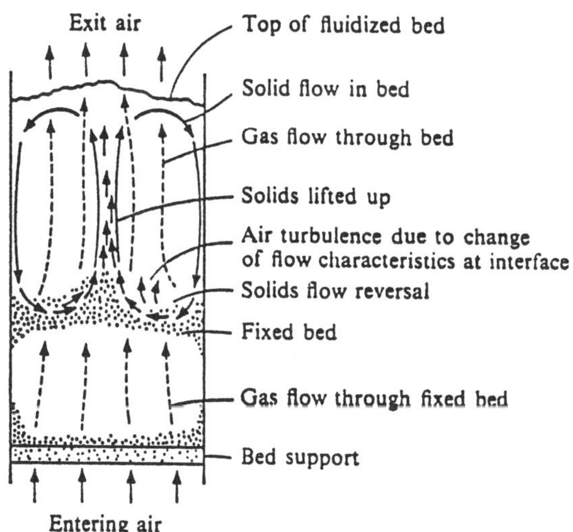

Figure 8.9   Flow model of fluidized bed [Leva 1962]

Using a radioactive tracer particle in a cylindrical bed, Lin et al. [1985] were able to determine the average particle motion which may take the form of a toroidal vortex or vortices in either directions and appears to be related to the ratio of superficial gas velocity and the minimum fluidizing velocity. The latter and other data on the transport properties on the particle clouds in a fluidized bed are summarized next. We note that these relations are highly empirical but are the only set of data base available to us for further development of knowledge. Details of empirical correlations are available in handbooks.

(a) Minimum fluidization velocity. From correlating data from various sources, Wen and Yu [1966] identified an Archimedes number (N_{Ar}) as a characteristic parameter:

$$N_{Ar} = \overline{\rho}_f(\overline{\rho}_p - \overline{\rho}_f)g(2a)^3/\overline{\mu}_f^2 \qquad (8.40)$$

and this minimum superficial (based on the cross-sectional area of the bed without considering presence of solids) velocity is given by:

$$U_{mf} = 7.5 \times 10^{-4}(\overline{\rho}_p - \overline{\rho}_f)g(2a)^2/\overline{\mu}_f \qquad (8.41)$$

for $N_{Ar} < 10^3$, and

$$U_{mf} = 0.202 \ [(\overline{\rho}_p - \overline{\rho}_f)g(2a)/\overline{\rho}_f]^{1/2} \qquad (8.42)$$

for $N_{Ar} > 10^7$. (Prob. 8.6).

(b) Voidage at minimum fluidization. The fraction voidage is equal to one minus the volume fraction solid. Wen and Yu [1966] correlated the voidage at minimum fluidization empirically to give:

$$\varepsilon_{mf} = (14\phi_s)^{-1/3} \qquad (8.43)$$

where ϕ_s is the ratio of surface area of spheres of equivalent volume to the actual surface area of the solids. In general the voidage of a stationary bed ε_b is less than 0.48, and $0.4 < \varepsilon_{mf} < 0.55$; $\varepsilon_{mf} =$

0.415 for spheres. The bed height h_{mf} at minimum fluidization is related to that of the stationary bed h_b according to:

$$h_{mf}/h_b = \varepsilon_{mf}/\varepsilon_b \tag{8.44}$$

(c) Bubble volume fraction and velocity. Bubbles occur in a gas-fluidized bed. Its origin will be treated in a separate section. A bubble in a fluidized bed does not have a well defined spherical shape. For a bubble diameter d_b, its velocity was given by Davidson and Harrison [1963] to be

$$U_B = 0.711 \sqrt{gd_B} + (U_s - U_{mf}) \tag{8.45}$$

where U_s is the superficial velocity of the fluid. The volume fraction of bubbles δ was given by Kunii and Levenspiel [1968] to be:

$$\overline{\varepsilon} = \delta\varepsilon_B + (1 - \delta)\varepsilon_e \tag{8.46}$$

where $\delta = (U - U_{mf})/U_B$, ε_e is the voidage of the gas-solid emulsion. The bubble size depends on a large number of factors and means for its prediction is limited.

(d) Expanded bed height. As the superficial velocity U_s is increased beyond the minimum fluidization velocity, the bed expands. Staub and Canada [1978] gave the overall bed voidage as:

$$\overline{\varepsilon} = \frac{U_s\varepsilon_{mf}}{1.05\, U_s\varepsilon_{mf} + (1 - \varepsilon_{mf})U_{mf}} = 1 - \overline{\alpha}_p \tag{8.47}$$

and the fraction of bed occupied by bubbles is then

$$\overline{\varepsilon}_B = \frac{\overline{\varepsilon} - \varepsilon_{mf}}{1 - \varepsilon_{mf}} \tag{8.48}$$

The expanded bed heighth is thus

$$h/h_{mf} = \overline{\varepsilon}/\varepsilon_{mf} \tag{8.49}$$

Other information including the terminal velocity of the particle cloud at minimum fluidization given in Sec. 3.1 and the dimensional correlations of viscosity of particle clouds of glass and quartz particles by Schuegerl [1971]. The latter is given in Fig. 8.10 with viscosities of the particle clouds vs. $\alpha_p/(\alpha_{po} - \alpha_p)$, where α_{po} is the volume fraction of particles at minimum fluidization. These two, together with the above items constitute a set of data base for the dynamic analysis of a fluidized bed.

8.5 Bubbles in Fluidized Beds

Beyond the minimum fluidization velocity, bed expansion occurs. Further increase in gas velocity will cause bubbles to appear in most cases depending on the properties of solid particles. Further increase will lead to slugging of large single bubbles in a deep bed. At still higher velocities than the latter, increased flow disturbance will lead to a turbulent regime and with

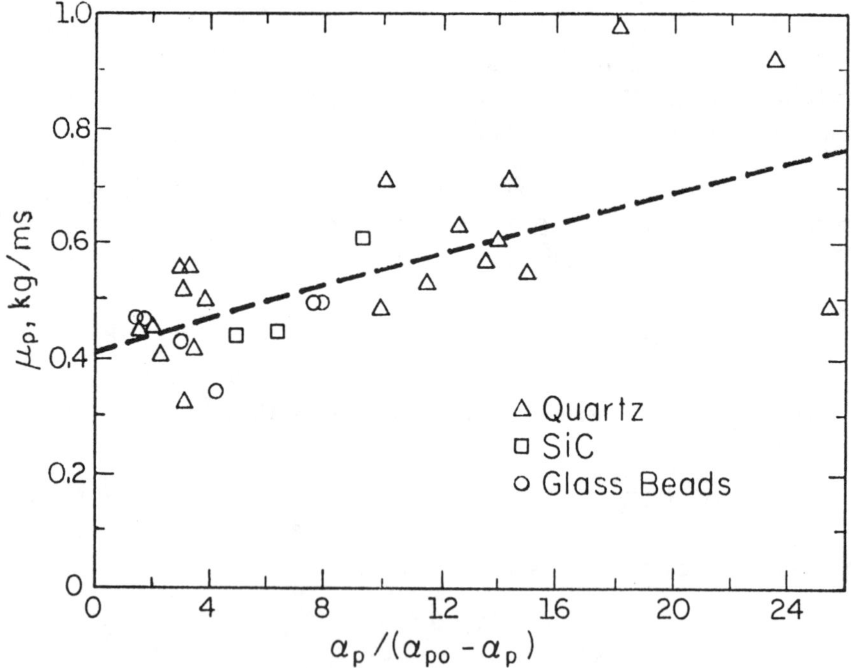

Figure 8.10 Shear viscosity measurements [Schuegerl 1971]

disappearance of the top surface, followed by a regime of fast fluidization.   Analogy can almost be drawn with the various flow regimes of gas-liquid flow (p. 196).   A classification of these regimes and materials is given in Hetsroni [1982].

Gas fluidized beds have been observed to be susceptible to bubble formation via bed inhomogeneity. Non-uniformity flow field in most liquid fluidized beds produce fluid channeling and bulk circulation of solids.   Observable bubbles were noted in water fluidized beds of particles of lead (sp. gr. 11.3) and tungsten (sp. gr. 19.3).   Murray [1965] treated instabilities in fluidized beds due to propagation of small disturbances and surface wave propagation in the gravitation field and concluded that a material density ratio of particle to gas greater than 10 will lead to unstable motion and growth of bubbles in a fluidized bed. This was confirmed in part by the experimental results of Rowe [1962].

Rowe suggested that at low values of $(\bar{\rho}_p - \bar{\rho})$ both particle and fluid flow are disturbed, bubbles cannot persist because a constant through flow of fluid does not exist. What appears to be a temporary space free of particles may be produced by centrifugal force of a turbulent vortex, but this is not a bubble in the sense of the present discussion. Liquid fluidized beds generally have low value of $(\bar{\rho}_p - \bar{\rho})$ When fluidized by water at rates that cause considerable expansion, temporarily empty vortices often resemble bubbles. Gas-fluidized beds are most susceptible to bubble formation.   The bubbling regime has received more attention than any other regime; many industrial gas-fluidized beds operate in this regime. Bubbles are the void regions that form at or near the gas distributor and propagate upward like that of pool-boiling liquid.   Some analogies are noted. Their formation is influenced by the grid configurations such as bubble caps or porous plates at the bottom of the bed which distribute the gas, but finite sized bubbles do evolve from porous plates.

A number of studies have been made, using a bubble injector, on the mechanism of formation and shape of bubbles in a fluidized bed. Details obtained by Rowe and Partridge [1962], outlining the stages of particle movement caused by a bubble as shown in Fig. 8.11 from observations made with an x-ray. The development of a bubble and its subsequent transitions are as follows. In Fig. 8.11 (a), the stable fluidized bed consists of a dense phase and a fluidized phase. In (b), a bubble is formed in the dense phase and rises in (c). It leaves the dense phase in (d), drawing a spout of particle behind it, and also forms a wake which it leaves behind. In (e), the wake rotates about its circular axis and grows in size as a result of the viscous forces until it becomes unstable; the particles flowing in streamline motion around the bubble and its attached wake "slice off" a portion of the rotating torous which remains behind in the now stationary particles. In (f), particles are stripped from the wake, and the latter appears to be a distorted torus filling the concave hollow in the rear of the bubble. Subsequently Rowe et al. [1964] recognized the cloud formation around bubbles in gas-fluidized

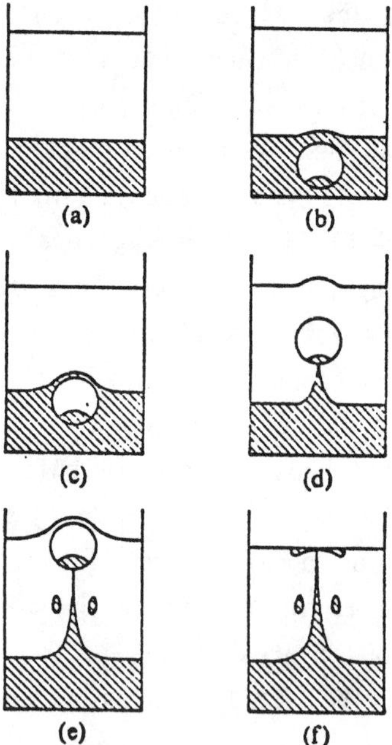

Figure 8.11   Stages of particle movement caused by a bubble in a
             fluidized bed [Rowe 1962]

beds. A bubble in a gas fluidized bed is a roughly spherical region
of space that rises by having particles flowing around it as if it
were a solid object moving through a liquid; gas flows continuously
through the bubble from the bottom and out at the top due to pressure
gradient in the bed and permeability of the bubble. At a high gas
velocity, the gas short circuits due to high permeability. At low
gas velocity, the gas circulates through the bubble due to the drag
of particles flowing around the bubble; the gas flowing out at the
top is dragged down again.

On the basis of various observed bubble shapes, Murray [1965]
treated the case of steady motion of the phases for the case of large
solid-to-fluid density ratio and neglected the inertia of the fluid
phase in the case of a two-dimensional bubble. He obtained the
streamlines of gas flow relative to the bubble as shown in Fig. 8.12,
with the stream function given by:

$$\psi = (1 - U^*) \sin\theta \left[r^* + \frac{U^*}{(1 - U^*)r^*} \right] + \frac{\sin 2\theta}{4r^*} \qquad (8.50)$$

where $U^* = U_B/U_O$, the ratio of the bubble velocity to gas velocity
(interstitial), $r^* = r/\ell$, ℓ is a characteristic length of the
fluidized bed and r is the radius from the center of the bubble, θ is

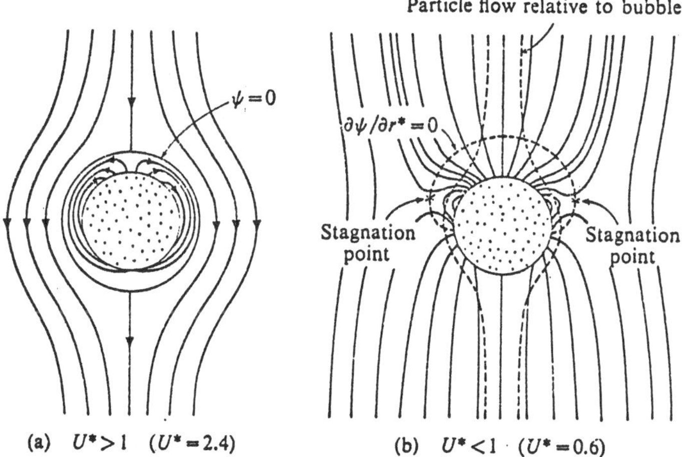

Figure 8.12 Gas flow relative to the bubble (two-dimensional)
            [Murray 1965]

the angle measured from the top of the bubble. $U^* > 1$ and $U^* < 1$ give different flow configurations as shown in Fig. 8.12.

The distinctive feature when the bubble rises faster than the free stream ($U^* > 1$) is shown by the streamline of $\psi = 0$. This circulating cloud between $\psi = 0$ streamline and the bubble moves with the bubble and the slightly ahead of it in reality [Davidson 1961]. Eq. (8.50) shows that the radius r_c of the cloud is given by, bubble radius r_B:

$$r_c \simeq r_B \ [(U^* + 1)/(U^* - 1)]^{1/2} \tag{8.51}$$

In the range $U^* < 1$, the effect of the particle flow on the gas amounts to increase the apparent region of void with flow through it evidenced by the dashed line in Fig. 8.12. For a spherical bubble, his result compares favorably with the experimental results of Rowe and Partridge [1962] for $U^* > 1$, and is close to the relation proposed by Davidson and Harrison [1963].

General information on bubble formation and stability is lacking. Bubbles help mixing in a fluidized bed. Rowe [1962] suggested a parameter that may distinguish between bubbling and non-bubbling fluidized beds as $(\bar{\rho}_p - \bar{\rho})/\bar{\mu}$, as evidenced by the following comparisons:

System	$(\bar{\rho}_p - \bar{\rho})/\bar{\mu}$, s/cm^2	
	Imminent bubbling	Vigorous bubbling
1/2 mm Copper shots in water	1000	1300
1/2 mm Lead shots in water	800	
Phenolic resin in CO_2 gas	800	1600
Other liquid-solid systems	500	

Note that for glass particles in air, the same parameter is nearly 10^4. Not only the characteristic parameter of the latter differs by an order of magnitude for inclusion in the general trend, visual observation may also tend to be subjective.

In the liquid-solid system, it appears that bubble formation may arise from surface force between the wetted particles and the Bernoulli force between adjacent particles due to interstitial liquid motion (attractive force between two spheres moving perpendicular to their line-of-centers which can be given from relations starting from p. 260 of Happel and Brenner [1965]), thus constituting a form of surface tension for a cavity. This is to be balanced by the increase in static pressure as liquid slows down as it enters the cavity. Since significant magnitude of Bernoulli force can only arise due to large relative motion, the suggestion that bubbling arises due to large difference in the density of phases appears reasonable.

In the case of a gas-solid system, a different mechanism is expected. First of all, the force due to electrostatic charges on particles is far greater than the close range surface force (Eq. 3.35) between the particles and the Bernoulli force between adjacent particles due to fluid motion in between. It is well known (see for instance, Page [1952]), that a spherical cavity of radius R in a material of dielectric constant ε_r, charge density ρ_e has its electric displacement D_e and electric field E given by:

$$\nabla \cdot D_e = \varepsilon_r \, \nabla \cdot E_e = \rho_e / \varepsilon_0 \qquad (8.52)$$

with $D_e = \varepsilon_0 E + P_e = \varepsilon_0 \varepsilon_r E$, $P_e = \varepsilon E$ being the polarization of the medium, and its permitivity. The force per unit volume is

$$f = P_e \cdot \nabla E_e = (1/2) \, \varepsilon_0 \, (\varepsilon_r - 1) \, \nabla E_e^2 \qquad (8.53)$$

giving rise to a tensile stress at the surface of the cavity of:

$$\sigma_e / R = (1/2) \, \varepsilon_0 (\varepsilon_r - 1) \, E_e^2 \qquad (8.54)$$

where σ_e denotes an electric surface tension which is opposed by the pressure difference created as the interstitial gas velocity slows down in the cavity. Eq. (8.54) is directly applicable to a continuum. In a fluidized bed of particle radius a and volume fraction α_p, we introduce a transmission coefficient K which has a limiting value of (a_c^2/a^2) $(\pi/4)$ for cubic piling; a_c is given as a_{12} in Eq. (3.58). The equivalent dielectric constant of the suspension is:

$$\varepsilon_r = 1 + \alpha_p \left[(\varepsilon_{pr} - 1)/(\varepsilon_{pr} + 1) \right] \tag{8.55}$$

For a given charge-to-mass ratio (q/m), $\rho_e = \rho_p(q/m)$. The Poisson equation can be solved for an average n_B bubbles per unit volume of the bed with the assumption of ρ_p being a constant outside of a bubble with the boundary conditions: $r = n_B^{-1/3} (3/4\pi)^{1/3} = R_b$, the electric potential $V_e = 0$, and $dV_e/dr = -E_e = 0$; at $r < R_B$, $\rho_p = 0$; to give at $r = R_B$,

$$E_B = - \left. \frac{\partial V}{\partial r} \right|_{R_B} = - \frac{1}{3 \, \varepsilon_r \varepsilon_o} \left(\frac{q}{m} \right) \rho_p \frac{R_B^3}{R_B^2} \left(1 - \frac{R_B^3}{R_b^3} \right) \tag{8.56}$$

for R_b much greater than the interparticle spacing; R_b is nearly the radius or characteristic dimension of the bed for a single bubble.

For an interstitial velocity $U_o = U_s/(1 - \alpha_p)$, U_s being the superficial velocity, with $U_o/U_{mf} \sim 6$ to 11, the maximum flow velocity through a bubble is nearly $3U_{mf}$ [Davidson and Harrison 1963]. Hence the static pressure in the bubble is higher than in the bulk by $\sigma_e/R_B = P_B - P \sim (1/2) \, \bar{\rho} \, U_o^2$. We get, from Eqs. (8.54) and (8.56):

$$\sigma_e/R_B = (1/2) \, K \, \varepsilon_o (\varepsilon_r - 1) \, E_B^2 \simeq (1/2) \, \bar{\rho} \, U_o^2 \tag{8.57}$$

Rearranging, for $R_b > 2R_B$, we get

$$\left(\frac{R_B}{R_b} \right)^4 = \left(\frac{2}{g} \right) K \, (\varepsilon_r - 1) \, \varepsilon_r^{-2} \, (N_{Re})_e^{-1} \tag{8.58}$$

where the electric Reynolds number $(N_{Re})_e$ is defined by (see Eq. 7.33):

$$(N_{Re})_e^{-1} = \frac{[R_b \, \alpha_p \, \bar{\rho}_p \, (q/m)]^2}{\varepsilon_o \, \bar{\rho} \, U_o^2} = [\alpha_p \, \bar{\rho}_p \, \left(\frac{q}{m} \right)^2 \frac{R_b^2}{\varepsilon_o \, U_o^2}] \, [\frac{\alpha_p \, \bar{\rho}_p}{\bar{\rho}}] \tag{8.59}$$

The corresponding electric surface tension is:

$$\sigma_e = \frac{R_b}{3(2^{5/4})}\ \bar{\rho}\ U_o^2\ [K(\epsilon_r - 1)\ \epsilon_r^{-2}]^{1/2}\ (N_{Re})_e^{-1/4} \tag{8.60}$$

(Prob. 8.7) The above points to a possible explanation of bubbles in a gas fluidized bed [Soo 1974]. Colver [1977] has demonstrated that bubbles in fluidized bed can be controlled or dissipated by a suitably applied electric field at moderate voltage and low current (a.c. or d.c.). It suggests a means for promoting mixing as well as bed stabilization.

8.6 Circulatory Motion in Fluidized Beds

Interests in predicting and correlating transport processes in a fluidized bed call for a knowledge of the average motion of phases in it. These transport processes include heat and mass transfer, mixing, chemical reactions, erosion of immersed solid surfaces, and method and location for introducing and removal of solid particles. The nature of average motion in a cylindrical bed in the form of toroidal vortices has already been introduced in Sec. 8.4. With the available data base (Sec. 8.4), though meager and empirical in some cases, the bed motion at various flow conditions can be predicted from solving the averaged multiphase equations (Ch. 2).

For a cylindrical bed of radius r_0 and expanded bed height h (Fig. 8.13) with a single species of solid particles in an isothermal gas. The radial and axial coordinates of the system are r and z with origin at the bottom center of the bed, with z aligned in the opposite direction of gravity of acceleration g. Gas enters at the bottom through a porous plate. The conjugate velocity components in the r, z directions are U, W of the gas and U_p, W_p of the particle cloud of density $\rho_p = \alpha_p \bar{\rho}_p$ which is taken to be uniform in this primary solution of one toroidal cell, although the solution can be extended to multiple cells. The averaged equations are now reduced to:

$$(1/r)\ [\partial(rU)/\partial r] + (\partial W/\partial z) = 0 \tag{8.61}$$

$$(1/r) \; [\partial(rU_p/\partial r] + (\partial W_p/\partial z) = 0 \tag{8.62}$$

the continuity equations; the momentum equation of the particle phase:

$$0 = \rho_p F \; [W - W_p) + \mu_p \; [\frac{1}{r} \frac{\partial}{\partial r} \, r \, \frac{\partial W_p}{\partial r} + \frac{\partial^2 W_p}{\partial z^2}] - \rho_p g \tag{8.63}$$

since the viscosity of the particle-particle interaction predominates in the conservation of momentum, and

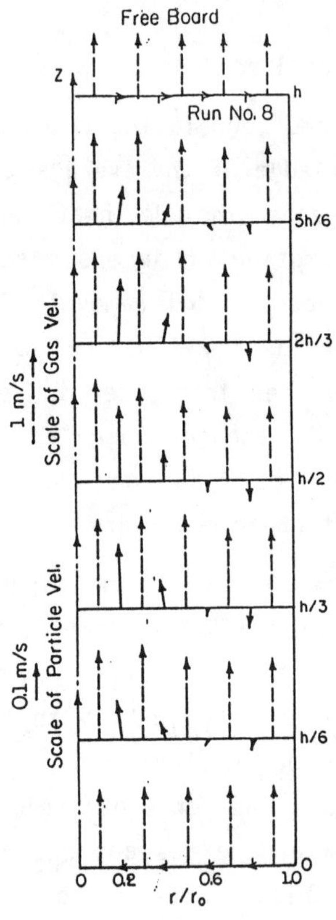

Figure 8.13   Motion of phases in the fluidized bed in the example in Fig. 8.14 as shown by velocity vectors (solids ACDW).

$$0 = -\frac{\partial P}{\partial z} - \rho_p F W_o - (\rho_p + \rho) g \qquad (8.64)$$

the latter momentum equation is for the mixture. The boundary conditions are: (a) At $r = 0$, $U = U_p = 0$, $\partial W/\partial r = \partial W_p/\partial r = 0$; (b) at $r = r_o - a$, a being the particle radius ($r_o \gg a$), $W_p = W_{pw}(z)$, $W = W_o + W_w(z)$, $U = U_p = 0$; and for the slip motion of particles at the wall (Eq. 3.122), $-\lambda_p(\partial W_p/\partial r)_w = W_{pw}(z)$, where $\lambda_p \sim a/3(2)^{1/2} \alpha_p$ is the mean free path of particle- particle interaction; and (c) at $z = 0$ and $z = h$, $W_p = 0$, and the net flow of particles is zero, or $\dot{m}_p/\alpha_p \bar{\rho}_p = 2\pi \int_0^{r_o} W_p \, r \, dr = 0$.

Eqs. (8.61) to (8.64) are satisfied by the solutions:

$$W - W_o = \sin(\pi z/n) \sum_n c_n(r/r_o)^n \qquad (8.65)$$

$$U = -(\pi r_o/h) \cos(\pi z/h) \sum_n [c_n/(n+2)](r/r_o)^{n+1} \qquad (8.66)$$

$$W_p = \sin(\pi z/h) \sum_n a_n (r/r_o)^n \qquad (8.67)$$

$$U_p = -(\pi r_o/n) \cos(\pi z/h) \sum_n [a_n/(n+2)] (r/r_o)^{n+1}$$

$$\equiv U_{pr} \cos(\pi z/h) \qquad (8.68)$$

Eq. (8.63) gives, at $r = 0$, $z = h/2$,

$$W_o + c_o - a_o = g/F \qquad (8.69)$$

which for $c_o - a_o \ll W_o$, which is physically valid, leading to:

$$0 = \sum_n c_n(\frac{r}{r_o})^n - [1 + \frac{\pi^2 \mu_p}{h^2 \rho_p F}] \sum_n a_n(\frac{r}{r_o})^n + (\frac{\mu_p}{\rho_p R \, r_o^2}) \sum_n n^2 a_n(\frac{r}{r_o})^{n-2}$$

$$(8.70)$$

The boundary conditions further give, $a_1 = 0$, and $c_1 = 0$, and $\sum_n c_n = w_w$, $\sum_n a_n = w_{pw}$; w_w and w_{pw} are now constants to be determined; we further have $U = \sum_n c_n/(n+2) = 0$, $U_p = \sum_n a_n/(n+2) = 0$. The series expansions in Eqs. (8.65) to (8.68) are now reduced to fifth degree polynomials with a_n's and c_n's given by:

$$a_o = -w_{pw}[(16/5) + (1/20)(\gamma/k)] + (13/140)(w_w/k)$$

$$a_1 = 0, \quad a_2 = w_{pw}[14 + (1/3)](\gamma/k)] - (26/63)(w_w/k)$$

$$a_3 = 0, \quad a_4 = -w_{pw}[21 + (3/4)(\gamma/k)] + (13/38)(w_w/k)$$

$$a_5 = w_{pw}[(56/5) + (7/15)(r/k)] - (13/90)(w_w/k)$$

and

$$c_o = \gamma\, a_o - 4\, k\, a_2$$

$$c_1 = 0, \quad c_2 = \gamma\, a_2 - 16\, k\, a_4$$

$$c_3 = -25\, k\, a_5, \quad c_4 = \gamma\, a_4, \quad c_5 = \gamma\, a_5;$$

where

$$\gamma = 1 + \frac{\pi^2}{h^2}\frac{\mu_p}{\rho_p\, F}, \qquad k = \frac{\mu_p}{r_o^2\, \rho_p\, F} \tag{8.71}$$

are the correlation parameters, relating viscous force to inertia force of phase interaction. The slip velocities at the wall w_w and w_{wp} are given by:

$$w_{pw}\left[\frac{16}{5} + \frac{1}{20}\left(\frac{\gamma}{k}\right) - \frac{1}{20}\left(\frac{\gamma^2}{k}\right) - 56\, k - \frac{68}{15}\gamma\right]$$

$$+ \frac{13}{20} w_w\left[\frac{1}{7}\left(\frac{\gamma}{k}\right) - \frac{1}{7}\left(\frac{1}{k}\right) + \frac{160}{63}\right] = \frac{g}{F} - W_o \tag{8.72}$$

and

$$w_{pw} = -\frac{13}{42}\frac{\lambda_p}{r_o}\left(\frac{1}{k}\right) w_w \tag{8.73}$$

A case with velocity vectors for motion ascending at the center and descending near the wall is illustrated in Fig. 8.13 with the polynomials representing the motions plotted in Fig. 8.14. They are for the physical parameters given by: r_0 = 0.069 m, h = 0.245 m, α_p = 0.24, W_0 = 0.99 m/s, $\bar{\rho}_p$ = 2500 kg/m^3, μ_p = 0.0235 kg/m-s, F = 9.85 s^{-1}, γ = 1.000654, k = 0.008356, with w_{pw} = -0.001975 m/s, and w_w = 0.001498. In this case we get a_0 = 0.2910, a_2 = -1.5557, a_4 = 2.6475, a_5 = -1.3847; c_0 = 0.2964, c_2 = -1.5921, c_3 = 0.0289, c_4 = 2.6540, c_5 = -1.3857. Also shown is the integrated flow rate from the axis to radius r:

$$\dot{m}_p(r) = 2\pi \, r_0^2 \, \rho_p \int_0^r w_p\left(\frac{r}{r_0}\right) \, d\left(\frac{r}{r_0}\right)$$

and

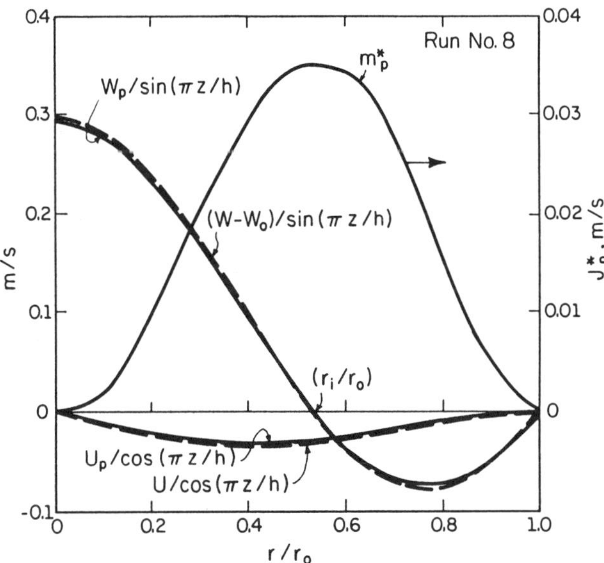

Figure 8.14 Distribution of components of circulating velocities of phases and integrated flow rate of solids in a cylindrical fluidized bed of a typical case of ascending near the center and descending near the wall (ACDW). (Based on calculation from Run No. 8 of Lin [1985])

$$J_p^\star = \frac{\dot{m}_p}{\pi\,r_o^2\,\rho_p\,\sin(\pi z/h)}$$

This case of ACDW conforms to the observations of Leva [1962] and models Test Run No. 8 of Lin [1985]. The average maximum mass flow is given by:

$$<J_p> = 2\,\alpha_p\,\bar{\rho}_p\,\left(\frac{r_o}{r_o}\right)^2 \int_0^{(r_o/r_o)} w_p\left(\frac{r}{r_o}\right)\,d\left(\frac{r}{r_o}\right) \qquad (8.74)$$

which has the value of 0.2334 $\alpha_p\,\bar{\rho}_p$ = 140 kg/m^{2}s. This quantity is significant in relating to mixing and the rate of circulation of solids. The motion in a two-dimensional fluidized bed is given in Soo [1986]; possibility of several vertical cells of various wave lengths and Benard cells in a shallow horizontal bed was identified.

Relation to experimental results. Validation of the above model was sought by evaluating the experimental results of Lin et al. [1985]. The latter were obtained by using a radioactive tracer particle in a fluidized bed. Typical data from their test runs are summarized in Table 8.2. The transitions from AWDC to ACDW are indicated from Runs No. 5 to No. 9. Listed in Table 8.2 also are the maximum average

Table 8.2 Experimental Results of Lin [1985]

2a = 0.5 mm, minimum (superficial) fluidization velocity = V_{mf} = 0.194 m/s, $\bar{\rho}_p$ = 2500 kg/m^3, bed radius = r_o = 0.069 m, static bed height = h_o = 0.113 m, volume fraction solid of static bed = α_{po} = 0.52.

Ref. Run No.	Superficial air velocity, m/s	Expanded bed height, h, m	Max. particle vel., W_{po} m/s	Flow pattern
5	0.320	0.138	-0.085	AWDC
6	0.458	0.170	-0.12	AWDC (small DW cell at top)
7	0.641	0.231	0.226	ACDW (small AW cell at bottom)
8	0.754	0.245	0.291	ACDW
9	0.892	0.246	0.438	ACDW

particle velocities, W_{po}, each corresponds to coefficient a_o in the polynomial profile, that is, the maximum particle velocity along the center of the cylindrical bed, upward or downward. Notable in the results of Run No. 6 is the small developing ACDW torous at the outer rim near the top of the bed, and the growth to ACDW in Run No. 7, leaving a small AWDC cell at the bottom (Fig. 8.15). Table 8.3 shows the computed average volume fraction solid α_p from the expanded bed height h, so also W_o for the interstitial air velocity and other parameters used in the computations from empirical relations in Sec. 8.4. The bubble velocity based on observed range of bubble sizes of 2 and 5 cm diameter were computed, as well as an average bubble volume fraction for later discussions. Note that the bubble velocity is close to the interstitial air velocity and the mean volume fraction of bubbles constitutes a part of the volume fraction of the gas. In all cases, the gas velocity is greater than the minimum fluidization velocity. However, the terminal velocity of particles in the bed (g/F) is not accurately given by the experiments of Ergun (Sec. 3.1), which were conducted at below minimum fluidization velocities. It was deemed desirable to determine the value of F from these experimental results. This is accomplished by solving the equation for a_o (= W_{po}) and Eq. (8.72) simultaneously with substitution of Eq. (8.73) for given W_o, W_{po}, λ_p/r_o, the particle Knudsen number; h/r_o, a dimensionless bed height, and the parameter:

$$\frac{gr_o^2 \, \rho_p}{\mu_p \, W_o} = \left[\frac{r_o \, \rho_p \, W_o}{\mu_p}\right] \left[\frac{W_o}{\sqrt{gr_o}}\right]^{-2} = \frac{\text{(Reynolds number)}}{\text{(Froude number)}^2}$$

to give w_{pw} and F from each test run. F and (g/F) - W_o are given in Table 8.3 and w_{pw}, k, and γ are given in Table 8.4. The values of F thus determined are, in general, approximately 3 times of those given by Eq. (3.3). It is also seen that (g/F) < W_o gives ACDW mode, while (g/F) > W_o gives ACDW circulation (Table 8.3). No net elutriation will occur as long as the gas velocity is lower than the terminal velocity of particles in the free board.

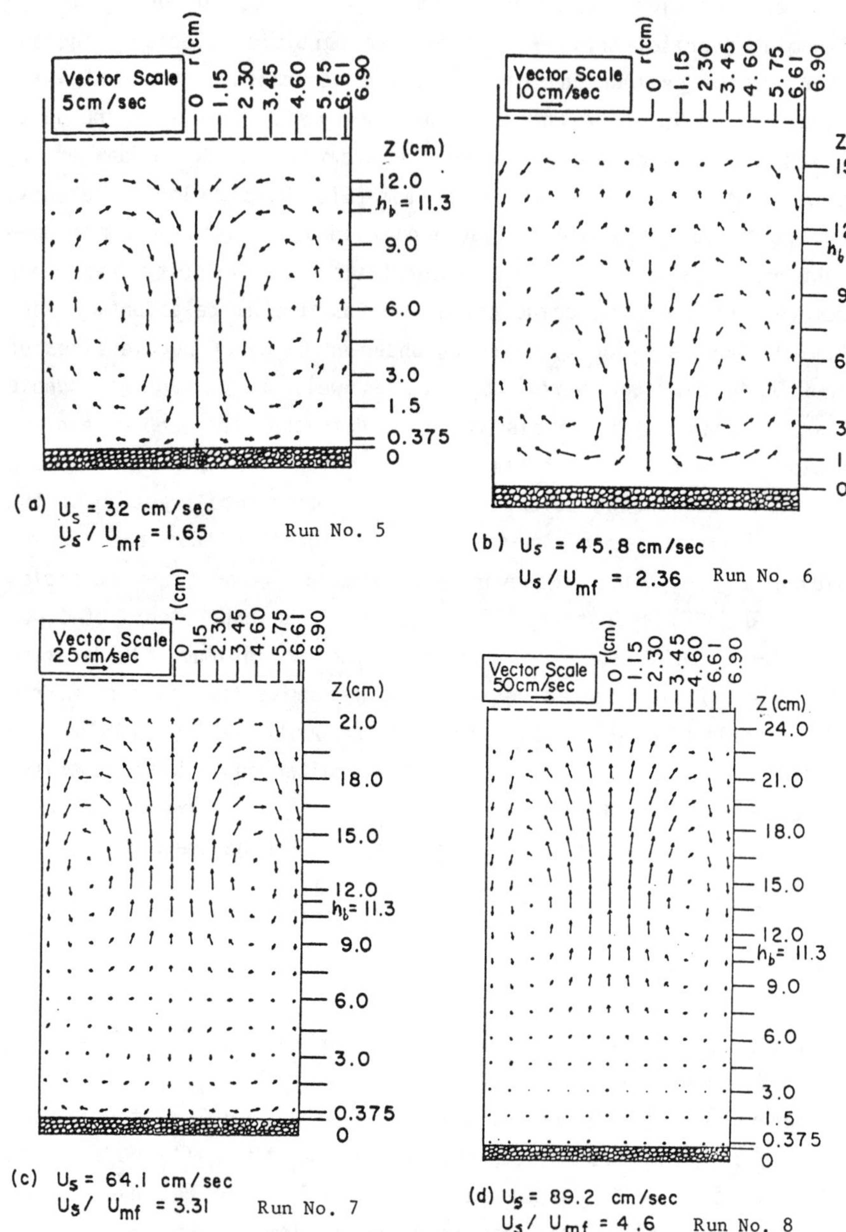

Figure 8.15   Particle circulation patterns at various fluidizing velocities for a gas fluidized bed of 0.42 - 0.6 mm glass beads; h_o denotes static bed height [Lin et al. 1985].

From the data in Table 8.3, except those of the bubbles, the polynomial coefficients were computed as shown in Table 8.4. Fig. 8.16 gives the corresponding distribution of velocity component W_p of cases of Run numbers 5 to 9. The maximum velocities of circulation W_{po} along the z-axis are given by a_o of these cases. These results do not conform to the experimental velocity distribution in details because of the assumption of uniform density and the available accuracy in the empirical correlations which are used as data base. Fig. 8.17 shows an overall comparison of the maximum particle

Table 8.3 Computed Data from Table 8.2

Ref. Run.No.	α_p α_{po} h_o/h	W_o m/s	W_{BO} $D_B=0.02m$	(average) $D_B=0.05m$	Est. α_B	F s^{-1}	μ_p kg/ms	$(g/F)-W_o$ m/s
5	0.42	0.55	0.44	0.62	0.25	17.84	0.358	-0.0017
6	0.35	0.71	0.58	0.76	0.40	13.83	0.160	-0.0014
7	0.25	0.86	0.76	0.95	0.53	11.34	0.0285	0.0042
8	0.24	0.99	0.87	1.06	0.58	9.85	0.0235	0.0049
9	0.24	1.17	1.01	1.20	0.64	8.31	0.0235	0.0093
(Method)			Eq. (8.45)		Eq. (8.48)		Fig. 8.10	

Table 8.4 Polynomial Coefficients of Velocity Distributions.
(Computed for cases with data in Tables 8.2 and 8.3)

Run No.	5	6	7	8	9
$\gamma-1$	0.009904	0.004514	0.0007438	0.0006540	0.0007687
k	0.004014	0.002776	0.0008446	0.0008356	0.0009900
λ_p/r	0.00203	0.00244	0.00342	0.00356	0.00356
w_{pw},m/s	0.0005197	0.0008319	-0.001505	-0.001975	-0.003174
w_w,m/s	-0.003320	-0.003058	0.001201	0.001498	0.002851
a_o	-0.08500	-0.12000	0.2260	0.2910	0.4380
a_1	0.39220	0.56657	-1.20222	-1.5557	-2.3027
a_2	-0.49300	-0.75466	2.82812	2.6475	3.8104
a_4	0.18631	0.30891	-1.5440	-1.3847	-1.9489
W_o	0.55	0.71	0.86	0.99	1.17

velocities given by this primary single cell solution to the experimental results including expanded bed heights for various interstitial velocities of the gas. The experimental results include a
continuous transition from AWDC to ACDW over a range of growth of an
ACDW cell at the bottom. The curves of the primary solution computed
from continuous variations of the transport parameters in Table 8.3,
however, show a discontinuous transition between a single AWDC cell
to a single ACDW cell for the same bed heights as in the experimental
data. Away from the transition region, the trend of the primary
solution agrees with that the experiments. The stable mode of bed
motion appears to be ACDW at the top of the bed. A bed starting with
AWDC may continue in that mode with increased speed of circulation as

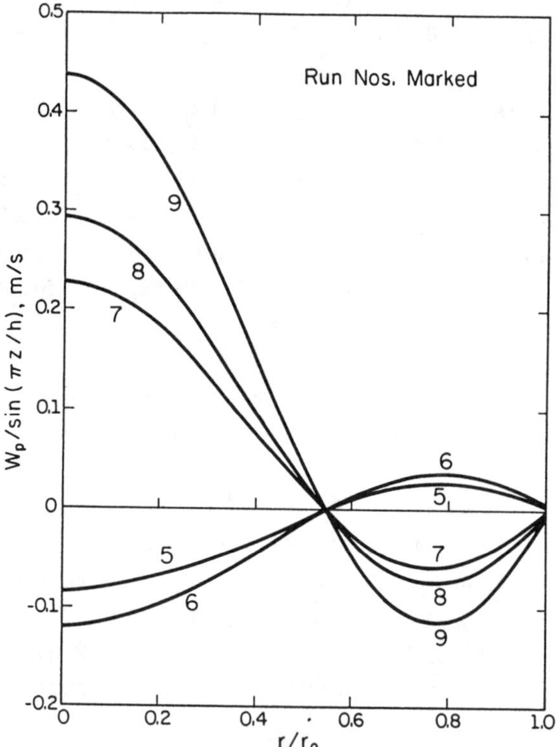

Figure 8.16   Computed axial component of particle velocities of Test
              Runs No. 5, 6 (AWDC) and No. 7,8,9 (ACDW) of [Lin et al.
              1985]

W_o increases, but instability will set in, switching it to an upper cell of ACDW.

Possible elaborations of the primary solution include the distributions in in particle density by gravity (Fig. 8.9) and electrostatic effects as illustrated by Eq. (6.36).

Motion of bubbles in a fluidized bed. Various conjectures have been made on the function of bubbles in a fluidized bed. They include the suggestion that the circulating motion in the bed is caused by the motion of the bubbles. Now the path of bubbles through a bed with a given circulatory motion can be determined. The radial component of bubble velocity can be represented by:

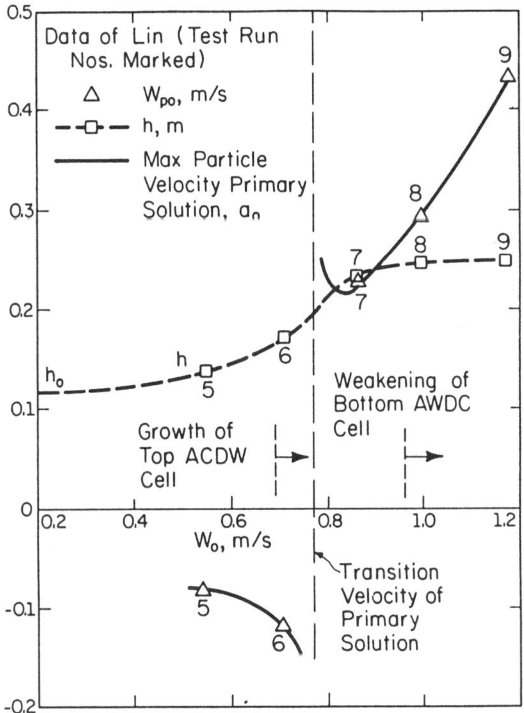

Figure 8.17 Relation of primary solution to experimental results of Lin and the trends of transition from AWDC to ACDW in the experimental system and the single cell solution.

$$d\ U_B/dt = F_{Bp}(U_p - U_B) \tag{8.75}$$

along the path of a bubble where F_{bp}, the inverse relaxation time for momentum transfer from the dense phase to the bubble; $F_{bp} = g/W_{Bo}$ (Table 8.3), t is the time, and subscript B denotes bubble properties. Since the bubble velocity is close to the interstitial gas velocity in a fluidized bed, W_o, to account for the acceleration of a bubble from the bottom of the bed, an approximation $dt = dz/W_B$ can be applied with W_B given by:

$$W_B = W_{Bo}(1 - e^{-F_{Bp}t}) \tag{8.76}$$

Eq. (8.75) can be integrated numerically to give the local radial velocities of the bubbles. The results are illustrated in Fig. 8.18 for the cases in Fig. 8.13 (ACDW) and Run No. 6 (AWDC). Fig. 8.18 gives the ratios of U_B/U_{pr} versus the height given by z/h, U_{pr} being

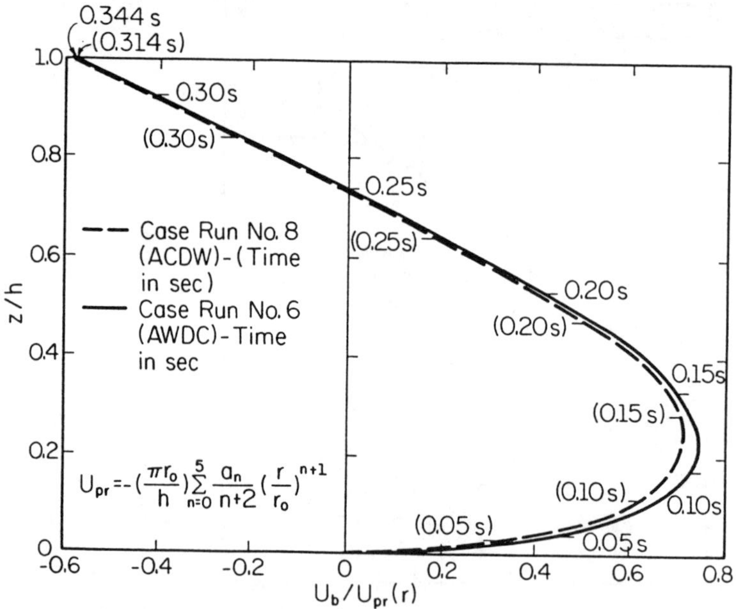

$$U_{pr} = -\left(\frac{\pi r_o}{h}\right)\sum_{n=0}^{5}\frac{a_n}{n+2}\left(\frac{r}{r_o}\right)^{n+1}$$

Figure 8.18   Local radial component of bubble velocities of cases in Test Runs No. 8 and No. 6.

the maximum radial component of velocity as a function of (r/r_0) at z = 0 given by Eq. (8.68). Fig. 8.18 shows that in the case of Run No. 6 (AWDC), the sign of U_{pr} is positive and the bubbles accelerating from the bottom of the bed will be displaced toward the wall over 80 percent of the time of rise (note the time in second marked parametrically on the curves) and over 70 percent of the height of rise because of a phase shift of the motion. For a case such as in Fig. 8.13 (ACDW), the bubbles are displaced toward the center of the bed $(U_{pr} < 0)$. The above estimation is conservative since the coalescing bubbles start from small sizes when first generated at the bottom porous plate and the rising time is longer than that illustrated in Fig. 8.18. It appears that the circulatory motion of particles in a fluidized bed is the cause of bubble migration rather than being a consequence of it as has been suggested by some. Therefore, bubbles tend to rise near the wall when particle motion is AWDC, and tend to rise near the center for ACDW motion. This is seen also from considering the relatively small inertia of the bubbles when compared to the dense phase of the particle suspension.

An interesting feature of a fluidized bed is that the fluctuating components of its velocity is of the same order of magnitude as the average velocity of the particle phase [Lin et al. 1985]. Agitation by bubble motion and bubble burst could be a factor. This has an important influence on the transport processes in the bed and erosion of heat transfer surfaces (Eqs. 3.84 and 3.85) immersed in the bed. (Prob. 8.8) A limiting case is seen in very fine powders (Prob. 8.9).

8.7 Transport Processes in Fluidized Beds.

One of the outstanding characteristics of a fluidized bed is that it tends to maintain uniform temperature even with nonuniform heat release. The relatively uniform temperature results from the rapid circulation of the solid in the bed. The magnitude of effective thermal conductivity of the batch fluidized system in the vertical direction was between 173 and 43,300 W/m K, depending on the condition and physical proportion of the gases and solids.

Heat transfer. Experiments by Mickley and Trilling [1951] have shown that the bed maintains essentially uniform temperature for an internally heated bed. They also showed that for particle size range of 70 μm to 4.52 mm and in 10.2 cm and 2.54 cm columns, the heat transfer coefficient for an externally heated bed is as shown in Fig. 8.19, correlated with $\rho_p G_0/(2a)^3$; G is the mass velocity of air flow, based on the cross-sectional rea of empty tube; for nonspherical lparticles, 2a = 6/(surface area per unit volume). Fig. 8.20 gives relations for internally heated bed (particle size range .041 to .45 mm). Correlations be Toomey and Johnstone [1952] and Mickley and Trilling [1951] for heat transfer coefficient h along an immersed axial surface is shown in Fig. 8.21. U_{mf} is the minimum fluidizing

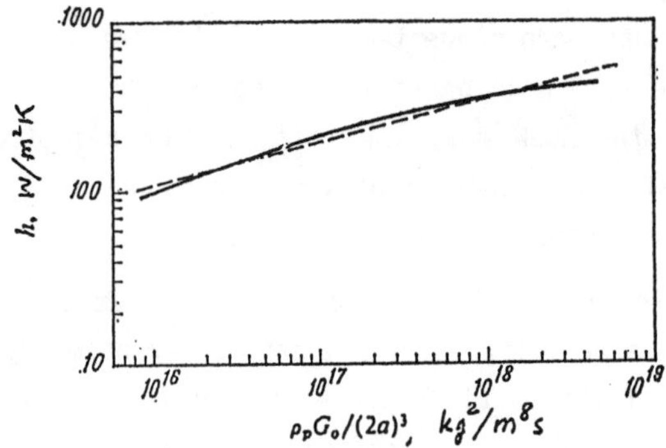

Figure 8.19   Correlation of heat-transfer coefficient in externally
heated fluidized bed [Mickley and Trilling 1951]

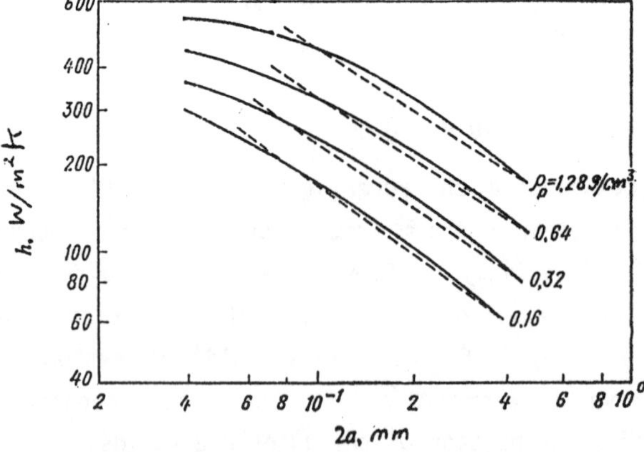

Figure 8.20   Heat-transfer coefficient with particle diameter and ρ_p
in internally heated fluidized bed [Mickley and Trilling
1951]

velocity and U is the actual superficial velocity. Lack of accurate description for the bed motion or momentum transfer in the bed prior to the procedure available in the previous section necessitated some form of logical approximation. A wide variation in experimental data was given by Chen and Withers [1978].

Lewis et al. [1962] made measurements on the radial and longitudinal heat transfer from a concentric 12.7 mm O. D. ($2r_2$) calrod heater in a fluidized bed of 74.7 mm I. D. ($2r_1$) with glass beads or petroleum cracking catalyst (microspheres of 74 to 149 μm) suspended in air or other gases (Refrigerant 12, He, CO_2, C_3H_8, H_2). The effective thermal conductivity in the longitudinal direction κ_e was estimated according to the temperature rise (ΔT) in the direction of the bed height L_b and heat flow $\dot{q}$ over area A:

$$\kappa_e = (\dot{q}/A)/(\Delta T/L_b) \tag{8.77}$$

and the effective thermal conductivity in the radial direction was estimated according to:

$$\kappa_r = \frac{\dot{q}}{L_b} \ln\left(\frac{r_1}{r_2}\right)/2\pi (T_2 - T_1) \tag{8.78}$$

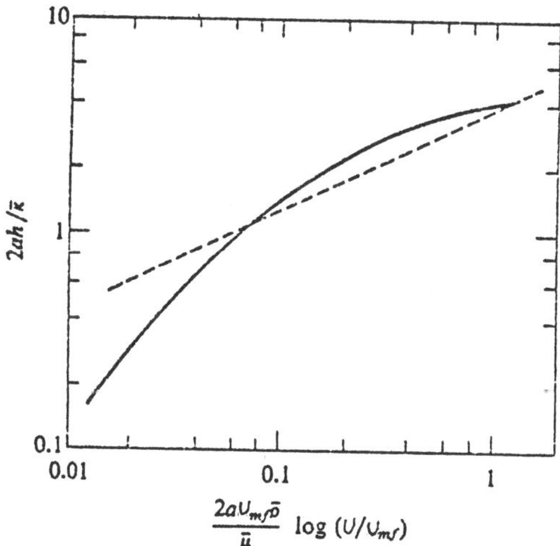

Figure 8.21 Correlation of h along immersed axial surface [Toomey and Johnstone 1952]

For superficial velocities ranging from 9 to 33 cm/s, κ_e ranging from 6.92 to 164 kW/m K were measured, for bed height to diameter ratios of 2.2 to 5.5. κ_r was nearly 2 percent of κ_e. Note that air at 450 K has a thermal conductivity of 0.042 kW/m K.

To determine the relation between κ_e and κ_r, they accounted for the fact that solids rise near the center and fall near the wall. Estimation of κ_r was refined as follows: taking a core radius r_c within which the mean velocity of the solids is W_{pc} and temperature T_c and average bed density ρ_m, over a height dz,

$$(W_{pc}\, \rho_m\, \pi\, r_c^2)\, c_p\, T_c = (W_{pc}\, \rho_m\, \pi\, r_c^2)\, c_p\left(T_c + \frac{dT_c}{dz}\, dz\right) + \dot{q}_w dz \quad (8.79)$$

where c_p is the specific heat of the solid. $\dot{q}_w$ is the heat flow at the boundary of T_w, or

$$\dot{q}_w = -\kappa_r\, 2\pi\, r_c\, (dT/dr)_{r_c} = \dot{q}/L_b \quad (8.80)$$

The net vertical heat flux is

$$W_{pc}\, \rho_m\, \pi\, r_c^2\, c_p(T_c - T_w) = -\kappa_e\, \pi\, r_0^2 (dT_c/dz) = \dot{q} \quad (8.81)$$

where r_0 is the outer radius of the bed. Hence

$$\kappa_r = \frac{\kappa_e r_0^2 (dT_c/dz)^2}{2\, r_c(T_c - T_w)(dT/dr)_{r_c}}$$

$$= \frac{\dot{q}(dT/dz)}{2\pi\, r_c(T_c - T_w)(dT/dr)_{r_c}} \quad (8.82)$$

For $r_c = 0.7r_0$, $r_0 = 38$ mm, $T_c - T_w = 2$ C, $dT_c/dz = 19$ C/m, $\rho_m = 400$ kg/m^3, $dT/dr = 109$ C/m, one gets $\kappa_r = 0.04\, \kappa_e$. (Prob. 8.10)

An approximate solution for simultaneous heat-mass transfer in two-phase flow through a packed bed was given by Tung and Dhir [1988].

Chemical reaction. Reaction time distributions in fluidized catalytic reactors were studied by Orcutt et al. [1962]. Measurements were made on a bed of iron oxide particles fluidized by a mixture of ozone and air, and the degree of decomposition of the ozone in passing through the bed was determined. Reaction persists only when

the gaseous mixture is in contact with the particles. The fraction
of reactants unconverted Φ is given by:

$$- \frac{d\Phi}{dt} = k_r \; \bar{R}T \; \left(\frac{1 - \varepsilon}{\varepsilon}\right) \; \Phi \tag{8.83}$$

where k_r is the rate constant (on the basis of mass of catalyst-
partial pressure), $\bar{R}$ is the universal gas constant, T is the absolute
temperature. Defining the dimensionless reaction time as:

$$\zeta = \frac{\bar{R} \; T \; \rho_p \; \dot{M}_g}{P \; M_p} \int_o^t \left(\frac{1 - \varepsilon}{\varepsilon}\right) \; dt' \tag{8.84}$$

where P is the overall pressure, M_p is the mass of catalyst,
and $\dot{M}_g$ is the molar feed rate. Eq. (8.83) integrates to

$$\Phi = \exp \left[- \frac{k_r \; \zeta \; P \; M_p}{\dot{M}_g} \right] \tag{8.85}$$

The random nature of the process was accounted for by assuming a
probability density function $f(\zeta)$; then the mass fraction of reactant
remaining at the exit of the bed is

$$\Phi(K) = \int_o^\infty e^{-K\zeta} \; f(\zeta) \; d\zeta \tag{8.86}$$

where $K = k_r PM_p/\dot{M}_g$; thus $\Phi(K)$ is the momentum-generating function
of $f(\zeta)$, with limits

$$\lim_{K \to \infty} \Phi(K) = \int_o^\infty f(\zeta) \; d\zeta = 1 \tag{8.87}$$

and

$$\lim_{K \to \infty} \frac{d \; \Phi(K)}{dK} = - \int_o^\infty \zeta \; f(\zeta) \; d\zeta = -1 \tag{8.88}$$

Since ζ was normalized with respect to $M_p/\dot{M}_g$, the average contact
time, $f(\zeta)$, satisfies the requirements for the density function of a

random variable with unit mean. Eq. (8.86) gives a basis for experimental determination of $f(\zeta)$ or an estimate of its lower moments. Numerical values of $f(\zeta)$ can be obtained from measuring the fraction of reactants remaining by varying K. The following cases were identified:

(a) The case of piston flow is given by $f(\zeta) = \delta(1-\zeta)$ and

$$\Phi(K) = \int_0^\infty e^{-K\zeta} \delta(1-\zeta) \, d\zeta = e^{-K} \tag{8.89}$$

(b) The case of perfect mixing is given by $f(\zeta) = e^{-\zeta}$ and

$$\Phi(K) = \int_0^\infty e^{-K\zeta} e^{-\zeta} \, d\zeta = (1+K)^{-1} \tag{8.90}$$

(c) When a fraction of the reactant bypasses the bed, such as in the case of bubble flow, a general representation is

$$f(\zeta) = f_B \, \delta(\zeta) + (1-f_B) \, g(\zeta) \tag{8.91}$$

where f_B is the fraction of fluid bypassing the bed; $0 < f_B < 1$, and $g(\zeta)$ is a continuous density function such that

$$\int_0^\infty g(\zeta) \, d\zeta = 1 \tag{8.92}$$

In this case,

$$\Phi(K) = f_B + (1-f_B) \int_0^\infty e^{-K\zeta} g(\zeta) \, d\zeta \tag{8.93}$$

For the bubble-flow model, they applied the relation given by Eq. (8.45) and gave, at the exit of the bed,

$$C_\ell = \left(1 - \frac{W_{fs}}{W_s}\right) C_{B1} + \frac{W_{fs}}{W_s} C_1 \tag{8.94}$$

where C_ℓ, C_{B1}, and C_1 are the concentrations in the mixed gas leaving the bed, in the bubble leaving the bed, and in the dense phase respectively. For uniformly distributed n_B bubbles per unit volume, each having volume v_B,

$$n_B \, v_B \, W_B = W_s - W_{fs} \tag{8.95}$$

and volumetric flow rate Q into a bubble

$$Q(C_1 - C_B) = V_B \frac{dC_B}{dt} = W_B \, V_B \frac{dC_B}{dz} \tag{8.96}$$

where C_B is the concentration in the bubble at height z, we have

$$C_B = C_1 + (C_0 - C_1) \, \exp\left(-\frac{Qz}{W_B \, V_B}\right) \tag{8.97}$$

where C_0 is the concentration entering the bed. Material balance of the dense, or particulate phase gives

$$n_B Q \int_0^{z_0} \left[C_1 + (C_0 - C_1) \, \exp\left(-\frac{Qz}{W_B \, V_B}\right) \right] \, dz + W_{fs} \, C_0$$

$$= n_B \, Q \, z_0 \, C_1 + W_{fs} \, C_1 + k_c \, z_0 \, (1 - n_B \, V_B) \, C_1 \tag{8.98}$$

where k_c is the rate constant on the volume-concentration basis, and z_0 is the bed height at which a fraction $(1 - n_B \, v_B)$ is occupied by the particulate cloud.

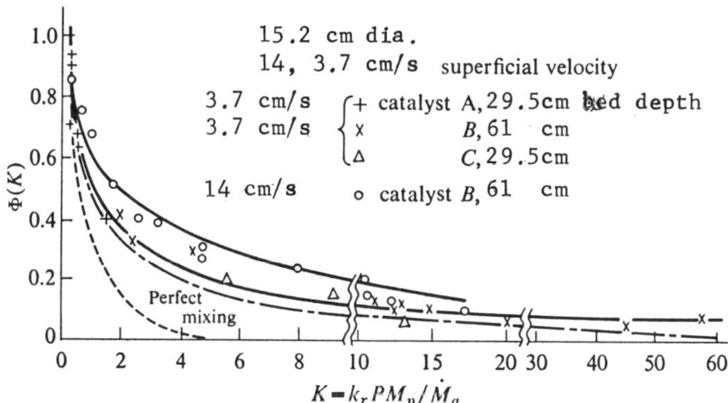

Figure 8.22   Reaction-time distribution in fluidized catalytic reactor [Orcutt, et al. 1962]

The bed expansion $z_0 - z_{mf}$ is assumed to be due to the bubbles; z_{mf} is the height at minimum fluidization. We have

$$n_B \, v_B \, z_0 = z_0 - z_{mf} \qquad (8.99)$$

Combining Eqs. (8.94), (8.98), and (8.99) gives

$$\Phi(K) = \beta \, e^{-n^*} + (1 - \beta \, e^{-n^*}) \left(\frac{K}{1 - \beta e^{-n^*}} + 1\right)^{-1} \qquad (8.100)$$

where $\beta = 1 - (W_{fs}/W_s)$, $n^* = Qz_0/W_B v_B$, and $K = k_r z_{mf}/W_s = k_r \cdot PM_p/M_g$. A comparison of these correlations is shown in Fig. 8.22. The above method is readily simplified to reactive flow through a catalytic packed bed. (Prob. 8.11)

Exercise Problems

8.1 Measurement gave the viscosity of a coal-water slurry as 0.045 kg/m-s for 0.6 volume fraction of 1 mm size coal flowing through a pipe of 0.46 m diameter at 1.5 m/s mean velocity. The specific gravity of coal is 1.3. Compute the coefficient C_μ in Eq. (8.7). Assume the velocity profile is given by: $W = W_0 [1 - (r/R)^{3/2}]$, W_0 being the core velocity.

Ans. 16.8

8.2 Show that the magnitudes of terms on the two sides of Eq. (8.19) are comparable for coal-water slurry of 1 mm size coal flowing through a pipe of 0.46 m diameter at a core velocity of 1.6 m/s; assume $W^*_{av} = 1$ and $|\partial W^*/\partial y^*|_{av} = 1.5$. Can this condition be achieved with a gas-solid suspension? (Sp. gr. of coal: 1.3)

Ans. No.

8.3 The measured pressure drop for flow of a coal-water slurry of 0.6 volume fraction of coal (sp. gr. 1.3) through a 0.46 m diameter pipe at a core velocity of 1.5 m/s is 64.2 Pa/m. Compute the velocity lags of the particle phase for $C\mu = 1$, for particles of $F_{pf} = 2.5$ s^{-1} (2a = 1 mm) and $F_{pf} = 250$ s^{-1} (2a = 100 μm).   Ans. 0.044, 4.4 x 10^{-4}.

8.4 For flow through a horizontal pipe of a suspension consisting of a layer of dense phase and a layer of dilute phase, show that in order that uniform density in the dilute phase be maintained through a balance of shear lift against the gravity settling (Eq. 6.24), the relation: $0.04\ W_o = (1 - \rho^*)g/F_{pf}$ must be satisfied. What does the relation mean?

8.5 Determine the permeability of a thin gap of width w for seepage of a liquid through it.   Ans. $w^2/12$

8.6 For minimum fluidization of a bed of 0.5 mm diameter glass particles in air (density 1.2 kg/m^3, viscosity 1.8 x 10^{-5} kg/m-s) a superficial velocity of 0.194 m/s was experimentally determined. Check this with the correlations given by Eqs. (8.40) to (8.42). Compute the case of 10 μm particles.
                    Ans. 0.201 m/s, 1.02 x 10^{-4} m/s

8.7 For solid particles of a dielectric constant of 4, density 1000 kg/m^3, charge to mass ratio of 10^{-6} C/kg, fluidized in a gas of density of 1 kg/m^3, interstitial velocity 0.3 m/s, bed radius of 0.06 m, and assume a transmission coefficient of 1.3x10^{-4}, compute the bubble radius and electrostatic surface tension. When compared to the surface tension of water of 0.07 N/m, what does the result mean?
                    Ans. Bubble radius 0.03 m, 4.2 x 10^{-4} N/m.

8.8 Modify the example in Figs. 8.13 and 8.14 by inserting tubes of radius R and horizontal pitch L_x and vertical pitch L_z to give a volumetric porosity of γ_v and an inverse relaxation time F_{wp} for the momentum transfer from particle to these internals. Assume $F_{wp} = F$, compute the maximum particle velocity when the other parameters are similar to the example in Fig. 8.13. Would this velocity be significantly increased if the same through flow of air is maintained as the case in Fig. 8.13?

Ans. 0.0228 m/s, 0.026 m/s.

8.9 Derive the polynomial coefficients for the circulatory motion of fine powders ($k^{-1} \gg 1$, $\gamma \sim 1$). Show that the relation will be universal for all expanded bed heights and for ACDW motion only for the primary solution.

8.10 Determine the radial effective thermal conductivity of the fluidized bed discussed in Sec. 8.7 for $r_c = 0.7\, r_0$, $r_0 = 38$ mm, $T_c - T_w = 2$ C, $dT_c/dz = 19$ C/m, $dT/dr = 109$ C/m, $\rho_p = 1$ kJ/kg C, $\rho_p = 400$ kg/m^3, $\kappa_r = 0.04\, \kappa_e$; when the mean particle velocity $W_{pc} = 10$ cm/s. Compute the heat transfer coefficient for $L_b = 0.2$ m.

Ans. 82.5 W/m K, 5.32 kW/m^{2}K.

8.11 Determine the fraction of reactant unconverted for flow through a packed bed of catalyst at superficial velocity W_s and reaction rate constant k_c and bed length z_0.

REFERENCES

Ailam, G., and Gallily, I., 1962, Phys. Fluids, Vol. 5, p. 575.

Allis, W. P., and Rose, D. J., 1954, Phys. Rev., Vol. 93, p. 83.

Anderson, T. B., and Jackson, R., 1967, Ind. & Eng. Chem. Fund., Vol. 6, p. 527.

Arastoopour, H., Wang, C. H., and Weil, S. A., 1982, Chem. Eng. Sci., Vol. 37, p. 1979.

Ardron, K. H., 1978, Int. J. Multiphase Flow, Vol. 4, p. 323.

Aris, R, 1962, Vectors, Tensors, and the Basic Equations of Fluid Mechanics, Prentice-Hall, Inc., Englewood Cliffs, NJ.

Baldwin, L. V., and Walsh, T. J., 1961, AIChE J., Vol. 7, p. 53.

Banister, R. R., and Davis, H. L., 1962, Phys., Fluids, Vol. 5, p. 136.

Barger, R. L., 1961, NASA TN D-740, p. 12.

Basset, A. B., 1888, Hydrodynamics, Deighton, Bell, Cambridge; Dover, New York (1961), p. 270.

Bingham, E. C., 1922, Fluidity and Plasticity, McGraw-Hill, NY.

Binnie, A. M., and Green, T. R., 1942, Proc. Roy. Soc. (London), Vol. 181, p. 134.

Bird, R. B., Stuart, W. E., and Lightfoot, E. N., 1960, Transport Phenomena, Wiley, NY, p. 101.

Blitz, J., 1963, Fundamentals of Ultrasonics, Butterworth, London.

Bobis, J. P., Porges, K. G. A., and Raptis, A. C., 1986, Gas Solid Flows, ASME, FED-Vol. 35, p.79.

Bogy, D. B., 1979, Ann. Rev. Fluid Mech., Vol. 11, p. 207.

Boussinesq, J., 1877, Mem. pres. par div.savants a l'acad. Sci. Paris, Vol. 23, p. 46.

Boussinesq, J., 1885, Comptes Rendu, p. 935.

Bowman, C. W., Ward, D. M., Johnson, A. I., and Trass, O, 1961, Can. J. Chem. Eng., Vol. 39, p. 1, 9.

Bowser, R. J., and Weinberg, F. J., 1974, Nature, Vol. 249, p.

Brandt, H. L., and Johnson, B. M., 1963, AIChE J., Vol. 9, p. 771.

Brodkey, R. S., 1967, The Phenomena of Fluid Motions, Addison-Wesley.

Brown, B., and McCarty, K. P., 1962, Proc. Eighth Symp. (International) on Combustion, Williams and Wilkins, Baltimore, p. 814.

Brunaur, S., 1943, The Adsorption of Gases and Vapors, Princeton University Press, Princeton, NJ.

Cadle, R. D., 1965, Particle Size, Reinhold, New York.

Cann, G. L., 1961, Energy Transfer Processes in a Partially Ionized Gas, Memo No. 61, Cal. Inst. of Tech., Army Ordance Contracts DA-04-495-Orb-1960 and 3231.

Carlson, D. J., 1962, ARS J., Vol. 32, p. 1107.

Carlson, D. J., 1962, Temperature, Its Measurement and Control in Science and Industry, Reinhold, NY, Vol. 3, Part II.

Carman, P. C., 1937, Trans. Inst. Chem. Engrs., Vol. 15, p. 150.

Carstensen, E. L., and Foldy, L. L., 1947, J. Acoust. Soc. Am., Vol. 19, p. 481.

Cawood, W., 1936, Trans. Faraday Soc., Vol. 32, p. 1068.

Chandrasekhar, S., 1960, Radiative Transfer, Dover, NY.

Chao, B. T., 1962, Phys. Fluids, Vol. 5, p. 69.

Chapman, S., and Cowling, T. G., 1958, The Mathematical Theory of Non-uniform Gases, Cambridge U. P. London.

Chen, J. C., and Withers, J. G., 1978, AIChE Symp. Ser. 174, Vol. 74, p. 327

Cheng, L., and Soo, S. L., 1970, J. Appl. Phys., Vol. 41, p. 585.

Cheng, L., Tung, S. K., and Soo, S. L., 1970, Trans. ASME, J. Eng. for Power, Vol. 92(A-2), p. 135.

Cheng, L., 1977, I/EC Process Design and Dev., Vol. 16, p. 192.

Cheremisinoff, N. P., and Gupta, R., 1983, Handbook of Fluids in Motion, Ann Arbor Science, Ch. 29, 36.

Chilton, E. G., Sauer, F. M., and Lundberg, R. E., 1963, Proc. of Multiphase Flow Symposium, ASME, p. 35.

Chong, Y. O., and Leung, L. S., 1986, Powder Tech., Vol. 47, p. 43.

Chopra, K. P., 1965, Astronautica Acta, Vol. 11, p. 157.

Chromey, F. C., 1960, J. Opt. Soc. Am., Vol. 50, p. 7.

Chu, J. C., Kalil, J., and Wetteroth, 1953, Chem. Eng. Progr., Vol. 49, p. 141.

Churchill, R. V., 1944, Modern Operational Method in Engineering, McGraw-Hill, New York, Appendix 4, Formula III.

Clift, R., Grace, J. R., and Weber, M. E., 1978, "Bubbles, Drops, and Particles," Academic Press.

Cobine, J. D., 1958, Gaseous Conductors, Dover, NY.

Cole, R., 1974, Adv. Heat Transf., Vol. 10, p. 85.

Colver, G. M., 1977, J. Powder Tech., Vol. 17, p. 9.

Corn, M., 1961, J. Air Pollution Control Assoc., Vol. 11, p. 523, 566.

Coronitti, S. C., 1965, Problems of Atmospheric and Space Electricity, Elsevier, NY, Chs. 1 and 2, p. 178, 410, 485.

Daily, J. W., and Bugliarello, G., 1959, Ind. Eng. Chem., Vol. 51, p. 887.

Daniel, J. H., and Brackett, F. S., 1951, J. Appl. Phys., Vol. 22, p. 542.

Davidson , J. F., 1961, Trans. Instn. Chem. Engrs., Vol. 39, p. 230.

Davidson, J. F., Clift, R., and Harrison, D., 1985, Fluidization, 2nd Edition, Academic Press, London.

Davidson, J. F., and Nedderman, R. M., 1973, Trans. Inst. Chem. Engr. Vol. 51, p. 29.

Delaplaine, J. W., 1956, AIChE J., Vol. 2, p. 127.

Delhaye, J. M., and Archard, J. L., 1976, Proc. CSNI Specialists' Meeting, Toronto (S. Bannerjee and K. R. Weaver, Eds.) AECL Vol. 1, p. 5.

Dimick, R. C., and Soo, S. L., 1964, Phys. Fluids, Vol. 7, p. 1638.

Dorsch, R. G., Saper, P. G., and Kadow, C. F., 1955, NACA TN 3587.

Drake, R. M., Jr., 1961, J. Heat Transfer, Trans. ASME, Vol. 83, p. 170.

Dryndrozhik, E. I., 1975, Fluid Mech., Soviet Research, Vol. 4, p. 33 (original Vol. 25, p. 103, 1974).

Duff, K. M., and Hill, P. G., 1964, Proc. Heat Transfer and Fluid Mechanics Institute, Stanford Univ. Press, p. 268.

Dunsky, V. D., Zabrodsky, S. S., and Tamarin, A. I., 1966, Proc. Third Int. Heat Transfer Conf., Vol. 4, p. 293.

Eddington, R. B., 1970, AIAA J., Vol. 8, p. 1.

Einstein, A., 1906, Ann. Phys., Leipzig, Vol. 19, p. 289.

Einstein, A., 1956, "The Theory of Brownian Movement," Dover, New York.

Elgobashi, S. E., Abou-Arab, T. W., Rizk, M., and Mostafa, A., 1984, Int. J. Multiphase Flow, Vol. 10, p. 697.

England, W. C., Firey, J. C., and Trapp, O. E., 1966, I/EC Process Des. Dev., Vol. 5, p. 168.

Elperin, I., Igra, O., and Ben-Cor, G., 1986, Trans. ASME, J. Fluid Eng., Vol. 108, p. 354.

Epstein, P. S., 1929, Z. Phys., Vol. 54, p. 537.

Epstein, P. S., and Carhart, R. R., 1953, J. Acoust. Soc. Am., Vol. 25, p. 553.

Ergun, D., and Orning, A. A., 1949, Ind. Eng. Chem., Vol. 41, p. 1179.

Faeth, G. M., 1967, Combustion and Flame, Vol. 11, p. 167.

Faeth, G. M., 1983, Prog. Energy Comb. Sci., Vol. 9, p. 1.

Farbar, L., and Morley, M. J., 1957, Ind. Eng. Chem., Vol. 49, p. 1143.

Finnie, I., Wolak, J., and Kabil, Y.H., 1967, ASTM J. Mater., Vol. 2, p. 682.

Forgacs, O. L., Robertson, A. A., and Mason, S. G., 1958, Brit. Paper and Board Makers' Assoc., Kenley, Surrey, England, p. 447.

Fraas, A. P., 1975, Survey of Turbine Bucket Erosion, Deposits, and Corrosion, ASME Paper No. 75GT-123.

Franklin, F. C., and L. N. Johanson, 1955, Chem. Eng. Sci., Vol. 4, p. 119.

Frenkel, J., 1946, Kinetic Theory of Liquids, Oxford Univ. Press; Dover, NY, 1955.

Friedlander, S. K., and Johnstone, H. F., 1957, Ind. Eng. Chem., Vol. 49, p. 1151

Friedlander, S. K., 1959, Chem. Eng. Prog. Symp. Ser., Vol. 55, p. 135.

Froessling, N., 1938, Gerlands Beitraze zur Geophysik, Vol. 52, p. 170.

Fuchs, N., 1959, Evaporation and Droplet Growth in Gaseous Media, Sec. 2, Pergamon, New York.

Goldberger, W. M., and Nack, H., 1964, Battelle Tech. Rev., Vol. 13, p. 11.

Goldstein, M. E., Yang, W-J, Clark, J. A., 1967, Trans. ASME, J. Heat Transfer, Vol. 89C, p. 185.

Gorring, R. L., and Churchill, S. W., 1961, Chem. Eng. Prog., Vol. 57, p. 53.

Govier, G. W., and Aziz, K., 1972, The Flow of Complex Mixtures in Pipes, van Nostrand Reinhold Co., NY

Graf, W. A., and Acaroglu, E. R., 1968, Bull. of Int. Assoc. of Sci. Hydrology, Vol. 23, p. 20, p. 123.

Green, H. L., and Lane, W. R., 1957, Particle Clouds, Dusts, Smokes, and Mists, E. and F. N. Spon Ltd., London.

Gugan, K., Lawton, J., and Weinberg, F. J., 1964, Proc. 10th Symp. (International) on Combustion, p. 709.

Haberman, W. L., and Morton, R. K., 1953, David Taylor Model Basin Report 802.

Hadamard, J., 1911, Compt. Ren., Vol. 152, 1735.

Halstrom, E. A. N., 1953, M. S. Thesis, Mechanical Eng., Princeton University, NJ.

Hammerton, D., and Garner, F. H., 1954, Trans. Instn. Chem. Eng. (London), Vol 32 (Supplement) S-18.

Hancher, G. W., and Jury, S., 1959, Chem. Eng. Progr. Symp. Ser., Vol. 55, p. 87.

Hanna, B. N., Raithby, G. D., and Nicoll, W. B., 1979, Two-phase Momentum, Heat and Mass Transfer, Hemisphere Pub. Corp., Washington, DC, p. 33.

Happel, J., and Brenner, H., 1965, Low Reynolds Number Hydrodynamics, Prentice Hall, Englewood Cliffs, NJ.

Hartman, G. K., and Focke, A. B., 1940, Phys. Rev. Vol. 57, p. 221.

Henry, R. E., Fauske, H. K., and McComas, S. T., 1970, Nucl. Sci. Eng., Vol. 41, p. N1

Hetsroni, G., 1981, Handbook of Multiphase Systems, Hemisphere Pub. Corp., DC.

Hettner, G., 1926, Z. Phys., Vol. 37, p. 179.

Hill, P. G., 1966, J. Fluid Mech., Vol. 25, p. 593.

Hinze, J. O., 1949, Appl. Sci. Res., Vol. A1. p. 263, 273.

Hirschkron, R., and Ehrich, F. F., 1964, ASME Paper 64-WA/FE-30.

Howarth, L., 1936, Proc. Roy. Soc., (London), Vol. A154, p. 3.

Hultberg, J. A., and Soo, S. L., 1965, Astronautica Acta, Vol. 11, p. 207.

Ishii, M., 1975, Thermo-fluid Dynamic Theory of Two-phase Flow, Eyrolles, Paris.

Jastrow, R., and Pearse, C. A., 1957, J. Geophys. Res., Vol. 62, p. 413.

Jenike, A. W., and Johanson, J. R., 1969, Trans. ASME, J. Eng. Ind., Ser. B, Vol. 91, p. 339.

Johanson, J. R., 1964, Trans. ASME, J. Appl. Mech., Ser. E., Vol. 86, p. 499.

Jokati, T., and Tomita, Y., 1973, Bull. of Japan SME, Vol. 16, p. 93.

Kada, H. and Hanratty, T. J., 1960, AIChE J., Vol. 6, p. 624.

Kaye, B. H., and Boardman, R. P., 1962, Proc. Symp. on Interaction between Fluids and Particles, Institute of Chem. Engrs., London, p.22

Keenan, J. H., and Keyes, F. G., 1950, Thermodynamic Properties of Steam, John Wiley and Sons, NY.

Kettenring, K. N., Manderfield, E. L., and Smith, J. M., 1950, Chem. Eng. Progr., Vol. 46, 139.

Kirkpatrick, R., and Lockett, M. J., 1978, Chem. Eng. Sci., Vol.

29, p. 2363.

Knudsen, V. O., Wilson, J. V., and Anderson, N. S., 1948, J. Acoust. Soc. Am., Vol. 20, p. 849.

Kraemer, H. R., and Johnstone, H. F., 1955, Ind. Eng. Chem., Vol. 47, p. 2426.

Kriebel, A. R., 1963, ASME Paper 63-WA-13.

Kronig, R., and Bruijsten, J., 1951, J. Appl. Sci. Res., Vol. A2, p. 439.

Krupp, H., 1967, Advances in Colloid and Interface Sci. (Amsterdam), Vol. 1, p. 111.

Kundt, K., and Warburg, E., 1875, Ann. Physik, Vol. 155, p. 337.

Kunii, D., and Levenspiel, O., 1968, Fluidization Engineering, Wiley, NY.

Laidler, T. J., and Richardson, E. G., 1938, J. Acoust. Soc. Am., Vol. 9, p. 217.

Lamb, H., 1932, Hydrodynamics, Dover, New York.

Landau, L. D., and Lifschitz, E. M., 1958, Statistical Physics, Addison-Wesley, Reading, MA, p. 322.

Lane, W. R., 1951, Ind. Eng. Chem., Vol. 43, p. 1312.

Langmuir, I., 1918, Phys. Rev., Vol. 12, p. 368.

Langmuir, I., and Blodgett, K., 1948, J. Meteorol., Vol. 5, p. 175.

Lapidus, L., and Elgin, J. C., 1957, AIChE J., Vol. 3, p. 63.

Laufer, J., 1951, NACA Report 1053.

Ledig, P. G., and Weaver, E. R., 1924, J. Am. Am. Chem. Soc., Vol. 46, p. 650.

Lee, Y. N., and Soo, S. L., 1970, Environ. Sci. and Tech., I/EC, Vol. 4, p. 678.

Leva, M., 1962, Proc. Symp. on Interaction between Fluids an Particles, inst. Chem. Engrs., London, p. 143.

Levich, V. G., 1962, Physicochemical Hydrodynamics, Prentice-Hall, Englewood cliffs, NJ, p. 209, 475.

Levy, A., Slamovich, E., and Jee, N., 1986, Wear, Vol. 110, p. 117.

Lewis, W. K., Gilliland, E. R., and Girouard, H., 1962, Fluidization (F. A. Zenz, Ed.) Chem. Eng. Prog. Symp. Ser. 38, Vol. 58, p. 87.

Liebermann, L., J., 1957, Appl. Phys., Vol. 28, p. 205.

Lin, J. S., Chen, M. M., and Chao, B. T., 1985, AIChE J., Vol. 31, p. 465.

Liu, B. Y. H., Whitby, K. T., and Yu, H. H. S., 1967, J. Coll. Interface Sci., Vol. 23, p. 367.

Loeb, L. B., 1955, Basic Processes of Gaseous Electronics, U. of Calif. Press, Berkeley, p. 421, 478, 560.

Loeffler, F., and Muhr, W., 1972, Chemie Ingenieur Technik, Vol 8, p. 510.

Loeffler, F., 1977, Proc. 4th International Clean Air Congress, p. 800.

Lorentz, H. A., 1896, Zittingsverl. Akad. Van Et., Vol. 5, p. 168.

Lorentz, H. A., 1915, The Theory of Electrons, Dover, NY, p. 147.

Love, T. J., Jr., and Grosh, R. J., 1964, Trans. ASME, J. Heat Transfer, Vol. 87c, p. 161.

Lumley, J. L., 1962, J. Math. Phys., Vol. 3, p. 309.

Marshall, T. J., 1962, Proc. Symp. on Interaction between Fluids and Particles, Inst. Chem. Engrs., London, p. 299.

Mason, B. J., 1957, The Physics of Clouds, Oxford, London, p. 369.

Maxwell, J. C., 1904, A Treatise on Electricity and Magnetism, Vol. 1, p. 231, Clarendon, Oxford.

Mazzone, D. N., Tardos, G. I., and Pfeffer, R., 1987, Powder Tech., Vol. 51, p. 71.

McDonald, D. A., 1960, Blood Flow in Arteries, Arnold, London.

McTigue, D. F., Givier, R. C., and Nunziato, J. W., 1986, J. Rheology, Vol. 30, p. 1053.

Metzner, A. B., and Whitlock, M., 1958, Trans. Soc. Rheol., Vol. II, p. 239.

Mickley, H. S., and Trilling, C. A., 1951, Ind. Eng. Chem., Vol. 43, p. 1220.

Miles, G. D., Schedlovsky, L., and Ross, J. J., 1945, J. Phys. Chem., Vol. 49, p. 93.

Mills, D., and Mason, J. S., 1978, Proc. 2nd Int. Powder & Bulk Solids Handling and Processing, Rosemont, IL.

Min, K., Chao, B. T., and Wyman, M. E., 1963, Rev. Sci. Instr., Vol. 34, p. 529.

Mitchell, R. I., and Engdahl, R. B., 1963, J. Air Pollution Control Assoc., Vol. 13, p. 11.

Monin, A. S., 1956, Adv. Geophys., Vo. 6, p. 435.

Montgomery, D. J., 1959, Solid State Phys., Vol. 9, p. 139.

Moore, D. W., 1959, J. Fluid Mech., Vol.6, p. 113.

Morr, G. F., and Soo, S. L., 1973, Fluid Studies in Air and Water Pollution, (R. A. Arndt et al., Eds.), ASME, p. 151.

Morrell, G., 1961, NASA TN D-677.

Mullin, J. W., and Treleaven, C. R., 1962, Proc. Symp. on Inter-action between Fluids and Particles, Institution of Chem. Engrs., London, p. 201.

Murphy, A. T., Adler, F. T., and Penny, G. W., 1959, AIEE Trans., Vol. 78, p. 318.

Murray, J. D., 1965, J. Fluid Mech., Vol. 21, p. 465, Vol. 22, p. 57.

Narayanan, S., Gossens, F. J., Dossen, N. W. F., 1974, Chem. Eng. Sci., Vol. 29, p. 2071.

National Research Development Corp. (London), 1971, Report No. 85, Interim No. 1, NTIS PB-210673.

Neilson, J.H., and Gilchrist, A. G., 1968, Wear, Vol. 11, p. 111.

Newitt, D. M., Richardson, J. F., and Shook, C. A., 1962, Proc. Symp. on Interaction between Fluids and Particles, Instn. Chem. Engrs. (London), p. 87.

Ni, S., 1986, personal communication.

Nieh, S., Chao, B. T., Soo, S. L., 1985, Particulate Sci. & Tech., Vol. 3, p. 127.

Orcutt, J. C., Davidson, J. F., and Pigford, R. L., 1962, Fluidization, (F. A. Zenz, Ed.) Chem. Eng. Prog. Sump. Ser. 38, Vol. 58, p. 1.

Ogawa, A, 1975, J. Coll. of Eng., Nihon Univ., Ser. A, Vol. 14, p.39.

Orr, C., Jr., 1966, Particulate Technology, MacMillan Co., New York.

Owen, P. R., 1960, Aerodynamic Capture of Particles (E. G. Richardson, Ed.), Macmillan, NY.

Page, L., 1952, Introduction to Theoretical Physics, Van Nostrand, Princeton, NJ.

Perez-Blanco, H., Chao, B. T., and Soo, S. L., 1980, Scaling in Two-phase Flows, ASME HTD-Vol. 14, p. 1.

Peskin, R. L., 1960, Proc. Heat Trnsfer & Fluid Mech. Institute, Stanford University Press, Stanford, CA.

Peskin, R. L., 1970, Analytical Study of Turbulent Diffusivity, ME-26E, Second International Clean Air Congress, Washington, DC.

Pfeffer, R., Rosetti, S., and Lieblein, S., 1966, NASA TN D-3603.

Pratt, H. R. C., Gaylor, R., Roberts, N. W., Murdock, K., and Thornton, J. D., 1953, Trans. Inst. Chem. Engrs., Vol. 31, p. 69.

Probert, R. P., 1946, Phil. Mag., Vol. 37, p. 94.

Ranz, W. E., and Wong, J. B., 1952, Arch. Ind. Hyg. and Occupational Med., Vol. 5, p. 464; Ind. Eng. Chem. Vol. 44, p. 1371.

Ranz, W. E., 1956, On Sprays and Spraying, Engineering Research Bulletin No. 65, Penn. State Univ., Univ. Park, PA.

Rayleigh, Lord (J. W. Strutt), 1917, Phil. Mag., Vol. 34, p. 94.

Rayleigh, Lord (J. W. Strutt), 1929, Theory of Sound, Dover, NY.

Reeks, M. W., 1977, J. Fluid Mech., Vol. 83, p. 529.

Reimann, A. L., 1934, Thermionic emission, Wiley, NY, p. 298.

Richardson, E. G., 1962, Ultrasonic Physics, (A. E. Brown, Ed.) Elsevier, NY, p. 259, 272.

Rietema, K., 1962, Proc. Symp. on Interaction between Fluid and Particles, Instn. Chem. Engrs., London, p. 275.

Ritter, J. E., Rosenfeld, L., and Jakus, K., 1986, Wear, Vol. 111, p. 335.

Roco, M. C., and Shook, C. A., 1981, "New Approach to Predict Concentration Distribution in Fine Particle Slurry Flow," Proc.

AIChE 91st National Meeting, Detroit, MI.

Rohsenow, W. M., and Choi, H. Y., 1961, Heat, Mass and Momentum Transfer, Prentice-Hall, Englewood Cliffs, NJ.

Rosseland, S., 1931, Astrophysik auf Atom-Theoretischer Grundlage, Springer-Verlag, Berlin, p. 41.

Rowe, P. N., 1962, Fluidization, Chem. Eng. Progr. Symp. Ser., Vol. 58, p. 42.

Rowe, P. N., and Partridge, B. A., 1962, Proc. Symp. on Interaction between Fluids and Particles, Inst. of Chem. Engrs, London, p. 135.

Rowe, P. N., Partridge, B. A., and Lyall, E., 1964, Chem. Eng. Sci., Vol. 19, p. 973.

Rubinow, S. I., and Keller, J. B., 1961, J. Fluid Mech., Vol. 11, p.447.

Rubinowitz, S. R., 1920, Ann. Physik, Vol. 62, p. 695.

Rudinger, G., and Chang, A., 1964, Phys. Fluids, Vol. 7, p. 1747.

Rumpf, H., 1972, Proc. Int. Symp. on Powder Tech. J. of Res. Assoc. of Powder Tech., Japan, p. 3.

Rybczynski, W., 1911, Bull. Acad. Sci. (Cracovie), Ser. A, p. 40.

Saffman, P. G., 1965, J. Fluid Mech., Vol. 22, p. 385.

Sauer, F. M., 1951, J. Aerosapce Sci., Vol. 18, p. 5.

Savage, S. B., 1965, Brit. J. Appl. Phys., Vol. 16, p. 1885.

Savage, S. B., 1983, Mechanics of Granular Materials and Constitutive Relations, (J. T. Jenkins and M. Satake, Eds.), Elsevier Sci. Pub. B. B. Amsterdam, p. 261.

Savic, P., 1953, Nat. Res. Council of Canada, Rept. No. MT-22.

Saxton, A. L., and Worley, A. C., 1970, The Oil and Gas Journal, Vol. 68, p. 82.

Schachman, H. K., 1959, Ultracentrifugation in Biochemistry, Academic Press, New York, p. 215-237.

Scheidegger, A. E., 1957, The Physics of Flow through Porous Media, Macmillan, NY.

Schlichting, H., 1979, Boundary Layer Theory, McGraw-Hill, New York.

Schuegerl, K., 1971, Fluidization (J. F. Davidson and D. Harrison, Eds.), Academic Press, NY, p. 261.

Segre, G., and Silberberg, A., 1962, J. Fluid Mech., Vol. 14, p. 115, 136.

Sevik, M., and Park, S. H., 1972, ASME Paper 72-WA/FE-32.

Sewell, C. J. T., 1910, Phil. Trans. Roy. Soc. (London), Vol. A120, p. 239.

Sha, W. T., and Soo, S. L., 1979, Multidomain Multiphase Fluid Mechanics in a Quasi-continuum, ANL-78-47, Argonne National Laboratory, NUREG-CR-0102.

Sha, W. T., Chao, B. T., and Soo, S. L., 1983, Proc. of Nuclear Thermal Hydraulics, Trans. ANS, p. 3.

Sha, W. T., Chao, B. T., and Soo, S. L., 1988, Report NUREG/CR-5065, ANL-87-51, Argonne National Laboratory, IL 60439.

Shapiro, A. S., 1953, Dynamics and Thermodynamics of Compressible Fluid Flow, Ronald, NY.

Sheen, S. H., Raptis, A. C., Bobis, J. P., Lee, S., and Simpson, T., 1984, Evaluation of Active Ultrasonic Cross-correlation Technique in Coal/liquid Pipe Flow Measurements, Report ANL/FE-84 -19, Argonne National Laboratory.

Sherwood, T. K., and Pigford, R. L., 1952, Absorption and Extraction, McGraw-Hill, New York, p. 237.

Shishkin, N. S., 1965, Problems of Atmospheric and Space Electricity (S. C. Coroniti, Ed.), Elsevier, NY, p. 268.

Skelland, A. H. P., 1967, Non-Newtonian Flow and Heat Transfer, Wiley, NY.

Slattery, J. C., 1967, Ind. Eng. Chem. Fund., Vol. 6, p. 108.

Sleicher, C. A., and Tribus, M., 1957, Trans. ASME, Vol 79, p. 789.

Smeltzer, C. E., Gulden, and Compton, W. A., 1970, Trans. ASME, J. Basic Eng., Vol. 92D, p. 639.

Soo, S. L., Tien, C. L., and Kadambi, V., 1959, Rev. Sci. Inst., Vol. 30, p. 821.

Soo, S. L., Ihrig, H. K., Jr., and El Kouh, A. F., 1960, Trans. ASME, J. Basic Eng., Vol. 82D, p. 609.

Soo, S. L., and Dimick, R. C., 1965, Proc. 10th Symp. (International) on Combustion, p. 699.

Soo, S. L., Trezek, G. J., Dimick, R. C., and Hohnstreiter, G. F., 1964, Ind. Eng. Chem. Fund., Vol. 3, p. 98.

Soo, S. L., 1965, Ind. Eng. Chem. Fund., Vol. 4, p. 426.

Soo, S. L., and Trezek, G. J., 1966, Ind. Eng. Chem. Fund., Vol. 5, p. 388.

Soo, S. L., 1967, Fluid Dynamics of Multiphase Systems, Blaisdell-Ginn, Waltham, MA.

Soo, S. L., 1968, ZAMP, Vol. 19, p. 545.

Soo, S. L., Wu, C., and Dimick, R. C., 1968, Intersociety Energy Conversion Engineering Conf. Record IEEE, p. 915.

Soo, S. L., 1969, Appl. Sci. Res., Vol. 21, p. 68.

Soo, S. L., Stukel, J.J., and Hughes, J. M., 1969, J. Environmental Sci. and Tech., I/EC, Vol. 3, p. 386.

Soo, S. L., 1970, ZAMP, Vol. 21, p. 125.

Soo, S. L., and Tung, S. K., 1971, Appl. Sci. Res., Vol. 24, p. 84.

Soo, S. L., 1971, Int. Rev. in Aerosol Phys. and Chem., Vol. 2, p. 61.

Soo, S. L., and Rodgers, L. W., 1971, Powder Tech., Vol. 5, p. 43.

Soo, S. L., and Tung, S. K., 1972, J. Powder Tech., Vol. 6, p. 283.

Soo, S. L., 1973, Powder Tech., Vol. 7, p. 267.

Soo, S. L., 1973a, Appl. Sci. Res., Vol. 28, p. 20.

Soo, S. L., 1973b, Int. J. Multiphase Flow, Vol. 1, p. 89.

Soo, S. L., 1974, J. Powder Tech., Vol. 10, p. 211.

Soo, S. L., and Tung, S. K., 1974, Proc. Symp. of Flow, Instr. Soc. of Am., p. 161.

Soo, S. L., 1975, Phys. Fluids, Vol. 18, p. 263.

Soo, S. L., 1976, Phys. Fluids, Vol. 19, p. 757.

Soo, S. L., 1977, Powder Tech., Vol. 17, p. 259.

Soo, S. L., 1978, Two-phase Transport and Reactor Safety,

Hemisphere Pub. Corp., DC, Vol. 1, p. 267.

Soo, S. L., 1980, Multiphase Transport, Hemisphere Pub. Corp., DC, Vol.1, p. 291.

Soo, S. L., 1981, J. Pipelines, Vol. 1, p. 57.

Soo, S. L., 1981a, AIChE Symp. Ser. 208, Vol. 77, Heat Transfer (R. Stein, Ed.), p. 152.

Soo, S. L., 1986, J. Powder Tech., Vol. 45, p. 169.

Soo, S. L., 1986a, Advances in Multiphase Flow and Related Problems (G. Papanicolaou, Ed.), SIAM, Philadelphia, 1986, p. 226.

Soo, S. L., 1987, J. Pipelines, Vol. 6, p. 193.

Sozzi, G. I., and Sutherland, W. A., 1975, G. E. Report HW-13418.

Staub, F. W., and Canada, G. S., 1978, Fluidization (J. F. Davidson and D. L. Keairns, Eds.) Camb. Univ. Press, p. 339.

Stern, A. D., 1968, Air Pollution, (Second Edition), Academic Press, Vol. 1, P. 240, 265.

Stokes, G. G., 1891, Mathematical and Physical Papers, Vol. 3, p. 55, Cambridge University Press.

Strauss, W., 1966, Industrial Gas Cleaning, Pergamon Press, London, p. 63.

Stuetzer, O. M., 1962, Phys. Fluids, Vol. 5, p. 534.

Stukel, J. J., and Soo, S. L., 1969, J. Powder Technology, Vol. 2, p. 278.

Sutton, O. G., 1953, Micrometeorology, McGraw-Hill, NY, p. 140.

Talbot, L., 1960, Phys. Fluids, Vol. 3, p. 289.

Taylor, G. I., 1922, Proc. London Mathematical Society, Vol. 20, p. 196.

Taylor, G. I., 1932, Proc. Roy. Soc. (London), Vol. A138, p. 41.

Taylor, G. I., 1937, Proc. Roy. Soc. (London), Vol. 164A, p.476.

Taylor, G. I., 1954, Proc. Roy. Soc. (London), Vol. A226, p. 34.

Taylor, G. I., 1956, The Shape and Acceleration of a Drop in a High Speed Air Stream, Ptn/6600/5278/49, Chemical Defense Experimental Establishment (Gt. Britain).

Tchen, C. M., 1947, Dissertation, Delft, Martinus Nijhoff, the Hague.

Temkin, S., and Dobbins, R. A., 1966, J. Acoust. Soc. Am., Vol. 40, p. 1016.

Ter Haar, D., 1956, Elements of Statistical Mechanics, Holt, Rinehart, and Winston, NY,

Tien, C. L., and Quan, V., 1962, ASME Paper 62-HT-15.

Timoshenko, S., 1934, Theory of Elasticity, McGraw-Hill, NY, p. 339.

Tomita, Y., Tashiro, H., Deguchi, K., and Jokati, T., 1980, Phys. Fluids, Vol. 23, P. 663.

Toomey, R. D., and Johnstone, H. F., 1953, Chem. Eng. Prog. Symp. Ser., Vol. 49, p. 51.

Towle, W. L., and Sherwood, T. K., 1939, Ind. Eng. Chem., Vol. 31, p. 457.

Trezek, G. J., and Soo, S. L., 1966, Proc. Heat Transfer and Fluid Mech. Inst., Stanford Univ. Press, p. 148.

Tsuji, Y., and Morikawa, Y., 1982, Int. J. Multiphase Flow, Vol. 8, pp. 657-667

Tung, S., and Soo, S. L., 1973, Trans. ASME, J. Appl. Mech., Vol. 40 (E2), p. 331.

Tung, V. X., and Dhir, V. K., 1988, Int. J. Multiphase Flow, Vol. 14, p. 47.

Turner, G. A., and Balasubramanian, 1976, J. Electrostatics, Vol. 2, p. 85.

Twomey, S., 1966, J. Atm. Aci., Vol. 23, p. 74.

Urick, R. J., 1948, J. Acoust. Soc. Am., Vol. 20, p. 283.

Vaitkus, D. R., Stukel, J. J., and Soo, S. L., 1971, "Measurements of Flwo Rate of Pulverized Coal in a Power Plant with an Electrostatic Probe," Symp. on Flow, ISA, Paper No. 2-6-179.

Vames, J. S., and Hanratty, T. J., 1986, Gas-Solid Flows, ASME FED-Vol. 35, p. 15.

Van der Hegge Zijnen, J. J., 1958, Appl. Sci. Res., Vol. A7, p. 207.

Van Driest, E. R., 1956, J. Aero. Sci., Vol. 23, p. 1007.

Viskanta, R., and Grosh, R. J., 1962, Trans. ASME, J. Heat Transfer, Vol. 84C, p. 63.

Waldmann, L., 1959, Z. Naturforsch., Vol. 14a, p.589.

Wallis, G. B., 1961, "Some Hydrodynamic Aspects of Two-phase Flow and Boiling," Int. Heat Transfer Conf., Boulder, CO.

Wallis, G. B., 1969, One-dimensional Two-phase Flow, McGraw-Hill, NY.

Walton, W. H., and Woolcock, A., 1960, Aerodynamic Capture of Particles, Pergamon, NY, p. 129.

Wamsley, W. W., and Johanson, L. N., 1954, Chem. Eng. Progr., Vol. 50, p. 347.

Weast, R. C., 1975, Handbook of Chemistry and Physics, 56th Ed., CRC Press, p. E47.

Wen, C. Y., 1966, Pneumatic Transport of Solids (Ed. S. D. Spencer, T. J. Joyce, and J. H. Farber), U. S. Dept. Interior, Washington, DC, Bureau of Mines Information Circular 8314.

Wen, C. Y., and Yu, Y. H., 1966, AIChE J., Vol. 12, p. 610.

Whitaker, S., 1969, Ind. Eng. Chem., Vol. 61, p. 4.

White, H. J., 1951, AIEE Trans., Vol. 70, p. 1189.

White, H. J., 1963, Industrial Electrostatic Precipitation, Addison-Wesley, Reading, MA.

Williams, J. C., and Jackson, R., 1962, Proc. Symp. Interaction Fluids and Particles, Inst. Chem. Eng., London, p. 282,291, 297.

Yao, K. M., Habibian, J. T., and O'Melia, C. R., 1971, Environmental Sci. and Tech., Vol. 5, p. 1105.

Zaloudek, F. R., 1964, The Critical Flow of Hot Water Through Short Tubes, G. E. Report HW-77594, Hanford Labs., WA.